Petroleum Economics
and Engineering

CHEMICAL PROCESSING AND ENGINEERING

An International Series of Monographs and Textbooks

EDITORS

Lyle F. Albright

Purdue University
West Lafayette, Indiana

R. N. Maddox

Oklahoma State University
Stillwater, Oklahoma

John J. McKetta

University of Texas
at Austin
Austin, Texas

IN PREPARATION

Petroleum Economics and Engineering

AN INTRODUCTION

Hussein K. Abdel-Aal

Department of Chemical Engineering
University of Petroleum and Minerals
Dhahran, Saudi Arabia

Robert Schmelzlee

Department of Economics and Business
Portland State University
Portland, Oregon
and
University of Petroleum and Minerals
Dhahran, Saudi Arabia

MARCEL DEKKER, INC.   New York and Basel

MARCEL DEKKER, INC.
270 Madison Avenue, New York, New York 10016

LIBRARY OF CONGRESS CATALOG CARD NUMBER: 75-213
ISBN: 0-8247-6293-2

Current printing (last digit):
10 9 8 7 6 5 4 3 2 1

PRINTED IN THE UNITED STATES OF AMERICA

CONTENTS

iii

Contents

PREFACE

In this book we have endeavored to fulfill a need for an introductory text in the economics of oil engineering. We have drawn from the general body of economic thought some parts that appear to be particularly relevant to oil engineering and the oil industry.

We do not, of course, believe that we have covered every aspect of economics that is relevant to the oil industry. For example, we have deliberately limited our material by avoiding discussion of the wholesaling and retailing of oil products.

Division of the text material is in relatively homogeneous chapters, although some topics are explored more thoroughly than others. Examples are given throughout the text to make the information presented more practical, as well as understandable. Care has been exercised to provide solutions in graphical and tabular form whenever graphs, charts, and tables are appropriate to visualization of the materials.

Owing to the scope of the subject, the economic "picture" of the industry had to be painted with relatively broad strokes, although detail is not entirely neglected. It is hoped that this book will be of value to students, to lay readers, and to all those interested in learning something about the oil industry.

The method used was to draw on pertinent published data, material from teaching courses in oil economics and oil engineering at the University of Petroleum and Minerals at Dhahran, Saudi Arabia, and information gathered directly from people in the industry. Use of the latter source was based on the sound assumption that one cannot ordinarily obtain a true picture of the oil industry from published data alone.

Most information derived from people in the industry was obtained through questions formed and asked by the authors, rather than from voluntary statements initiated by industry representatives. Cooperation of industry members was anything but uniform; sometimes the absence of complete cooperation handicapped the collection of data, but this absence was often offset by enthusiastic aid from others. No actual oil company records have been used, so sample data do not involve any company operational secrets or policies. Information compiled and used has been carefully reviewed and prepared. A conscious attempt, therefore, has been made to avoid identifying particular companies or company officials.

We wish to thank all of the people who have had a hand in making this book possible. We particularly wish to acknowledge the aid of Dr. Bakr Abdullah Bakr, Rector of the University of Petroleum and Minerals, for his encouragement and financial aid in making the presentation of this manuscript possible. Also, our sincere thanks to Drs. Nasser Rashid, Ronald Scott, Marwan Kamal, and James Corey, all of the University of Petroleum and Minerals, for their assistance and many valuable suggestions.

In addition, we would also like to thank Mr. Carlo Trampini of Agip, the Saudi Arabian Oil Company, the British Petroleum Company, Ltd., and the American Arabian Oil Company for their assistance.

Despite all this assistance, errors of exposition and inelegancies of expression undoubtedly remain. These are our responsibility. Additional suggestions and comments for the improvement of this text will always be welcome.

This has been a unique experience as well as opportunity for the authors. Writing this book has been rewarding to us, and it is hoped that universities and colleges, particularly in the Middle East, will be able to use it in informing their students of the great natural resource and industry that is responsible, in most cases and to a great degree, for the forward movement of their economies.

Hussein K. Abdel-Aal

Robert Schmelzlee

Chapter 1

INTRODUCTION

PLAN OF THE BOOK

This book provides an overall picture of the importance of engineering economics to the petroleum industry. The outline of the book follows the operations of production, processing, and marketing, including some general aspects of the petroleum industry as they pertain to individual operations rather than to operations on a world-wide basis. The object of this study has been to prepare a fairly comprehensive report on the treatment of some oil engineering problems from an economic point of view, yet to encompass an easily understandable exposition of the oil industry and its associated procedures, from the oil fields to tankers and pipelines.

The book includes material on each area of the oil industry, mainly from the viewpoint of the Middle East, although some material is applicable to North America and the Caribbean as well. This is an introductory text, intended for students and lay readers who may be exposed to the oil industry for the first time. Since the distribution channels through which crude oil and oil products move form such a vast and complex network, the authors of this book have left the distributive network of wholesaling and retailing for another place. Therefore, the marketing discussion, as presented, go only as far as tanker and pipeline liftings.

Despite these limitations, the schematic presentation provides a general but accurate picture of the industry and some of the economics involved. Since this book was designed mainly as an introductory and explanatory exposition of oil operations particularly oil operations of the Middle East, the following outline is used.

Chapter 2 explains where world supplies of oil are located, and where oil demand is to be found. It is interesting to discover in this chapter that,

1

with the exception of the United States, the world's suppliers of oil are not
necessarily the world's demanders. In other words, supplies or reserves
of oil are at sites other than those of consumers of oil and oil products.
For example, the Middle East, North Africa, West Africa, Southeast Asia,
the Caribbean, and lately the North Sea are the areas where most oil reserves are found. The developed countries of Western Europe, Japan, the
United States, Canada, and Mexico are the principal consumers of oil. The
United States, formerly the leading supplier and user of oil, is in fact fast
becoming the largest buyer of oil and oil products, and what is more important is moving more toward the position of buyer rather than a supplier.
Using the current statistical review of the British Petroleum Company,
Chapter 2 also analyzes current statistics on reserves, production, consumption, and refinery capacity. Much interesting information is presented
here. For example, an Arab Oil Supplement, derived from the new annual
publication, The Arab Petroleum Directory, 1972 edition, is presented,
giving a short summary of contributions of all Arab countries relative to the
oil industry of the world in terms of reserves, production, consumption,
and refinery capacities.

Chapter 3 describes the various types of crude oil and the numerous
oil products derived from these crudes. This chapter also discusses the
various uses to which oil products have been put. In addition, oil demand
and supply are analyzed, as well as some of the basic economic considerations underlying prices of crude oil and oil products including their elasticities, or their sensitivities to demand and supply with changes in prices.

The aim of Chapters 4 and 5 is to present an abbreviated analytical approach to financial management: Some of the practical tools available to
oil management for guidance in making decisions involving investments are
presented. In Chapter 4, such important aspects of returns on investment
as the use of compound interest, capital cost, net cash flows, present values, payouts, etc., are offered. Furthermore, in Chapter 5 the subject of
amortization is considered. This involves the depreciation of oil equipment
and the depletion of oil resource properties as they are exhausted.

Chapter 6 discusses, in an economic and engineering context, some of
the aspects of an oil company's operation in the oil fields. Particular emphasis is placed on the determination of reserves, well drilling, and spacing, and the factors involved in efficiency of drilling, such as selection of
drilling bits, penetration rates of drilling, etc.

Economies of scale are also included in Chapter 6. In the production of
crude oil, for instance, the emphasis is on large-scale operations, as in the
Middle East, where the big integrated oil companies assume risks that would
be beyond the capacity and resources of small firms. These large companies
are able to apply more advanced technological methods and research to exploration and development, and so are able to export large amounts of crude
resources and oil products more efficiently.

Some economic concepts related to pipelines, including economies of scale, are also discussed in Chapter 6. For instance, an increase in the diameter of pipe will usually create a much larger proportionate increase in volume, with a consequent sharp fall in cost per barrel of throughput. For example, in 1960 it was determined by a number of U.S. companies that a 1,000-mile, 10-inch pipeline with a throughput of 45,000 barrels of crude per day would have to charge at least 37¢ per barrel to obtain a 7% return on investment. But a larger-diameter pipeline would bring a sharp fall in cost per barrel throughput, and thus a 7% return on investment could be obtained at a much lower charge per barrel.

From the oil field in Chapter 6, we move into the refinery in Chapter 7. In this chapter we consider such economic and engineering concepts as economic yield and recovery, economics of design in a minimum sense, as this can be a volume in itself, costs and refinery margins, etc. The principles discussed are also given some quantitative content, particularly with reference to economics of balance in yield. Problems are used liberally to illustrate the economics presented.

There are economies of scale in the refinery as well as in the oil field. Investment per unit of capacity falls sharply with increase in scale, up to a fairly high level. For instance, even in 1950, it was estimated that investment per barrel of capacity in the United States fell from $1,300 for a 10,000 bbl/day refinery to $833 for a 30,000 bbl/day refinery, and to $545 for a 200,000 bbl/day refinery. This is also true today, mainly because total costs of operation per barrel of refined oil fall sharply with a rise in scale because of a drop in fixed unit costs.

Having produced the oil products from the crude oil, we next consider marketing aspects in Chapters 8 and 9. Since some crudes are not processed in the Middle East, distribution is discussed from the standpoint of both crude oil and refined oil products.

Marketing, or the distribution stage, is discussed in two parts: Chapter 8 discusses prices and price making, and Chapter 9 covers the transporting of oil and oil products by sea tankers and pipeline carriers. The exposition on price and price making is based mostly on Middle East operations. Monopoly is assumed, at the crude oil production level, thus prices of oil products to the ultimate consumer are not as dependent on cost as they are on value of oil products to the consumer. Costs of producing oil in the Middle East oil fields are extremely low relative to prices for crudes and oil products, basically because of the enormous capacity of the oil fields in the Middle East. With such immense oil fields and such huge levels of production per day, economies of scale are easily achieved in Middle East fields and refineries.

In Chapter 9, a thorough discussion of the two main carriers in the movement of crude oil and oil products, sea tankers and pipelines, is given.

The principle of economies of scale is especially apparent in transportation by its application to the use of increasingly larger tankers and larger diameter pipelines. Actually, the larger tankers are less expensive to build and to operate on a per-ton basis, because building costs per ton decrease with size, and power requirements and fuel consumption increase less than proportionately as size increase. Also, operating crews on tankers are not generally increased to any appreciable extent as the dwt (deadweight ton) size of the tanker increases. Crews are generally increased only as more maintenance is needed.

Since the text follows largely a Middle East point of view, only movement of oil and oil products by sea and pipeline, as found principally in the Middle East, are discussed. Further distribution of oil and oil products after the tanker and pipeline stages are not discussed in this book. This will be reserved for another text by these authors. But rates and rate structure in the tanker market are discussed in detail, in order to show how they affect costs of movement of oil and oil products.

Chapter 10 is a summary chapter in which conclusions are made about the oil industry. This chapter, thus, summarizes this introductory study of the oil industry by an analysis of costs and outputs, again referring specifically to the Middle East.

SUMMARY AND CONCLUSIONS

The reader will find that the petroleum industry requires large amounts of initial investments because of (1) the economies of scale involved once production starts, (2) the capital-intensive nature of the oil industry, and (3) the uncertainties encountered in oil exploration, where tremendous amounts of capital must be invested before discovery and then, in addition, after discovery and before production can start.

Generally, fixed costs, rather than running or variable costs, account for a large proportion of total costs in the petroleum industry. This is due to the capital-intensive nature of the industry in general, because of the high initial investment, and because of economies of scale in the oil fields, refineries, pipelines, sea tankers, etc., all of which dictate the use of fixed equipment rather than labor, which is usually regarded as a variable cost.

Since variable costs, which basically include only materials (crude oils) because all refining processes are low on labor and high on mechanization, are utilized directly in the refining of products and constitute the only changing portion of total costs, a high ratio of fixed costs to total costs becomes extremely meaningful to the oil industry. For instance, with large fixed investments or costs to total costs in the refineries (and also in the oil fields in the exploration and development stages) spread over a high amount

of throughput, lower costs of refining per barrel result. The Middle East
is noted for its lower costs per barrel of production, due mainly to its large
reserves and high production levels even in the face of high exploration and
development costs and the high costs of building and outfitting refineries.
With lower costs per barrel, the Middle East has developed a competitive
advantage over other oil-producing areas of the world.

The reader will also find that the additional cost incurred by a small
increase in output, as barrels produced in the oil fields or gallons and tons
produced in the refineries, is very low. Thus, the short-term average cost
curve falls rather sharply; also, the marginal cost curve drops to a very
low level after a certain volume of output is reached. In the Middle East,
output has continued to increase as unit average costs and marginal costs
are kept under control due to improved techniques, even though additional
capacity is added. Some oilmen say that some developed fields in the Middle East are so large that very great increases in output are still possible
at constant costs, or even at slightly falling marginal costs.

The Economics Department in the Ministry of Petroleum and Mineral
Resources of the Kingdom of Saudi Arabia, in its Petroleum Statistical
Bulletin on world petroleum statistics, presented some interesting figures in
in 1967 on production in average barrels per well per day. These are given,
in abbreviated form, in Table 1.1.

TABLE 1.1

Production of Crude Oil in 1967

Country	Reserves (bbl)		Number of producing wells	Total oil production (bbl/day)	Average barrels per day per well
Abu Dhabi	15	billion	54	382,500	7,090
Iran	43.8	billion	163	2,575,000	15,800
Iraq	23.5	billion	118	1,229,500	10,420
Kuwait	70	billion	503	2,276.700	4,525
Qatar	3.75	billion	63	323,000	5,125
Saudi Arabia	77	billion	362	2,597,600	7,200
Indonesia	9	billion	2,244	504,800	225
Venezuela	17	billion	10,547	3,537,000	300
Total Western Hemisphere	73.24	billion	748,078	14,629,000	20

Source: Petroleum Statistical Bulletin, 1968, Department of Economics,
Ministry of Petroleum and Natural Resources, Kingdom of Saudi Arabia.

Figures are not given for capital investment; it is obvious, however, that with more wells being necessary to maximize production, capital investment will also increase. The Middle East, predominantly Abu Dhabi, Saudi Arabia, Iran, Iraq, and Qatar, as well as Kuwait, indicated a high average production per well per day in 1967. This has also increased notably today. Undoubtedly, with fewer wells and higher average barrels per day per well in the Middle East, return on investment and payback are exceedingly more favorable to oil investors as compared to oil investments in Indonesia, Venezuela, and other Western Hemisphere areas, such as the United States and Canada.

However, the oil industry is an inherently unstable one, since it cannot operate under competitive conditions. It is monopolistically competitive in nature because results of exploration for oil for uncertain, overhead costs at all stages of production are high, demand for oil and oil products is highly inelastic in the short run, and the industry is not self-adjusting since a fall in prices may significantly choke off supplies or stimulate demand for crude and refined oil products.

Because crude oil production, transportation, refining, storage, and distribution all require large amounts of capital, heavy fixed charges result. This forces competitive producers to produce as much as they can as quickly as possible in order to recoup their "sunk costs," and to cover at least part of their overhead. With this kind of operation, the oil industry is subject to continuous crises in the absence of reasonably strong control over supply. Because of inelastic demand or the fact that consumption is not readily expanded when prices of oil products fall, prices can be pushed below costs of production by relatively small surpluses. To plug these bottlenecks, the large companies and oil-producing nations have attempted to provide the control needed of supply and demand in order to preserve prices of oil products and of crude oils.

Prices are determined largely by costs of developing further production from the oil fields and not by the low operating costs of developed capacity. Also, the cost of developing discovered reserves is a prime factor in preventing unprofitable flooding of markets with oil supplies. The reader will become aware of refinery margins, or the differences between the costs of barrels of crude oil input and receipts from the refined products of the crudes, or output. The "product mix" is the key to total profit and return on investment in a refinery, since each refined product produced has a different margin of profit, with some products having a bigger margin than others. The product mix in the refineries of the Middle East usually has a lower proportion of the more valuable products due to the nature of demand for their refined products, the less complex types of refineries in the Middle East regions, the lower proportion of catalytic or reforming plants, and the relatively low price of crude oil.

Refinery costs in the Middle East are relatively high because all refinery equipment, as well as spare parts, is imported, and Middle East refineries carry greater costs for economic and social overheads. On the other hand, Middle East refineries are much larger in capacity than the world average, so their capital costs per barrel, and operating costs per barrel or ton, are reduced substantially. Furthermore, Middle East refineries are less complex than those in other regions. This has the effect of reducing the value of their output but also of cutting capital and operating costs.

One conclusion that may be drawn from the presentation to follow is that in the future the capital structure of the industry will change in the direction of an even higher ratio of capital to operating costs, as prices continue to rise, oil becomes harder to find, and demand increases at a rate exceeding the present rate of consumption.

Current and prospective crude prices are and will be sufficiently high in Western Europe and the United States to support high-cost production close to market; but if exploration and development costs rise sharply at the margin, a substantially higher level of prices will be needed. It will then be difficult to reconcile such a higher level of prices with the continuance of moderate-cost oil from the Middle East.

Chapter 2

WORLD OIL: SUPPLY AND DEMAND

INTRODUCTION

The pattern of oil consumption and production has changed remarkably since
World War II, when petroleum became important in the tremendous expansion and rebuilding of industry, particularly in the Eastern Hemisphere.
The United States is the largest user of petroleum and has retained its position as the largest single producer, but it is not the largest producer when
the Middle East is considered collectively either with or without North Africa
as a supply region. Oil is not concentrated in any one part of the world, but
is found in many places including North America, the Caribbean area (including Venezuela, Colombia, and Trinidad-Tobago), the Middle East, North and
West Africa, Southeast Asia, and now even in Alaska and in the North Sea
region of Europe.

Outstanding in postwar growth of the world oil industry has been the
regional shift in the pattern of oil consumption and supply. Demand has
grown, especially Western Europe and Japan, for more gasoline for automobile travel, more electric power, more heating fuels, and more light.
Those countries and areas with firm industrial bases and a supply of technically trained workers have rapidly expanded oil product consumption.

On the other hand, although the rate of oil consumption in parts of the
world other than Western Europe and Japan, such as the Middle East and
Asia, rose much more rapidly than that of the United States, the relative
proportion of world oil production which they absorbed did not change materially. All areas outside North America and Western Europe and, to a
limited degree, Eastern Europe, fall into this group, since they entered the
postwar era with little industry, few motor vehicles and inadequate roads,
and populations with low incomes living near subsistence levels.

WHERE THE CRUDE OIL IS (RESERVES)

World "published proven" oil reserves at the end of 1973 were 634.7 billion bbl, or 38.0 billion bbl less than the 672.7 billion bbl reported at the end of 1972. The United States had 41.8 billion bbl, or 6.3% of the world total, while the Caribbean area, mostly Venezuela, had 17.6 billion bbl, representing 2.9% of the world total. Canada and other unidentified Western Hemisphere areas had 23.3 billion bbl, making the Western Hemisphere's share of the world's oil 82.7 billion bbl. Table 2.1 summarizes the distribution of these oil reserves, and Figure 2.1 shows the location of all oil discovered between 1859 and 1973.

With total reserves in the Eastern Hemisphere of 552.0 billion bbl, or 87.2% of the world's total, the Middle East claimed 349.7 billion bbl, or 63.4% of the Eastern Hemisphere's total. Actually, the Middle East's share represented 55.4% of the world's total. The USSR-Eastern Europe-China complex, the accuracy of whose figures may be questionable, were reported to have reserves of 103.0 billion bbl, or 16.3% of the world's total. North and West Africa in the Eastern Hemisphere had reserves of 67.3 billion bbl, or 10.4% of the world's total. Other Eastern Hemisphere countries accounted for 15.6 billion bbl, or 2.5%, making the Eastern Hemisphere's total reserves 552.0 billion bbl or 87.2% of the world's reserves.

Subtracting the USSR-Eastern Europe-China complex, total "Free World" reserves at the end of 1973 were estimated at 531.7 billion bbl, which represents 83.7% of the world's total. The remainder of 103.0 billion bbl, 16.3%, is thus the Eastern bloc's total reserves.

Proven Reserves and Price

The market price of crude oil affects the amount of proven reserves that is discovered, because more effort is made to find reserves when the price of crude is up. When the market price of crude rises, more crude can and will possibly be recovered. The increased price makes new exploration and new recovery techniques feasible for the oil companies. And the converse is also true: As the price of crude goes down, exploration and recovery costs make it unprofitable to explore and to develop, hence the quantity of proven reserves is more apt to diminish. This is, basically, the economic concept of the workings of supply and demand. Demand is reflected by price; as long as it is profitable, the industry will do anything feasible to increase supply.

A high price for crude oil may be necessary if the financial requirements facing the petroleum industry of $1.3 trillion for the period of 1970-1985 are to be met [1]. The largest share of these capital requirements and

TABLE 2.1

World Published Proven Oil Reserves at the End of 1973

Country/Area	Billions of barrels	Percent share of total
United States	41.8	6.3
Other Western Hemisphere countries, excluding the Caribbean	23.3	3.6
Venezuela and other Caribbean countries	17.6	2.9
Western Hemisphere total	82.7	12.8
Western Europe	16.4	2.6
African—North and West	67.3	10.4
Middle East	349.7	55.4
USSR, Eastern Europe, China	103.0	16.3
Other Eastern Hemisphere countries	15.6	2.5
Eastern Hemisphere total	552.0	87.2
World total	634.7	100.0
Less USSR, Eastern Europe, China	103.0	16.3
"Free World" total	531.7	83.7

Source: Statistical Review, 1974 (for 1973), British Petroleum Company, London.

exploration expenditures, $810 billion, must come from earnings [2]. John Winger, a leading financial observer of the industry, recently stated:

> For the past decade and a half, the net income of the petroleum industry has been too small relative to its capital needs. And, as a consequence of the underinvestment during that time, the discovery of new petroleum reserves throughout the non-Communist world has not kept pace with expanding market requirements. In no sense can the shortfall of new discoveries be construed as reflecting a lack of petroleum to be found. There were far more opportunities throughout the world to employ capital in the search for petroleum than the available funds could possibly accommodate. And there still are [3].

Quantity of proven reserves can be increased through better use of existing reserves or through the addition of new reserves. Also, rising prices for petroleum should make secondary and tertiary recovery activities more

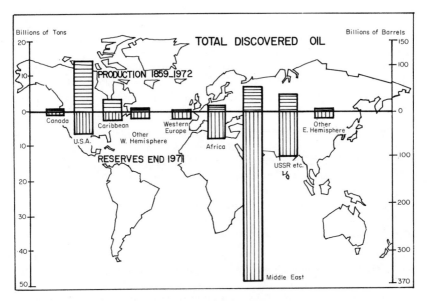

FIG. 2.1. Total oil discovered between 1859 and 1973.

attractive from an economic standpoint. This should serve measurably to
stimulate increases in production on the part of domestic industry. "This
is because numerous old, previously uneconomical fields will be revived.
At old price levels, only about 30% of the oil in place was economically recoverable. At today's market prices, recoverable oil should be raised by
10% at least" [4].

The increase of market price for crude should also stimulate new technology for both exploration and development. Currently, the costs of exploration and development are being eased by the employment of new methods and
more sophisticated facilities for selecting drilling locations and testing wells
being drilled. However, 90% of exploratory wells still turn out to be dry
holes [5]. New marine equipment, such as submersible drilling rigs, will
make drilling in offshore areas increasingly accessible. Improvements in
recovery techniques would make it possible to produce increasing proportions of petroleum in place in underground reservoirs. A doubling of world
reserves on this basis would appear entirely feasible [6]. New technology
could increase proven reserves through the development of completely new
sources of petroleum. High market prices of crude may make economical
the use of oil shale, tar sands, and coal to produce petroleum and petroleum
products.

Exploration for new reserves is increased with increases in market
prices. Areas that were formerly considered uneconomical will possibly

be probed for the first time now. Exploration in the more difficult and costly offshore areas may be possible to an increasing extent with advancing crude oil prices. For example, the cost of North Sea wells, believed to average about $4 million, including all support services, would indicate annual expenditures in the area of $750 to $800 million. Shell Oil Company estimates the cost of the new generation of drilling rigs at $25 to $30 million each, and operating costs at about $32,500/day [7].

The primary limiting factor in the relationship between market price and quantity of proven reserves is government intervention. Intervention may be in the form of price control or in the limiting of exploration or production. Exploration is limited primarily by the use of leases specifying when and where exploration may be undertaken. Production may be limited artificially by quotas or environmental regulations. Another factor limiting increases in proven reserves has been materials shortages, that is, shortages of casings, rigs, drilling crews, etc.

Thus the market price of crude oil has a great influence on the quantity of the world's proven oil reserves. Increases in market prices make it more feasible economically, and encourage more efficient methods of exploration and production. This in turn increases the quantitites of proven reserves.

Additional Probable Reserves

The search for new reserves of petroleum appears to be gathering momentum every day. Higher prices for crudes, increasing worldwide demand for crudes, and the concentration of crude reserves by the Organization of Petroleum Exporting Countries (OPEC) are major factors behind the world's continuous search for new oil reserves.

The majority of new "field finds" in the future will be offshore finds such as the big discovery in the North Sea. But offshore locations complicate the problem because of many unresolved questions, such as the ownership of the petroleum, how much petroleum there may be, how to get to the petroleum, and how to get the petroleum out of the water when we do find it.

Most offshore oil comes from waters less than 100 ft deep. So far, the deepest development well with subsea production facilities lies in 374 ft of water, but submarine platforms are in the planning stage for depths below 984 ft [8]. Maximum drilling depths at this time are 1,640 and 1,804 ft. But soon, wells are expected to be drilled in waters deeper than 3,231 ft, or beyond the continental shelf. The industry is already studying technical and economic problems involved in working on the continental rise in waters up to 9,843 ft deep [9].

Offshore drilling first began in 1948. According to a report prepared for the United Nations, more than 100 billion bbl of oil have already been discovered below the seas [10]. Currently, recoverable offshore reserves and offshore oil rigs can be found all over the world (Figs. 2.2 and 2.3). "At current levels, the offshore is yielding 19% of total world output, but in the years ahead, probably the late 1980s, it is expected to provide at least half of the world's production" [11].

North America. The greatest potential for reserves in the United States is in offshore areas and in the Arctic. The most important offshore developments should be in the Northern Gulf of Mexico off Texas, Alabama, Mississippi, and northwest Florida. Other offshore areas that show good potential are offshore eastern United States, California, and Bristol Bay, Alaska [12].

Onshore potential U.S. reserves are in southern Alabama, northwestern Florida, western Texas, western Oklahoma, western Pennsylvania, northern Michigan, central Illinois, northeastern Utah, southwestern Wyoming, and western North Dakota [13].

The National Petroleum Council Energy Study estimates ultimate discoverable oil in place for the United States at about 800 billion bbl. Of this total, only about 50% has been discovered to date, leaving some 400 billion bbl yet to be discovered. Over half of the new discoveries are expected to be found in Alaska and offshore areas [14].

In Canada, concern about government oil policies has reduced exploration. However, potential exists off the shores of Nova Scotia and Labrador. Onshore possibilities lie in the Northwest Territory, specifically the Mackenzie Valley and Delta [15].

The potential for self-sufficiency of petroleum needs for Mexico exists in potential fields in southern Mexico [16].

South America. The greatest possibilities for major new reserves in South America lie in the offshore areas of Trinidad. Other potential offshore areas are Brazil and Peru [17].

Europe. The major potential in Europe is the North Sea between the British Isles, Norway, and Denmark. According to Peter Odell, Director of the Economic Geography Institute in Rotterdam, full exploitation of the North Sea can provide Western Europe with more than half of its total requirements by 1980 [18].

The Mediterranean, the Adriatic, and the Aegean Seas also hold some offshore potentials. The main interest is in the area between Greece and Turkey [19]. Currently, the situation is severely complicated because of the Cyprus political problem and the existing boundary conflicts between these two long-time enemies.

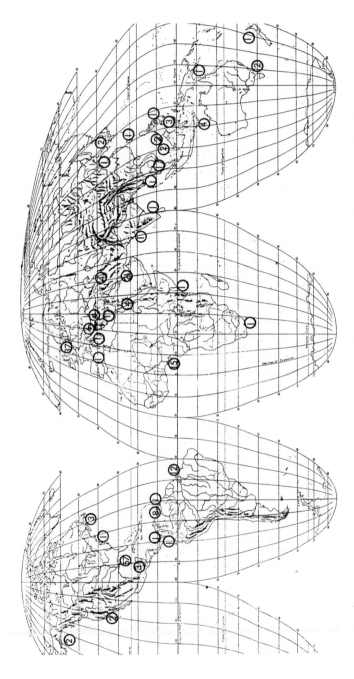

FIG. 2.2.   Where the offshore oil rigs are—worldwide.   Circled numbers indicate the number of oil rigs at each location.   [From Gardner, Frank, "Offshore Oil—Only the Beginning," Oil and Gas Journal, vol. 72, no. 8, p. 124 (1974).]

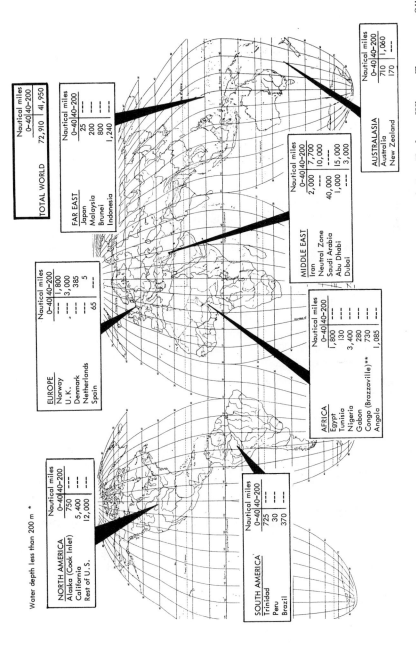

FIG. 2.3. Offshore potentials—worldwide. [From Gardner, Frank, "North Sea Today: Where Tomorrow?" Oil and Gas Journal, vol. 71, no. 50, p. 77 (1973).]

Water depth less than 200 m *

TOTAL WORLD	Nautical miles	
	0-40	40-200
	72,910	41,950

FAR EAST	Nautical miles	
	0-40	40-200
Japan	25	---
Malaysia	200	---
Brunei	800	---
Indonesia	1,240	---

AUSTRALASIA	Nautical miles	
	0-40	40-200
Australia	710	1,060
New Zealand	170	---

MIDDLE EAST	Nautical miles	
	0-40	40-200
Iran	2,000	7,700
Neutral Zone	---	10,000
Saudi Arabia	40,000	---
Abu Dhabi	1,000	15,000
Dubai	---	3,000

EUROPE	Nautical miles	
	0-40	40-200
Norway	---	1,800
U.K.	-	3,000
Denmark	---	385
Netherlands	---	5
Spain	65	---

AFRICA	Nautical miles	
	0-40	40-200
Egypt	1,800	---
Tunisia	130	---
Nigeria	3,400	---
Gabon	280	---
Congo (Brazzaville)**	730	---
Angola	1,085	---

NORTH AMERICA	Nautical miles	
	0-40	40-200
Alaska (Cook Inlet)	750	---
California	5,400	---
Rest of U.S.	12,000	---

SOUTH AMERICA	Nautical miles	
	0-40	40-200
Trinidad	725	---
Peru	30	---
Brazil	370	---

Africa. Offshore Nigeria holds the spotlight as far as oil potential in Africa is concerned. Egypt, Angola, and Gabon are also areas with offshore oil potentials [20].

Middle East. Over half of potential offshore oil in the Middle East is within 40 nautical miles of the shores of Saudi Arabia. Potential further offshore has been indicated for the Neutral Zone (above Saudi Arabia and next to Kuwait) and Abu Dhabi of the United Arab Emirate States [21].

Far East. Southeast Asia has the greatest crude oil potential of any area in the Far East. This potential is offshore Indonesia, offshore Malaysia, the Gulf of Thailand area, and offshore South Vietnam. At this writing three U.S. firms are involved in bidding for offshore South Vietnam search rights. The firms involved are Exxon, Shell, and Mobil. The prospects are good for discovery of important quantities of low-sulfur oil. However, the overriding problem of political stability and border disputes in these areas, particularly in Vietnam, still exists [22]. Vietnamese oil would of course be non-Arab oil. It would be close to the Japanese market and to Singapore, which is expected to become the world's third largest refining center by 1975.

Conclusion. The world demand situation requires that new sources of energy be developed. Until that is done, however, most of our energy must come from petroleum. This requires the exploration and development of new petroleum reserves. The greatest potential appears to be in offshore regions. Today we finally have the capabilities and technology to find oil offshore.

WHERE THE CRUDE OIL IS PRODUCED

Over the past 100 years, the United States has led in the production of crude oil, producing about 16 billion tons; by the end of 1973, its reserves were still estimated to be around 5.7 billion tons (43 billion bbl). In contrast, production of oil for the past 100 years in the Middle East, regarded as "the infant producer," was roughly 50 billion bbl, along with estimated reserves in the Middle East at the end of 1973 of close to 360 billion bbl. The Eastern Hemisphere, particularly the Middle East, has led the Western Hemisphere in reserve discoveries and production by an amount that has more than doubled.

Total world oil production in 1973 was 57,710,000 bbl/day (21,064,150,000 bbl/year). This represented an 8.3% increase over 1972's total of 52,925,000 bbl. The Eastern Hemisphere produced 68.5% of this total in 1973, a 12.2% increase over 1972, with a daily total of 39,360,000 bbl. At the same time, the Western Hemisphere produced 18,350,000 bbl/day, with its 31.5% of world total, but gained only 2.0% when compared to 1972. Average annual percent increases in daily production for the years 1968 through 1973 indicate that the Eastern Hemisphere increased +10.9%, whereas the Western Hemisphere lagged far behind with a +1.5% increase. Table 2.2 summarizes these results.

It is obvious that the Eastern Hemisphere is experiencing a rapid rise in oil production that will outdistance the Western Hemisphere in years to come. The Western Hemisphere appears to be settling down to a modest growth rate in building reserves. Its reserves are clearly dwarfed by the Eastern Hemisphere's total reserves.

WHERE THE CRUDE OIL IS CONSUMED

In 1973, world daily consumption of crude was 56,425,000 bbl (20,595,125,000) bbl/year). This is quite comparable to world daily production of 57,710,000 bbl. Consumption in 1973 represented an 8.3% increase over 1972 on a world basis (Table 2.3).

The Eastern Hemisphere had 61.4% of the daily crude oil consumption, or 34,365,000 bbl, while total consumption in the Western Hemisphere was 22,060,000 bbl/day or 38.6% of world total consumption.

The United States was the largest individual nation consumer of crude oil, with consumption of 16,815,000 bbl/day. This amount represented 29.5% of the 1973 world total. Western Europe, including both the EEC countries and other Western European nations, consumed 15,155,000 bbl/day, or 27.0% of total world consumption. Actually, the United States, Japan, and Western Europe together accounted for 66.2% of total world consumption in 1973. The USSR, Eastern Europe, and China together, with

TABLE 2.2

World Crude Oil Production, 1973

Country/area	1973 (thousand) bbl/day)	Share of total, 1973 (%)	Increase over 1972 (%)	Average Annual increase, 1968–1973 (%)
World total	57,710	100	+8.7	+7.4
Eastern Hemisphere total	39,360	68.5	+12.2	+10.9
Western Hemisphere total	18,350	31.5	+2.0	+1.5

Source: Statistical Review, 1974 (for 1973), British Petroleum Company, London.

TABLE 2.3

World Crude Oil Consumption by Area, 1973

Country/area	(thousand bbl/day)	Share of world trade, 1973 (%)	Increase over 1972 (%)	Average annual increase, 1968–1973 (%)
World total	56,425	100	+7.3	+8.0
Eastern Hemisphere	34,365	61.4	+8.6	+9.5
Western Hemisphere	22,060	38.6	+5.3	+5.8
Some major countries and areas with high consumption figures				
United States	16,815	29.5	+5.0	+5.1
USSR, Eastern Europe, China	8,775	15.7	+10.2	+8.9
Japan	5,425	9.7	+14.0	+13.4
Western Europe	15,155	27.0	+6.7	+8.2

Source: Statistical Review, 1974 (for 1973), British Petroleum Company, London.

8,775,000 bbl/day (15.7% of the world total), followed by Japan with 5,425,000 barrels of crude per day (9.7% of the world total), complete the list of "biggest" consumers.

The largest percent gains in crude oil consumption in 1973 over the previous year were seen in Spain, with a 20.0% gain, followed by the USSR-Eastern Europe-China complex and France, with a reported 10.2% increase for each. Not far behind was Japan, with a 14% increase. All four country/areas met or surpassed the average increase of 7.3% for the world. The total percent increase of Western Hemisphere consumption was rather far behind the consumption increase of the Eastern Hemisphere, with 5.3% to 8.6% in 1973 over 1972. The four areas of Spain, Japan, France, and the Eastern bloc, however, individually increased crude oil consumption more rapidly over the 5-year period 1968-1973, with increases of 14.9%, 13.4%, 11.9%, and 8.9%, respectively, than the United States showed with its average annual increase of 5.1% for the period 1968-1973 (Table 2.3).

Table 2.4 gives the figures for the years 1957-1970, showing demand and percent of increase of demand for crude and liquefied natural gas (LNG) by the United States and consuming areas outside the United States. As this

TABLE 2.4

Growth of Oil Use Within and Outside of the United States
(thousand bbl/day)

Year	United States Demand	United States percent up	Outside United States Demand	Outside United States percent up	World Demand	World percent up
1957	8,797	0.2	9,410	7.6	18,207	3.9
1958	9,065	3.0	10,155	7.9	19,220	5.6
1959	9,451	4.3	11,015	8.5	20,466	6.5
1960	9,661	2.2	12,332	12.0	21,993	7.5
1961	9,806	1.8	13,533	9.7	23,339	6.1
1962	10,235	4.4	15,067	11.3	25,302	8.4
1963	10,551	3.1	16,652	10.5	27,203	7.5
1964	10,816	2.5	18,512	11.2	29,328	7.8
1965	11,304	4.5	20,392	10.2	31,696	8.1
1966	11,831	4.6	22,153	8.6	33,984	7.2
1967	122,273	3.7	23,572	9.5	35,845	7.5
1968	13,050	6.3	26,171	11.0	39,221	6.3
1969	14,137	8.3	29,098	11.2	43,235	10.2
1970	14,716	4.1	32,169	10.5	46,885	8.4

Note: Includes crude and liquefied natural gas.
Source: World Oil, August 15, 1971.

table illustrates, U.S. demand for crudes and LNG is up, but there is an
even faster rate of increase in demand in countries outside the United States,
notably the many industrialized countries in Western Europe and Japan. In
most of the years the ratio of increase in demand by "outside-the-U.S."
countries has been almost 2 to 1. In some years, the ratio has been more
than 2 to 1.

Figure 2.4 shows trends in world oil consumption by major consuming
areas from 1950 through 1970. In order of consumption, the major consumers
were as follows:

1. United States

2. Western Europe

3. Rest of World

4. USSR-Eastern Europe-China

5. Japan

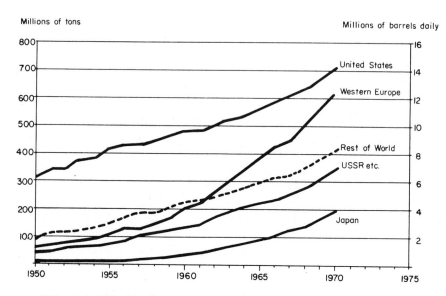

Millions of tons

Millions of barrels daily

FIG. 2.4. World oil consumption. [From Statistical Review, 1974 (for 1973), British Petroleum Company, London.]

All five show upward consumption curves, indicating a steady increase in oil consumption. Western Europe had the most abrupt turn upward, beginning in 1961 as it vaulted into second position ahead of "Rest of World," and then continued to rise rapidly.

Figure 2.5 illustrates both oil production and consumption in 1973 by areas around the world. If we think of consumption as a hat and production as a face on a head, it is obvious from Figure 2.5 that consumption does not fit production in most areas of the world: Either the hat is too large for the face, or the face is too large for the hat. Only in the United States does consumption come close to matching production, and even there, as indicated in Figure 2.5, consumption is rising beyond production. Western Europe is low on production and heavy on consumption. This situation may soon change, however, since new reserves have been discovered and are being developed in this area, particularly in the North Sea.

Japan's position is particularly unusual. Consumption completely dominates the hat/face concept of consumption/production for Japan. Japan must rely entirely on outside sources for its consumption requirements; it has no oil reserves. The Middle East is the biggest source of Japan's oil, followed closely by Southeast Asia.

As Figure 2.5 also shows, the Middle East, North and West Africa, and the Caribbean are heavily one-sided in favor of production over consumption.

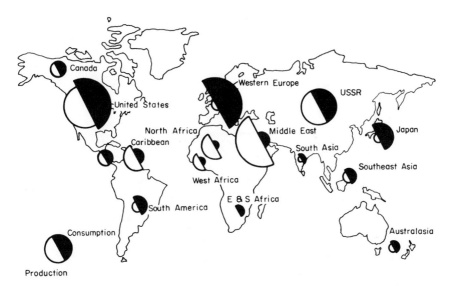

FIG. 2.5. World oil production and consumption, 1973. [From Statistical Review, 1974 (for 1973), British Petroleum Company, London.]

PRODUCTION OF REFINED OIL PRODUCTS

As of 1973, refinery capacity was located mainly in the developed consuming countries, excepting the United States which is both a consuming and producing country. Western Europe had 28.8% of total capacity, followed closely by the United States with 22.1%, then the USSR-Eastern Europe-China complex with 16.1%. Figure 2.6 gives a breakdown of location and capacity of refineries around the world. The Middle East is conspicuously last, with only 4.0% of total world refinery capacity; sales to developed consuming countries are largely crudes.

The Middle East, easily the world's largest producer of crudes, is low in refinery capacity because refineries, operated mostly by large, integrated oil companies, have been located on the basis of market orientation rather than raw material orientation. This means that the refineries have been situated close to the market for refined oil products rather than near the oil fields. Today, with oil producing countries participating in oil ownership and assuming more "direct say" in the management and distribution of oil and oil products, some increase in refinery capacity in the Middle East is a probability for the future.

Total world oil refining capacity in 1973 was 64,510,000 bbl, or 22,766,150,000 bbl/year (Table 2.5). The Eastern Hemisphere produced 41,420,000 bbl/day, or 64.2% of the world total. Western European

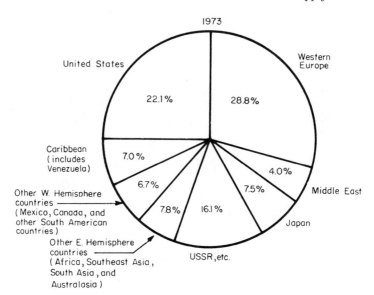

FIG. 2.6. Location of refinery capacity, 1973. [From Statistical Review, 1974 (for 1973), British Petroleum Company, London.]

refineries in the Eastern Hemisphere accounted for almost half of the total of 41,420,000 bbl/day, or 18,615,000 bbl.

At the same time in 1973, the Western Hemisphere produced a total of 23,090,000 bbl of refined oil products, or 35.8% of world refinery capacity. The United States, of course, led all others in refinery processing with a total of 14,250,000 bbl/day, or roughly 62% of the refinery capacity attributed to the Western Hemisphere in 1973.

Table 2.5 also indicates refinery capacity in terms of annual average in barrels per day from 1968 through 1973. Figures show that the Eastern Hemisphere increased in annual average barrels per day by 8.9% in 1973 over 1972 and by 9.6% from 1968 through 1973 as against the Western Hemisphere's increase of only 5.6% in 1973 over 1972, and 5.5% for 1968 through 1973.

Of the crude oil refined in Western Europe in 1973, which accounted for 28.8% of total world refinery capacity, Italy ranked first with almost 3,830,000 bbl of crude refined per day. France was second with 3,160,000 bbl/day, followed closely by West Germany, England, and the Benelux countries in that order.

A "pattern" of demand for refined oil products processed by refineries in 1973 is given in Table 2.6.

TABLE 2.5

World Oil Refining Capacities at the End of 1973

Country/area	1973 (thousand bbl/day)	Share of total, 1973 (%)	Increase over 1972 (%)	Average annual increase, 1968-1973 (%)
World total	64,150	100	+7.7	+8.0
Eastern Hemisphere total	41,420	64.2	+8.9	+9.6
Western Hemisphere total	23,090	35.8	+5.6	+5.5
United States	14,250	62% of Western Hemisphere's 35.8% total share of world (or 23,090,000)		

Source: Statistical Review, 1974 (for 1973), British Petroleum Company, London.

TABLE 2.6

Pattern of Demand for Refined Oil Products, 1973

Country/area	First choice	Second choice	Third choice
United States	Gasolines	Fuel oils	Miscellaneous products, including diesel oils
Western Europe	Fuel oils	Gasolines/ diesel oils	Gasolines
Japan	Fuel oils	Gasolines	Gasolines/ diesel oils

Source: Statistical Review, 1974 (for 1973), British Petroleum Company, London.

As shown in Table 2.6, the United States leads the world in demand for gasolines, and its demand is increasing sharply each year. U.S. refineries, therefore, are geared for gasoline processing. Refinery yields from crude in the United States in percent by weight over the last 7 years are given in Table 2.7.

Western Europe requires more fuel oils and middle distillates, so Western European refineries are geared more to these products. Refinery

yields on crude in Western Europe in percent by weight over the past 7 years
are given in Table 2.8.

SUPPLY AND DEMAND

Table 2.9 shows the oil needs of various countries and geographic areas as
percentages of their total energy requirements. The USSR and the Eastern
bloc countries are least dependent on oil for energy or power purposes,
with oil fulfilling only 25% of their total energy needs in million ton equivalents;

TABLE 2.7

Yields from Crudes in U.S. Refineries
in Percent by Weight, 1967-1973

Oil products	1973	1972	1971	1970	1969	1968	1967
Gasolines	42	43	43	42	42	42	42
Middle distillates	29	29	28	29	29	29	29
Fuel oils	7	6	6	6	6	7	7
Other products, including refinery fuel and loss	22	22	23	23	23	22	22

Source: Statistical Review, 1974 (for 1973), British Petroleum Company,
London.

TABLE 2.8

Yields from Crudes in Western European Refineries
in Percent by Weight, 1967-1973

Oil products	1973	1972	1971	1970	1969	1968	1967
Gasoline	17	17	17	17	18	18	18
Middle distillates	33	33	32	31	30	31	30
Fuel oils	37	37	37	38	38	37	38
Other products, including refinery fuel and loss	13	13	14	14	14	14	14

Source: Statistical Review, 1974 (for 1973), British Petroleum Company,
London.

TABLE 2.9

Percentage of Oil to Total Energy (Power) Needs of Leading
Western Countries and the Eastern Block in 1973
(in million tons of oil equivalent)

Country/area	Total energy needs	Oil needs	Oil as % of total energy needs
United States	1726.2	814.7	47.2
Canada	162.1	83.8	51.7
Total Western Hemisphere	2128.4	1068.3	50.2
Belgium and Luxembourg	52.3	32.5	62.1
Netherlands	76.2	41.3	54.2
France	173.4	125.7	72.5
West Germany	255.2	149.5	59.1
Italy	132.8	104.5	78.7
England	216.3	112.6	52.1
Scandinavia	72.3	56.0	77.4
Spain	51.0	35.7	70.0
Other Western European countries	141.0	89.9	63.7
Total Western Europe	1170.5	747.7	63.9
Japan	332.2	267.2	80.4
Australasia	63.4	32.2	50.8
USSR, Eastern Europe, China	1639.3	434.0	26.5
Total Eastern Hemisphere	3576.9	1697.6	49.8
Total world	5705.3	2765.9	46.7

Source: Statistical Review, 1974 (for 1973), British Petroleum Company, London.

U.S. energy requirements, relative to all the "Western" areas of the world, is least dependent on oil as a power source, with oil fulfilling only 46% of its total needs. The United States, however, is the largest consumer of oil as a power base, indicating the vast needs of the United States for energy purposes. Actually, a large part of the total 1,726,200,000 tons of oil equivalent energy needs in the United States is supplied by natural gas, solid fuels including coal, and hydroelectric power, as indicated in Table 2.10. Nuclear fuels, at this time, account for only a small percentage of U.S. energy consumption, as it does in all other countries. (The United States leads in use of nuclear fuels.)

TABLE 2.10

World Primary Energy Consumption 1973
(in million tons of oil equivalent)

Country/area	Oil	Natural gas	Solid fuels	Water power	Nuclear fuels	Total energy needs
United States	814.7	566.9	307.6	23.8	13.2	1726.2
Canada	83.8	48.3	16.7	12.3	1.0	162.1
Total Western Hemisphere	1068.3	662.7	338.1	45.1	14.2	2128.4
Belgium and Luxembourg	32.5	7.2	12.4	0.2	—	52.3
Netherlands	41.3	32.2	2.6	—	—	76.2
France	125.7	14.1	28.0	4.4	1.2	173.4
West Germany	149.5	25.7	76.8	2.1	1.1	255.2
Italy	104.5	13.3	10.8	3.9	0.3	132.8
England	112.6	28.5	72.7	0.4	2.1	216.3
Scandinavia	56.0	—	6.1	9.8	0.4	72.3
Spain	35.7	1.0	11.1	2.8	0.4	51.0
Other Western European countries	89.9	5.0	38.2	7.5	0.4	141.0
Total Western Europe	747.7	127.0	258.7	31.1	6.0	1170.5
Japan	267.2	5.3	51.2	7.3	1.2	332.2
Australasia	32.2	3.5	25.5	2.2	—	63.4
USSR, Eastern Europe, China	434.0	237.5	949.6	17.4	0.8	1639.3
Total Eastern Hemisphere	1697.6	400.6	1406.0	64.3	8.3	3576.9
Total world	2765.9	1063.3	1744.2	109.4	22.5	5705.3

Source: Statistical Review, 1974 (for 1973), British Petroleum Company,
London.

Italy and Japan are especially dependent on oil for their energy requirements. This partly explains why Italy has the largest oil refinery capacity
in Europe and why Japan's refineries are producing to near capacity. Japanese purchases from the Middle East are mainly crude oils, since it re-
fines most of its own oil products.

On a per capita basis, petroleum's share of the energy market is misleading, as indicated in Table 2.11. The Middle East, with smaller populations, shows an 80% share of oil to their energy requirements, while the
United States, with a large percentage of the world population, shows only

TABLE 2.11

World Energy and Petroleum Demand on Per Capita Basis in 1970

Country	Energy demand, oil equivalent (bbl per capita)	Petroleum demand (bbl per capita)	Petroleum's share (%) of energy market on per capita basis (petroleum demand ÷ energy demand)
United States	56.0	22.5	40
Canada	44.0	23.5	54
USSR	24.0	7.0	29
Australia	21.0	13.0	62
Western Europe	20.0	11.0	55
Japan	13.0	9.0	70
Central and South America	5.0	3.5	70
Middle East	4.0	3.5	80
Africa	1.5	1.0	63

Source: Statistical Review, 1971 (for 1970), British Petroleum Company, London.

a 40% share of oil to their energy needs. But the United States was high in energy demand equivalent (barrels) of oil, or 56 bbl per capita in 1970, followed closely by Canada with 44 bbl per capita.

A recapitulation of reserves, production, consumption, and refinery capacity in millions of barrels is given in Table 2.12. As the recapitulation shows, world proven reserves fell by 5.6% from 1972 to 1973 by 5.6%, with the Eastern Hemisphere showing a decrease of 35.2 billion bbl (587.2 - 552.0) or -6.0%. There was also a decrease in the Western Hemisphere of 2.8 billion bbl, equal to a -3.3% (85.5 in 1972 - 82.7 in 1973). Thus, there was an overall decrease in reserves in 1973 over 1972 by the Eastern Hemisphere over the Western Hemisphere reserves of 2.7% (-6.0 + 3.3).

On production of crude, production has continued to increase since 1969, with the rate of increase in 1973 having spurted ahead in both the Eastern and Western Hemispheres. In 1973, the Eastern Hemisphere actually produced 12.2% more than in 1972. This was due mostly to the United States increasing imports in place of domestic supplies, as that nation embarked on a program of prorating reserves.

Consumption, following production, showed a slight decrease in 1973 due to the oil crisis in the last three months of the year. There was a

TABLE 2.12

Summary of Reserves, Production, Consumption, and Refinery Capacity

(thousands of barrels per day, except reserves which is in thousands of barrels per year)

	1973	1972	1971	1970	1969	1973 change over 1972 (%)	1972 change over 1971 (%)	1971 change over 1970 (%)	1970 change over 1969 (%)	Average annual %, 1968–1973
Reserves										
World	634.7	672.7	641.8	620.7		−5.6	+4.8	+3.4		
Eastern Hemisphere	552.0	587.2	554.6	537.4		−6.0	+5.8	+3.2		
Western Hemisphere	82.7	85.5	87.2	83.3		−3.3	−1.9	+3.9		
Production										
World	57,710	52,925	50,370	47,740	43,680	+8.7	+5.4	+5.6	+9.6	+7.4
Eastern Hemisphere	39,360	34,935	32,345	29,675	26,400	+12.2	+8.1	+9.1	+12.7	+11.1
Western Hemisphere	18,350	17,990*	18,025	18,065	17,280	+1.8	+0.5	+0.4	+4.8	+1.9
Consumption										
World	56,425	52,695	48,770	46,260	42,910	+7.3	+8.0	+5.4	+8.7	+8.0
Eastern Hemisphere	34,365	31,750	29,390	27,560	25,050	+8.6	+8.3	+6.6	+11.6	+9.5
Western Hemisphere	22,060	20,945*	19,380	18,700	17,860	+5.3	+7.6	+3.6	+4.3	+5.8
Refinery Capacity										
World	64,510	59,425	55,600	51,020	46,860	+7.7	+7.1	+9.0	+7.3	+8.0
Eastern Hemisphere	41,420	37,755	34,670	31,530	28,720	+8.9	+9.0	+10.0	+6.9	+9.8
Western Hemisphere	23,090	21,670	20,930	19,490	18,140	+5.6	+3.9	+7.4	+7.9	+5.4

Note: The Western Hemisphere's position is more consumption over production, due largely to the United States, indicating the strong position of imports or exports from the Eastern Hemisphere, particularly the Middle East, to the Western Hemisphere.

Source: Statistical Review, 1974, 1973, 1972, 1971, 1970, respectively), British Petroleum Company, London.

decrease of 0.7% for the world, but an increase in consumption of 0.3% in the Eastern Hemisphere, and a large decrease of 2.3% in the Western Hemisphere in 1973 over 1972.

Refinery capacity continued to increase in total daily barrel capacities in 1973. The percentage, or rate, of increase shows a greater upward trend in the Western Hemisphere and a slightly lower trend downward in the Eastern Hemisphere in 1973 over 1972. In fact, in this period both hemispheres showed a slower rate of change than from 1971 to 1972, but the Western Hemisphere increased more sharply, to +5.6% from +3.9%. This indicates that the Western Hemisphere may be increasing still further its share of world refinery capacity as the world increases its total capacity.

Finally, in comparing world production and consumption of crude in 1973 over 1972, production went up 8.7% but consumption slipped 0.7% to 7.3%. Thus, consumption "lost" 0.7% to production. In addition, proven reserves at the end of 1972 were reputed to be 672.7 billion bbl and at the end of 1973 to be 634.7 billion bbl, or a decrease in reserves in 1973 over 1972 of 38 billion bbl, which represents a decrease of 5.6%. Refinery capacity for the world increased 0.6% (7.7% from 7.1%) in 1973 over 1972. Thus, the oil industry did not keep up with increases in demand in 1973 over 1972, with consumption gains of 7.3%, production gains of 8.7%, refinery capacity gains of 7.7%, but reserve losses of 5.6%.

In summary:

1. The rate of increase of consumption or demand is less than the rates of increase of refinery capacity and production, but greater than the rate of increase of reserves.

2. The rate of increase of production is greater than the rate of increase of reserves. This is a trend to be watched closely in the future.

3. The rate of increase of refinery capacity is less than the rate of increase of production. This trend will also be carefully observed.

4. Reserves, with a rate of decrease, after lower rate of increase in 1972, indicates considerable concern for the future if this low rate continued and the rate of increase of consumption continues to rise.

THE ARAB OIL SUPPLEMENT

Because of the predominant position of the Arab countries, particularly in North Africa and the Arabian Gulf region (Iran is not an Arab country, although it is an important oil producing country and a member of the Organization of Petroleum Exporting Countries), the balance of this chapter is devoted to oil and gas data and figures on the Arab states' contribution to the world's requirements for petroleum and natural gas. This special treatment indicates the importance that the authors attach to Middle East and North African oil.

The Arab oil supplies far "outshine" all supplies, or reserves, of other regions in the world. With demand constantly increasing, especially in terms of power requirements for growing industrial complexes, Arab oil reserves will increase more and more in value, possibly even more than gold reserves.

The following data, extracted from The Arab Petroleum Director, published annually in Kuwaiti City, Kuwait, are intended as an abbreviated reference on Arab oil producing countries and a source of comparison data. Figures given below are mainly those of 1970 and 1971.

Abu Dhabi

Area—80,000 square kilometers.

Population—90,000, located mostly in capital of Abu Dhabi, and big towns, such as El-Teeh.

Reserves—2.135 million tons as of 1970, or about 16 billion bbl of crude.

Production of crude—32.8 million tons in 1970, equivalent to 252 million bbl at a rate of 694,000 bbl/day.

Refineries—None at present. The Japanese are presently conducting studies in Abu Dhabi.

Petrochemical plants—None at present. The government is interested in establishing a plant, but one which will produce products that do not compete with neighboring petrochemical plants in Saudi Arabia and Kuwait.

Natural gas—Estimated proven reserves, as of 1971, are 10 billion cubic feet. Gas is separated from crude at Hobashan and Shamis terminals.

—Production in 1970 was 266.205 billion cubic feet.

—Consumption amounted to 22 to 42 million cubic feet, mostly by industry and the government.

—Liquid gas production: none at present.

Algeria

Area—About 919,591 square miles.

Population—Estimated in 1970 at 13.2 million.

Reserves—Not announced.

Production of crude—In 1969 produced 9,804,100 tons of oil, of which 8,049,100 was exported and national consumption was 1.8 million tons.

Refineries—In 1969, Societe Reffinerie d'Alger refined 2 million tons of oil; the Hassir Refinery at Messaoud refined 100,000 tons, and a new refinery, Arzew Refinery, began production in August 1972. (Algerian oil is noted for its low sulfur content of 0.30%; therefore, there is a heavy demand for it on the international market.)

Petrochemical plants—Fertilizer plant at Arzew has been operating since 1969. The range of products makes the plant an industrial complex of world importance. In the eastern section of Algeria there are also several plants to develop the region and to utilize the oil and gas.

Natural gas—National reserves now estimated at about 3 trillion cubic meters or roughly 15% of world reserves.

　　　　—Total production in 1969 was 4.325 trillion cubic meters, of which 66% was exported and 34% produced for internal consumption.

　　　　—Has liquid natural gas, but no figures given.

Bahrein

Area—Total 255 square miles, which includes 33 islands.

Population—205,000.

Reserves—In 1969, reputed to have 46.8 million barrels.

Production of crude—16,639 bbl/day in 1970, a record for Bahrein, mostly from the Awali field with its 332 productive wells, each producing an average of 329 bbl/day from depths of 1,850 to 4,600 ft.

Refineries—A huge refinery built in 1936. In 1970, it had a capacity of 9.4 million barrels, or about 180,000 bbl/day. In addition to most of its own crudes, the Bahrein refinery also uses crudes from Saudi Arabian fields, which amounts to about 35 to 50% of its crude feed. Furthermore, the Bahrein refinery produces gasolines, kerosenes, aviation gasoline, fuel gas, fuel oil, and liquefied petroleum gases as well as naphthas and asphalt. Most of the refined products are exported to East Africa and Asia.

Petrochemical plants—None.

Natural gas—Reserves are estimated to be 1 trillion cubic feet. Domestic uses or consumption are 70 million cubic feet daily for generating electricity and several million cubic feet daily for the aluminum factory in Bahrein.

Dubai

<u>Area</u>—1,500 square miles.

<u>Population</u>—100,000 (60,000 living in capital of Dubai).

<u>Reserves</u>—As of 1970, proven reserves 133 million tons, or 1 billion bbl.

<u>Production of crude</u>—El-Fateh, the main field, reached 110,000 bbl/day in 1971. Target for 1972 was 300,000 bbl/day.

<u>Refineries</u>—None at present.

<u>Petrochemical plants</u>—None at present.

<u>Natural gas</u>—Reserves estimated in 1971 at 500 billion cubic feet. No figures on production given.

Dubai is currently building one of the biggest ports in the Gulf area. It began in 1968 and is to be completed in 1975. It will be able to accommodate 30 large-size dwt tankers. Cost of venture is estimated at £124 million sterling.

Egypt Arab Republic

<u>Area</u>—386,198 square miles (1 million square kilometers).

<u>Population</u>—At end of April 1971, estimated at 34 million.

<u>Reserves</u>—Figures not given.

<u>Production of crude</u>—Figures not given.

<u>Refineries</u>—As of 1971, refinery capacity was 9.5 million tons annually, with two refineries at Musturud (near Cairo), and two at Alexandria.

<u>Petrochemical plants</u>—Plans have been made for construction of a £100 million (Egyptian) complex.

<u>Natural gas</u>—No figures given.

Iraq

<u>Area</u>—444,442 square kilometers, or 169,240 square miles.

<u>Population</u>—In mid-1970, estimated at 9.7 million.

<u>Reserves</u>—23.5 billion bbl as of 1970, placing Iraq fifth in oil reserves among the Middle East oil producers.

<u>Production of crude</u>—76.4 million tons in 1970.

Refineries—There are seven refineries, with the principal one being the
Daura refinery, near Baghdad, the capital, with 75,000 bbl/day capacity.
Total annual output for refined products for the seven refineries together is
over 2 billion tons with installed refining capacity of 100,800 bbl/day at the
end of 1970.

Petrochemical plants—Key petrochemical plant is Abu Al-Khusail fertilizer
plant, begun in 1971 with capacities of 66,000 tons/year of ammonia, 56,000
tons/year of urea, 11,000 tons/year of sulfuric acid, and 140,000 tons/year
of ammonium sulfate.

Natural gas—North Rumeilah field production should be 100,000 bbl/day.
No figures are given on reserves.

Jordan

Area—37,000 square miles.

Population—2,103,000 as of a 1968 census.

Reserves—None.

Production of crude—None.

Refineries—One refinery in Al-Zarga region. Total production in 1970 was
464,000 tons of oil products.

        —Refining capacity is 15,683 bbl/day with a catalytic cracking
capacity of 1,400 bbl/day; crude needs, purchased, were 10,000 bbl of
crude per day in 1970.

Petrochemical plants—None.

Natural gas—None.

Kuwait

Area—7,000 square miles.

Population—733,200, as of 1970 census.

Reserves—Current estimate, 1970, was 74.5 billion bbl.

Production of crude—In 1969 was 1.1 billion bbl (including their share of the
Neutral Zone) or 2.77 million bbl/day.

Refineries—First in total refinery capacity in the Middle East with 489,000
bbl/day of oil product production.

        —Three refineries are in operation. One is at Mina Al-Ahmadi,
with a 250,000 bbl/day capacity; one is at Mina Abdullah with a 110,000

bbl/day capacity, using Neutral Zone crude, and one is in the Shuaiba area with a capacity of 95,000 bbl/day.

Petrochemical plants—Kuwait Chemical Fertilizer Company, which used 38.168 billion cubic feet of natural gas in the first quarter of 1970.

Natural gas—Kuwait Oil Company is responsible for production of LPG (liquefied petroleum gas). This amounted to 15.9 million barrels in 1969, a 33.6% increase over 1968. Principal market for LPG is Japan. Cost of transporting LPG is about three times that of transporting crude oil.

—Production of natural gas in 1969 was 513.094 billion cubic feet, or about 1.405 billion cubic feet a day. There are no independent gas fields as such, but all oil fields yield gas. No reserve figures thus given. Gas utilization as a percentage of production was 35.2% in 1969 as compared to only 16.5% in 1965.

—Principal uses of Kuwaiti natural gas are (1) for reinjection in oilfields to maintain pressure, (2) production of energy for generating electricity and water distillation capacities, and (3) as a raw material for petrochemicals and fertilizers.

Lebanon

Area—4,015 square miles.

Population—2.6 million estimated (1971). Most of the population is in Beirut, one of the largest commercial and financial centers in the world, as well as in the Arab regions.

Reserves—At this writing, none.

Production of crude—None, at this writing.

Refineries—Two refineries, one at Tripoli and the other at Zahrani. The Tripoli refinery has a capacity of 20,000 bbl/day of oil products, which meets 60% of Lebanon's domestic needs. Crudes come from the Kirkuk oil fields of Iraq by an Iraq Petroleum Company pipeline. The Zahrani refinery has a capacity of 16,500 bbl of oil products a day, and makes up the other 40% of Lebanon's oil product consumption. The Zahrani refinery gets its crude requirements from Tapline, the pipeline of Saudi Arabia.

—Both refineries together have a total capacity of 1.8 million tons of crude per year, or 36,500 bbl/day of oil products.

Natural gas—None.

Libya

Area—679,358 square miles (1,759,540 square kilometers).

Population—1,870,000, according to 1969 estimates.

Reserves—As of July 1969 approximately (estimated) 35 billion barrels.

Production of crude—Production reached 3.5 million bbl/day at mid-1970, but was reduced to 3.1 million bbl/day in early 1971 by the government.

—Libya has a strong crude oil position, with high-quality sulfur-free crude and a natural location on the west side of the Suez, just across the Mediterranean from its Western European customers.

Refineries—One refinery at Moroa Al Brega has a capacity of 9,500 bbl/day of oil products, and another at Al-Merisa has a daily capacity of 5,000 bbl of oil products. A new refinery at Zawia, with a capacity of 60,000 bbl/day of oil products, is in the planning stage at this writing.

—A LPG plant is also under consideration.

Petrochemical plants—A gigantic petrochemical complex as of 1971 should have been completed by now. It will have a floating terminal to receive crudes and to export surplus petrochemical production.

—An ammonia plant with a 220,000 tons/year capacity is contemplated.

Natural gas—No figures are given on natural gas reserves or production.

—In 1970, LPG production reached 345 million cubic feet per day.

Morocco

Area—172,000 square miles.

Population—About 15 million as of this writing. Rabat, the capital has 500,000.

Reserves—920,000 bbl as of this writing.

Production of crude—50,000 tons in 1970, with most of the crude production coming from Rahareb Basin. There is great future possibility for reserves and production offshore.

Refineries—Two, with a total capacity of 1.8 million tons/year, or 35,000 bbl of oil products a day. The two refineries are the Sidi Qassem Refinery, with a refining capacity of 9,000 bbl/day, and the Al-Mohammadieh Refinery with a capacity of 26,000 bbl of oil products per day.

Petrochemical plants—Plans are to build one with a capacity of 50,000 tons/ year in the near future near the Al-Mohammadieh Refinery; there are none at the present time.

Natural gas—Proven reserves are 18 million cubic feet.

Oman Sultanate

Area—130,000 square miles.

Population—700,000.

Reserves—5 billion bbl (estimated in December 1969), or 667 million tons.

Production of crude—In 1969, the latest figures, 328,000 bbl/day.

Refineries—None at present.

Petrochemical plants—None at present.

Natural gas—1.5 trillion cubic feet in reserves estimated at this time.

Qatar

Area—11,400 square kilometers, including all the islands.

Population—130,000 estimated.

Reserves—Not given.

Production of crude—9 million tons of crude a year. Its oil, with a specific gravity of 41.5 °API standard, is regarded as the best in the Gulf area.

Refineries—No refineries, but a small petroleum purifying installation at Um Saeed, the principal port. Its capacity is 24,000 of crude oil. Its products are gasoline, kerosene, and diesel, as well as super-gasoline. A new refinery, with a capacity of 6,000 bbl/day in oil products for domestic needs, is being considered at this time.

Petrochemical plants—None at present. A plant to produce around 1,000 tons of ammonia and 1,100 tons of urea fertilizer is being planned. Also, a huge complex costing $74 million (U.S.), which would produce 200 commodities, is being studied, with Um Saeed as the site.

Natural gas—No figures on reserves or production. Plans are being made for liquefying natural gas in Qatar.

Saudi Arabia

Area—927,000 square miles, or 2.4 million square kilometers.

Population—Estimated between 5 and 7 million.

Reserves—Estimated proven reserves, including 50% of reserves of the Neutral Zone, by a 1970 estimate, are 146.56 million bbl, which is close to 30% of "Free World" reserves (see Table 2.1).

Production of crude—In 1970, the latest reported year's figures, had 1.4 billion bbl; today, over 6 million bbl/day.

Refineries—Saudi Arabia has one of the largest refineries at Ras Tanura (Aramco), and rates with those of Kuwait, the refinery in Abadan, Iran, and in Bahrein in capacity and in products produced and shipped. Also, Getty Oil Company, inshore in the Neutral Zone part of Saudi Arabia, and the Arabian Oil Company (Saudi-Japanese), offshore near the Neutral Zone, each have a refinery with 50,000 bbl/day capacity at Mina Saud (Getty) and about 30,000 bbl/day at Khafgi (Arabian Oil). Furthermore, there is now a refinery at Jeddah on the Red Sea for Eastern domestic needs where, in 1970, the latest figures showed that capacity was 33,000 bbl/day in oil products. There are also plans for a refinery in the near future in Riyadh with a capacity of 15,000 bbl/day in oil products.

Petrochemical plants—Fertilizer plant at Dammam, in the Eastern Province, with a daily production capacity of 1,100 tons of urea, 600 tons of ammonia, and up to 50 tons of sulfur. An oil lubricating plant for Jeddah and a sulfur plant at Abqaiq are planned for the future.

Natural gas—In 1969, reserves estimated at 45.7 trillion cubic feet.

　　　　—Production of natural gas used for Dhahran Electric Power Company, for SAFCO, the petrochemical plant, and local consumption of gas, all supplied by Aramco.

Sharjah

Area—Not declared.

Population—Not declared.

Reserves—Not declared, only began exploring for oil in 1965.

Production of crude—No figures.

Refineries—None.

Petrochemical plants—None.

Natural gas—No figures.

Sudan

Area—1 million square miles.

Population—15 million estimated in 1969. Capital is Khartoum.

Reserves—No figures given.

Production of crude—No figures given.

Refineries—One, built at Port Sudan on the Red Sea with a 20,000 bbl/day capacity in oil products, supplied by light Iranian crude. Products refined are gasoline, kerosene, gas oil, and LPG, all for Sudan's own needs. Only the surplus of fuel oil after domestic needs is exported.

Petrochemical plants—None.

Natural gas—No figures given.

Syria

Area—Figures not given.

Population—1970 census showed 5.8 million. Damascus is the capital.

Reserves—Up to December 31, 1970, were estimated at 162 million tons.

Production of crude—In 1970, was 3.557 million tons total. Expected to be 4.5 million tons in 1971.

      —Syrian crude is heavy, with average density of 25 °API for Suwaidiah field and 19.1 °API for Kratchuk field. Proportion of sulfur in Suwaidiah crude is 3.6% and for Kratchuk crude is 4.5%. In addition, the proportion of fuel gas in Syrian crude is 62% of the total derivatives produced after refining the crude. But Syria also has one field of light crude at Jabsa, where crude is 40.5 °API.

Refineries—One at Homs with a capacity of 59,000 bbl/day in oil products, and a catalytic reform capacity of 2,500 bbl/day. A new refinery at Al-Riqqa on the Euphrates river and expansion of the refinery at Homs is planned.

Petrochemical plants—There are two. One petrochemical plant, at Homs, manufactures fertilizer, sulfur, and other organic products. The other plant, also at Homs, produces 140 tons of ammonia per day.

Natural gas—In 1971, reserves estimated at 900 billion cubic feet, or 25 billion cubic meters.

Tunisia

Area—48,200 square miles.

Population—5 million. Capital is Tunis.

Reserves—40 million tons of crude estimated.

Production of crude—Average today is 3.3 million tons/year. Most crude produced comes from the major Burmah field.

Refineries—The country's only refinery is located at Benzert, with a capacity of 30,000 bbl/day of oil products.

Petrochemical plants—One phosphate plant.

Natural gas—Reserves estimated at 200 million cubic meters; average production per year is 10 million cubic meters, mostly from Abdul Rahman area in Ras Bon. Production of natural gas in 1971 estimated at 280 million cubic meters.

Yemen (People's Republic of South Yemen)

Area—Over 1,600 square miles.

Population—1.5 million estimated with 100,000 in Aden, the capital.

Production of crude—No figures.

Refineries—A big refinery at Aden, with a yearly capacity of 5 million tons of crude oil. The refinery supplies fuel to ships visiting the port of Aden, also supplies gasoline, kerosene, and gas to Red Sea markets. The loading terminal at the refinery can provide tankers with crude oil or refined oil products.

　　　—Daily refining capacity is estimated at 200,000 bbl of oil products.

Petrochemical plants—None.

Natural gas—Nothing declared on natural gas.

North Yemen

No oil figures.

NOTES

1. "The Interdependence of Profits and Energy," Public Affairs/Briefings, Standard Oil Company of California, March 1974, p. 3.

2. "Petroleum Industry Competition: It Has Been Verified Time and Time Again," Public Affairs/Briefings, Standard Oil Company of California, March 1974, p. 4.

3. "The Interdependence of Profits and Energy," p. 1.

4. "Oil Current Analysis," Standard and Poor's Industry Survey, sec. II, April 4, 1974, p. 43.

5. "Oil Basic Analysis," Standard and Poor's Industry Survey, sec. II, July 5, 1973, p. 70.

6. "Oil Basic Analysis," sec. II, p. 58.

7. Gardner, Frank, "Offshore Oil—Only the Beginning," Oil and Gas Journal, vol. 72, no. 18, p. 125 (1974).

8. Gardner, Frank, "North Sea Today: Where Tomorrow?" Oil and Gas Journal, vol. 71, no. 50, p. 78 (1973).

9. McCaslin, John, "Offshore Oil Production Soars," Oil and Gas Journal, vol. 72, no. 18, p. 77 (1974).

10. Gardner, Frank, "Offshore Oil—Only the Beginning," p. 125.

11. Ibid., p. 125.

12. King, Robert E., "Big Reserve Boost Foreseen in Gulf of Mexico in 1974," World Oil, vol. 178, no. 5, p. 72 (1974).

13. Ibid., p. 74.

14. Speech by M. A. Wright, Chairman of Exxon Corporation, Wall Street Transcripts, vol. 42, no. 9, p. 35128 (1973).

15. King, Robert E., "Big Reserve Boost Foreseen in Gulf of Mexico in 1974," p. 75.

16. King, Robert E., "Big Reserve Boost Foreseen in Gulf of Mexico in 1974," p. 76.

17. Gardner, Frank, "North Sea Today: Where Tomorrow?" p. 77.

18. Rose, Standford, "Our Vast Hidden Oil Reserves," Fortune, vol. 84, no. 4, p. 107 (1974).

19. Gardner, Frank, "Offshore Oil—Only the Beginning," p. 124.

20. Gardner, Frank, "North Sea Today: Where Tomorrow?" p. 77.

21. Gardner, Frank, "North Sea Today: Where Tomorrow?" p. 77.

22. "Going Fishing in South Vietnam," Forbes, vol. 113, no. 2, p. 39 (1974).

Chapter 3

CHARACTERISTICS OF CRUDE OIL
AND REFINED PRODUCTS

COMPOSITION OF CRUDE OIL

Crude oil is a complex mixture in which hydrogen and carbon are combined in various ways to form many different compounds. Because of the complex chemical nature of crude oil, elaborate techniques have been developed by the refining industry in order to achieve higher yields of those products of increasing demand.

Petroleum crude oil is made up essentially of carbon and hydrogen atoms. The combination of these two elements is almost limitless because of the capacity of the carbon atom to combine with other elements and with itself. Compounds composed of only carbon and hydrogen, commonly known as petroleum hydrocarbons, belong to the paraffin class. The simplest paraffinic hydrocarbon is methane, CH_4. The next simplest compound is ethane, C_2H_6, followed by propane, C_3H_8, and butane, C_4H_{10}.

In addition to the hydrocarbons, compounds containing other elements in addition to carbon and hydrogen, such as sulfur, oxygen, and nitrogen, are found in petroleum crude.

Table 3.1 shows the composition of petroleum crude by element. By volume, petroleum crude oil is composed of 84 to 87% carbon, 11 to 13% hydrogen, and 1 to 4% impurities [1]. These impurities consist largely of sulfur, nitrogen, oxygen, and helium [2]. Of these impurities, sulfur is the most difficult and costly to remove.

TYPES OF CRUDE OIL

Oils from different sources differ widely in properties due to the varying nature of the hydrocarbon series which they may contain. The two most

41

TABLE 3.1

Composition of Petroleum Crude

Element	Percent by weight
C (carbon)	83–87
H_2 (hydrogen)	11–14
S (sulfur)	0.05–2*
N_2 (nitrogen)	0.1–2*
O_2 (oxygen)	0–2*

*Regarded as impurities.

useful series of hydrocarbons are called paraffins and naphthenes (asphalts). Consequently, crudes are classified roughly into the following types:

1. Paraffinic base crudes: These are rich in paraffin, and contain very little of the naphthenes and asphaltenes.

2. Asphaltic base crudes: These consist largely of aromatic and/or naphthenic hydrocarbons.

3. Mixed or intermediate-type base crudes: These have a higher paraffinic content than asphaltic base crudes, but a lower paraffinic content than paraffinic base crudes.

A typical example of the paraffinic type as found in the United States is Pennsylvania oil. Similarly, oils found in Iran and Saudi Arabia (Abqaiq and Ain Dar) are generally considered mixed but more on the paraffinic side.

A typical example of asphaltic oil in the United States is California crude oil.

The mixed or intermediate type is best illustrated by the oil found in Saudi Arabia (Fadhili), Iran, and Iraq.

It is obvious that the physical operations involved in the refining of oil, such as flash vaporization, fractional distillation, and vacuum distillation, are governed to a large extent by the properties of the hydrocarbons themselves, since they constitute the bulk of the oil. The same reasoning also applies to the chemical conversion processes such as reforming, cracking, etc. Similarly, chemical treatment operations, such as sulfur removal, are governed by the presence of sulfur, oxygen, and nitrogen compounds.

The types and quantities of products made from particular crude oils are largely determined by technical factors involving the quality of the crude oil and the type of equipment available, and by economic factors involving market demand and accessibility to these products. Accordingly, useful yields of lubricating oils and waxes may be obtained from one type of oil, while little lubricating oil but high-octane gasolines may be obtained from another type of oil.

QUALITY OF CRUDE OIL

Some of the physical properties used in determining the quality of crude oil are listed in Table 3.2.

The price of petroleum crude oil is basically determined by four factors: API (American Petroleum Institute) gravity, sulfur content, viscosity, and capillarity. API gravity is the density, or specific gravity, of crude oil as measured against a common denominator, in this case an equal amount of water at 60° F. Sulfur content is described as the percentage of sulfur impurity in a sample of petroleum crude oil. Viscosity is a measure of the fluidity or resistance-to-flow characteristics of crude oil. Viscosity, like API gravity, is measured against water as a standard reference. And lastly, capillarity is a measure of the adherence properties of crude oil.

Of these four factors which affect crude oil prices, only API gravity and sulfur content are of concern to the refiner. API gravity, essentially, tells the refiner how much crude oil he is getting for his money. The higher the API gravity, the greater the potential value of the crude oil. The most common formula for gravity quality differentials is 2¢ per °API [3]. For example, in the mid-continent fields of the United States, the price is set at 40 °API with a differential of 2¢ per °API per barrel [4]. Therefore, the price of crude oil is adjusted through a differential to equalize the quality in density to the price.

Sulfur content, on the other hand, tells the refiner the amount of basic impurity that is in the crude oil. Even though the API gravity figure might be attractively high, this might reflect a high sulfur content. The sulfur content for most crude oils falls between 1 and 2.5% [5], where 1% sulfur content is considered "sweet" crude and 2.5% sulfur content is considered "sour" crude.

Presently there is no universally accepted differential, similar to the quality differential system used for API gravity adjustments, for adjusting the price of crude oil to the amount of sulfur content. Up until the late 1960s, refineries were set up to handle a specific type of crude. Some refineries would simply not purchase "sour" crudes (high sulfur content) at any price because they were not equipped to process them [6]. Recently, however,

TABLE 3.2

Considerations Affecting Quality of Crudes

Property	Comment	Range or limitation
API gravity, °API	$°API = \dfrac{141.5}{sp.\ gravity} - 131.5$	From 20 to 45; for Middle East oils it is around 35.
Sulfur content	High-sulfur oils require extensive processing.	Maximum of 0.5% by weight.
Carbon residue	Related to the asphalt content in oil.	A lower carbon residue is a better quality.
Salt content	Severe corrosion takes place if the salt content is high.	Up to 15 lb per 1,000 bbl of oil.
Nitrogen content	Not desirable if high; it causes poisoning of catalysts.	Up to 0.25% by weight.
Pour point	The temperature in °F at which an oil will no longer flow from a standard test tube.	The lower the pour point, the lower the paraffin content of the oil.

the ecology push has forced refiners to reduce the amount of sulfur in their finished products, and the energy crisis has forced them to refine whatever they can get to go through their fractionating tower. As a result of this dilemma and the uncertainty of the future, the refineries have not been able to establish a price differential system for sulfur impurities.

Although API gravity and sulfur content are major factors in determining the value of petroleum crude oil at the refinery, viscosity and capillarity are not of particular concern to the refiner. These latter properties describe the resistance properties of crude oil which affect the rate at which it will flow through pipelines, and they are therefore of primary importance to the producer and transporter of petroleum crude oil in determining value. All four factors, however, would be used in determining the value of oil at the wellhead.

IMPURITIES IN CRUDE OIL

Separation of Impurities

Of the significant impurities found in crude oil, sulfur is the only element which is retrieved as a solid. The other impurities, mainly nitrogen and helium, are recovered as gases and are easier to separate from the crude oil. Also, the commercial value of these gases makes their recovery a profitable venture.

Until as recently as 1969, the recovery of sulfur was strictly an economic question [7]. A refinery was set up to handle a certain type of crude and would realize a normal profit from its sulfur recovery unit. Now things are different. In the United States, for example, the Environmental Protection Agency has forced the users of petroleum fuels (gas oils, fuel oils, and gasolines) to purchase fuels which have a restricted amount of sulfur. The refinery can still make fuels with greater sulfur content, but they have a federally limited market.

The cost of reducing the sulfur content reflects many factors, such as size of plant, heating requirements, labor, and maintenance [8]. A good discussion of the mechanics of sulfur recovery systems (as well as nitrogen and helium recovery systems) is presented by H. S. Bryant, a research scientist for Mobil Oil Co. [9]. The most important factor, however, in determining the final cost of removing sulfur impurities from crude oil is the commercial price of sulfur. As can be seen from Table 3.3, the price of sulfur is extremely volatile. Therefore any attempt by a refiner to formulate an equation to adjust the price of crude to the amount of sulfur impurity

TABLE 3.3

Price of Sulfur Since 1947

Year	Price/long ton
1948–1949	$27.30
1950	28.60
1951	31.80
1952	22.00
1953	24.60
1954–1957	26.50
1958–1964	23.50
1965–1966	25.50
Jan. 1967	28.00
May–Sept. 1967	33.50
Oct. 1967–Feb. 1968	39.00
Mar. 1968–Jan. 1969	42.00
July 1969	40.00

Source: Nelson, W. L., "Net Cost for Desulfurizing of Residues," Oil and Gas Journal, vol. 68, no. 3, p. 60 (1970).

would be very complex and misleading. With a quality-related adjustment already in effect, namely the API gravity adjustment, another quality related adjustment would simply complicate the buying and selling of crude oil. Generally, crudes purchased from certain geographic areas are assumed to have impurity concentrations within certain parameters. As a result, the aggregate cost of removing sulfur from crude oil fluctuates considerably. In 1973, for example, in 30,000 bbl/day refining and with the price of sulfur fluctuating between $38.00 and $44.00 a long ton, the aggregate cost of removing the sulfur was between 54¢/bbl and 73¢/bbl [10].

Environmental Aspects of Separating Impurities

Essentially, the recovery of helium and nitrogen from crude oil is economically profitable and environmentally stable. This is certainly not the case, however, for sulfur. The ecology movement has played a large role in setting the climate for regulation governing all phases of sulfur use and production.

This aspect of petroleum refining cannot be deemphasized as an important factor in crude oil impurity recovery. It is also an issue that will probably intensify and become increasingly important in the future.

Economic Implications of Impurities

In most cases, paraffinic base crudes are easiest and least expensive to refine, with the mixed base and then naphthenic base (or asphaltic base) crudes following in that order. This applies to the light distillates such as gasoline as well as to the heavy lubricating oils. Many naphthenic base oils contain more sulfur than mixed base crudes, and paraffinic base crudes may contain scarcely any sulfur. But, paraffinic and mixed base crudes contain troublesome wax. The fact that a true naphthenic base crude contains no wax enhances the ease of manufacture of lubricants from these crudes.

The physical operations of refining, such as vaporization, fractionation, and cooling, are governed to a large extent by the properties of the hydrocarbons (hydrogen and carbon) because they make up the bulk of the crudes. But the chemical operations, such as treating and filtering, are governed by the extent of the presence of sulfur, oxygen, and nitrogen compounds.

So the refiner is interested in the base and in the general properties of the crudes he uses, in the presence of impurities, such as sulfur, salt, and emulsions which cause general difficulties in processing, and in the rising costs of processing, which lead to higher-cost refined products. Desulfurization processes, whereby sulfur is eliminated in the form of hydrogen sulfide, are particularly important in this regard.

Further, refiners seek to avoid sulfide and acid corrosion. The cost of corrosion due to high sulfur content or "sourness" varies with the amount of sulfur and the degree of sourness found in the crudes. For example, hydrogen sulfide is often found in natural gas as well as dissolved in crude oil, and it may also be formed by the decomposition of organic sulfur compounds at high temperatures. Hydrogen sulfide gas rapidly attacks steel parts, such as storage tanks and pipelines, that may be exposed to it. Refineries that handle high-sulfur crudes must use expensive chromium steels, which are resistant to hydrogen sulfide, in tanks and pipelines. Also, refinery equipment is very complex, and large quantities of chromium are used in vessel linings, bubble caps and trays, centrifugal pumps, pump liners, wear rings, pump rods, connections to vessels, pipestill bends, etc. Surprisingly, however, it has been reported by many refineries that the corrosiveness of sulfur-bearing crudes is not directly proportional to the amount of the sulfur content. For example, crudes that contain less than 0.7% sulfur most often cause more sulfide corrosion than crudes with, say a 2% sulfur content. Chromium steel, although expensive, is thus used extensively to reduce sulfide corrosion.

The combination of sulfide and hydrochloric acid corrosion in the processing of sour crudes is also exceedingly troublesome. To neutralize the hydrochloric acid, caustics are injected into the crude when reducing the salt content by 5 to 10 lb/1,000 bbl of crude oil, which then results in a reduction of sulfide corrosion. This process, of course, adds to refinery costs. Because of the wide differences in crudes, refinery processing methods also differ. What is an economical method of handling one crude may be inadequate for another crude. Thus, the refining of crude oil involves numerous economic problems.

Many factors must be considered by refineries in determining their processing plans and schedules: the characteristics of the crudes and their accessibilities, the yields of refined products that can be expected from these crudes, the marketing possibilities of these refined products including their values (prices) and potential profits, the cost of processing, etc.

Impurities present in crudes, and those impurities produced during refining distillation and cracking operations, must be removed from nearly all commercial refined products. Improvements in refined products, such as color, stability to light, odor, sulfur content, amount of gumlike or asphaltic substances, corrosiveness, and composition, are accomplished by treatment operations.

Sulfuric acid was once widely used in treatment because sulfuric acid partly removes sulfur, precipitates asphaltic or gum-like materials, improves color and stability, and to some extent improves the odor. Sweetening of "sour" distillates is always necessary, but sulfuric acid today has been supplanted by mercaptan-removal sweetening processes, including processes that oxidize mercaptans and processes that destroy and remove other sulfur compounds along with mercaptans, hydrogen sulfide, and sulfur. Also, acid and absorbent clays are used to improve the color of lubricating oils.

OIL PRODUCTS FROM CRUDE (PETROLEUM PRODUCTS)

In order to understand the oil industry and the economics applied to oil products, one must know something of the number, nature, characteristics, and economic importance of these products.

A survey by the API of the petroleum industry in the United States in 1965 revealed over 2,000 different refined oil products made from crude oils (Table 3.4).

TABLE 3.4

Oil Products Made by the U.S. Petroleum Industry, 1965

Class	Number of products
Fuel gas	1
Liquefied gases	13
Gasolines	40
Motor	19
Aviation	9
Other (tractor, marine, etc.)	12
Gas turbine (jet) fuels	5
Kerosenes	10
Distillates (diesel fuels and light fuel oils)	27
Residual fuel oils	16
Lubricating oils	1,156
White oils	100
Rust preventatives	65
Transformer and cable oils	12
Greases	271
Waxes	113
Asphalts	209
Cokes	4
Carbon blacks	5
Chemicals, solvents, misc.	300
Total number of products	2,347

Source: American Petroleum Institute, Facts About Oil, 1966.

The principal oil products and fractions resulting from crude oil are classified as follows:

Gasolines—mainly aviation, premium, and regular grades.

Kerosenes—as many as eight kerosene products are refined from light distillates.

Distillate fuel oils, such as diesel fuel oil and gas oil.

Residual fuel oils, such as diesel type and others.

Lubricating oil base stocks.

Asphalts.

Petroleum waxes.

Naphtha solvents.

Liquefied gases.

From the point of view of the oil industry, the refined oil products which are of the greatest value are those products which have the highest profit potential. Thus, most refineries want to sell these oil products in greater volume than the less profitable ones. However, there are exceptions, because of the type of crude available, or for some other reason such as specialization in a few products for the asphalt market, the lubricating oil market, etc.

Three products—gasoline, middle distillates (including diesel fuels), and residual fuels—make up the bulk of the output of the oil industry, in terms of both value and volume. Gasoline is most important, not only because of its high value (profit potential) to the industry, but because problems of oil companies largely revolve around gasoline.

Actually gasoline, one of the derivatives of petroleum, does not compete with coal or nuclear fuels as a source of energy. Other petroleum derivatives, such as diesel and natural gas, however, do compete with coal and nuclear fuels. Heating oil, another derivative of petroleum, competes with coal as a source of heat.

Table 3.5 shows typical composition and boiling point ranges used in refining petroleum, as well as uses of various oil products obtained by the multistep process of fractional distillation for separating crude petroleum.

Actually, in the straight distillation of petroleum, an average of less than 20% of the crude oil appears in the gasoline fraction. This amount, even from all crude oil sources in the world, would not meet the tremendous world demand for gasoline. Also, this straight-run gasoline possesses poor "antiknock" properties. Thus, some further processing of crude is necessary to improve both the quantity and the quality of gasoline. For example,

TABLE 3.5

Oil Products Obtained by Fractional Distillation of Crude Petroleum[a]

Oil product	Approximate composition	Boiling point range (°C)	Uses
Gas	C_1-C_4	-164 to +30	Fuel, carbon black, polymerized gasoline, and casing head gasoline
Petroleum ether, or ligroin	C_5-C_7	+30 to +90	Solvent, dry cleaning, refrigerant
Straight-run gasoline[b]	C_5-C_{12}	+40 to +220	Motor fuel
Kerosene	C_{12}-C_{16}	+200 to +315	Lighting, fuel for oil stoves, fuel for diesel engines, fuel for tractors, rockets, and jets
Gas oil, or fuel oil	C_{15}-C_{18}	up to +375	Furnace oils, fuel for diesel engines, cracking stock
Lubricating oils	C_{16}-C_{20}	+350 and up	Lubrication
Greases, petroleum (oily) or petrolatum	C_{18} and up	Semisolid	Lubrication, sizing paper, ingredient of medicines
Paraffin ("wax")	C_{20} and up	Melts at 51-55	Candles, impregnating matchsticks, waterproofing fabrics, household canning
Pitch and tar (artificial asphalt)	—	Residue	Roofing, paving, protective paints, manufacture of rubber
Petroleum coke	—	Residue	Fuel, carbon electrodes

[a]Data from Nelson, W. L., Petroleum Refining Engineering (New York: McGraw-Hill Book Company, 1973), pp. 80-106.
[b]From Aramco Oil Company, Chemical Engineering Department.

cracking, by which large and less volatile hydrocarbon molecules are broken down into a variety of smaller and more volatile ones, is used to boost production and improve the properties of gasoline. Alkylation is another refining process that serves the same purpose.

MARKET CHARACTERISTICS

The characteristics of refined products such as gasoline, middle distillates (heating oil and diesel fuel), and residual fuel have both a direct and an indirect influence on the price policies and selling arrangements of those who market these products.

For example, demand for such principal products as gasoline and heating oil obviously depends on the need for transportation services and heat. A particular consumer will want a relatively constant amount of gasoline and heating oil, so sales expansion is limited except to what can be achieved by advertising and sales promotion (e.g., persuading consumers to use more gasoline by taking more motor holidays).

There is also, of course, the possibility of substitution of products. Substitutes for some petroleum products are provided by the same oil industry which provides the original product. For example, diesel fuel, a petroleum product, is rapidly replacing another petroleum product, heavy fuel oil, for use in powering trains. Diesel fuel, as well as liquefied petroleum gas, is being substituted for gasoline in trucks and tractors. But some of these substitutes are not immediately interchangeable; there are time lags before substitution can be made, because facilities must be altered to use the substitute. For example, fuel oil can replace anthracite coal only as new furnace facilities become available.

The petroleum industry is difficult to classify as being monopolistic, oligopolistic, or competitive. Actually, the industry is more competitive at the refinery and distribution stages than at the oil field or crude oil stage. As an industry of a few sellers (an oligopoly), the crude production level is oligopolistic, partly because competition is usually restricted by law in any case. The number of suppliers of crude in a particular market area, such as the Arabian Gulf, the Caribbean, or the North Mediterranean area, is small, enough so that the selling efforts of one petroleum company may have an appreciable effect on one or more of the other oil companies in that area.

The refinery and distribution levels of the industry are regarded as monopolistically competitive, since there are many more suppliers at these levels than at the crude production level. Furthermore, ease of entry by new competitors is greater at these levels, especially at the distribution level, than at the crude oil stage. Suppliers at these levels are constantly introducing a much more intensive type of competition than at the crude oil stage, yet an element of monopoly still prevails to a certain extent. This is particularly evident in the area of pricing, where prices of refined oil products to both consumers and industrial consumers by competitors at the refinery and distribution levels are fairly uniform, due to some "understanding" between competitors.

Generally speaking, some types of buyers of refined oil products have some influence in price making. The industrial buyer knows more about prices and qualities of oil products on the market than does the household consumer. The industrial buyer also buys in much larger quantities. Hence, prices to industrial consumers are generally lower and more uniform than prices for the same products sold to household consumers. Gasoline is an example of a product sold to both types of buyers for which the price to the industrial buyer is lower than the price to the household buyer.

Along with the increase in consumption of oil products and changes in patterns of consumption and supply areas, there has also been a shift in the location of refineries. Refineries today are generally located more near the market for their products than near the source of crude oil. Before World War II the tendency was to locate refineries near the source of crude oil and then ship the refined products to markets. Some of the reasons for the shift in refinery locations are as follows:

1. It is generally cheaper to transport crude oil than refined products, at least over long distances.

2. Nations can buy more crude oil than refined oil products with the same available foreign exchange.

3. Users in industrial centers now obtain feed stocks formerly considered petroleum wastage from nearby refineries, and use them in making thousands of petrochemical products.

4. The consuming nations, rather than the foreign producing countries, get the economic benefits of refinery construction and operation.

5. Large-diameter pipelines handle crude oil better than oil products.

Crude oil is also refined at the source of origin, mainly to supply nearby local markets, as well as to provide bunker fuel for vessels in foreign trade and to supplement output of consumer-located refineries during periods of peak demand.

PRICE ELASTICITY OF SUPPLY

Supply of crude oil is generally regarded as inelastic. Demand for crude must be derived, because demand for crude depends on the demand for refined oil products. In most cases, demand for refined oil products is inelastic, therefore demand for crude becomes inelastic; and this, in turn, affects the elasticity of crude supply.

But there are also other reasons for the inelasticity of crude supply. For one thing, each production stage, including exploration, development, and production of oil on a commercial basis, is dominated by high fixed "sunk" costs relative to the variable costs that might be saved by contracting output. Also, the major supply of crude oil offered to the market reflects neither the level of demand nor the level of costs, but rather the discoveries of oil.

Exploration and development of crude means large capital expenditures, which may be unrelated to short-term fluctuations in the volume of output. The search for oil involves large outlays for scientific equipment and personnel and for test drilling. Drilling costs increase at least geometrically with depth. Thus, a large production of development costs become sunk costs whether oil is discovered or not. Yet despite large investments with heavy fixed costs, and the desire to obtain a quick return on investment, a big price increase is poor policy whether supplies are scarce or not. If price increases resulting from the desire for a quick return on investment are too big, there may be a quick contraction in amounts of supplies (barrels of oil) that buyers will take. The result could be a big price drop, bringing surplus stocks of crude.

The degree of elasticity of supply, that is, the effect on supply of oil products of expansion and contraction of prices of oil products, is high, at least in terms of increasing production of most oil products. Usually, only a slight increase in price will bring additional quantities of production of an oil product. But elasticity of supply is conditioned; that is, supply is conditioned to some extent by time. Thus, time may limit expansion of production of an oil product.

For example, when demand (D) is greater than supply (S), the price of oil products will go up. This condition should remain relatively constant. Over a period of time, supply should maintain a close relationship with demand. This can be achieved by allowing supply to lag behind demand, yet allowing supply to be sufficiently flexible that a quick upsurge in demand, as we see today because of rapidly expanding energy needs, will bring a proportionate increase in supply and thereby safeguard against large price increases. The position of the market should remain, however, one in which D > S, and not S > D, so as to guarantee against big price drops that could discourage investments in the oil industry.

Normally, supply should act inelastically. For instance, with some price upsurges due to increase in demand, supply should be somewhat controlled. This indicates that the oil industry should be responsible for stability in price and output by efficient matching of supplies with consumption.

An inelastic supply means that products change production plans upward more slowly with increasing price changes. An elastic supply indicates that producers of oil products should change their supply plans upward more quickly with increase in price.

The formula for price elasticity, or sensitivity, of supply which determines whether supply is elastic or inelastic relative to price changes is as follows:

$$\text{Elasticity of supply} = \frac{\% \text{ of change (increase) in quantities supplied}}{\% \text{ of change (increase in price)}}$$

where any figure over 1 is elastic and any figure under 1 is inelastic. Therefore, if 200,000 bbl/day of gasoline had been produced at $2.00/bbl, and price went up to $2.60/bbl and production of gasoline was thus raised to 300,000 bbl/day, elasticity of supply for this refinery relative to gasoline would be 1.7, or elastic:

$$\frac{50\%}{30\%} = 1.7$$

A 1.7 price elasticity means that a price increase will increase quantities of gasoline supplied by refineries by 1.7 times, in terms of percent, to the amount of the increase in price per barrel.

If production was increased to 240,000 bbl/day, price elasticity of supply would have been

$$\frac{20\%}{30\%} = 0.67$$

This would indicate an inelastic price. In other words, refiners would not be responding with additional supplies to a price increase as they were when elasticity was elastic.

The difference between the two examples of price elasticity of supply is that in the inelastic supply case, the refiner may restrict output for various reasons. High prices need not necessarily increase supply of refined petroleum products. Also, low prices will sometimes not curb output. The small refiner may produce more, not less, merely to make a living, regardless of price, but only up to a point. Large producers may produce more, not less, even at heavy losses. For more refiners, thus, it may appear to be more profitable to maintain turnover of supplies, even at the risk of some oversupply and price reductions in order to maintain standing, or position in the industry. This seems to be fundamental for most petroleum refineries.

Limiting of Supply

Demand for oil and oil products is fairly inelastic. Knowing this, the oligopolistic oil producer is strongly tempted to limit supply. In the exercise of control over supply, the producing oligopolist encounters inelastic conditions of both production and demand.

Since the supply of oil is fixed to a large extent, the problem presents itself as to how the producing oligopolist can dispose of his supply of oil. Since the oil oligopolist has no control over demand, market forces will operate naturally. Because the demand for oil is generally regarded as being inelastic, the oil oligopolist knows that selling only part of his supply of oil at any time brings more in profits to him than if he were to put his entire output on the market and sell the whole amount at one time.

If demand for crude oil or oil products was elastic, the oil oligopolist might find his highest profits accruing from the sale of the entire amount at one time, for any attempt to raise oil prices if demand for oil was elastic would cause a falling off in amount of oil purchased, and so a loss in gross returns received.

Thus, limitation of supply of oil is likely to be found today in many countries, as it is in Kuwait and Libya, for example. Limitation of supply is a possibility when demand is inelastic, as in the case of oil and oil products.

PRICE ELASTICITY OF DEMAND

Generally, elasticity of demand, or the sensitivity of demand for petroleum and petroleum products, is tied in with price. Changes in selling price affect demand for oil and oil products; when barrels of oil and/or oil products demanded by buyers exceeds price reductions, on a percentage basis, the demand is referred to as an elastic demand. When quantities demanded do not exceed price reductions, on a percentage basis, the demand is known as an inelastic demand.

Demand can be plotted on a graph, as illustrated in Figure 3.1, to give a "picture" of elasticity of demand. Price elasticity of demand, or the relationship between changes in amounts bought and changes in prices per unit, is expressed as a ratio, and assumes the shape of a curve sloping from left to right and down. The wider, or more horizontal, the sweep of the curve, the more elastic demand is; the less the sweep, or the more vertical the more inelastic demand is.

Differences between elastic and inelastic demand, however, are only a matter of degree. There is no sharp line dividing demand for one type of

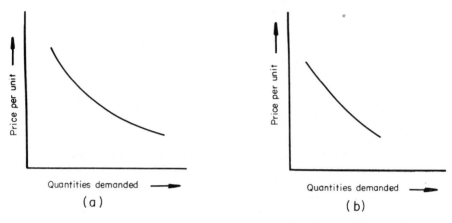

FIG. 3.1.   Typical demand curves.   (a) Elastic type.   (b) Inelastic type.

product from that for another type.   Generally, demand for the necessities of life is inelastic, and demand for the luxuries of life is elastic.   As an example of a necessity, consider a product such as bread.   The amount of bread purchased will vary comparatively little with changes in price.   Consumers may economize slightly on bread when price is high, but the amount of increase in consumption of bread when the price drops is relatively small. When prices of luxury items such as automobiles or color television sets drop, however, demand for these items increases considerably.

Elasticity of Demand for Crude Oil

As stated earlier, demand for crude is a derived demand, because demand for crude is wholly dependent on demand for oil products which use the raw material.   Therefore the elasticity of crude can be either elastic or inelastic depending on the elasticity of the oil products which it serves.   Thus, prices of crude are governed greatly by prices of refined oil products.

Determining Elasticity of Demand for Gasoline

Since gasoline is the most important petroleum product, both in terms of amounts produced and sold and in terms of profits earned for the industry, its demand pattern is worth examining.   For one thing, demand for gasoline seems to be highly inelastic with respect to price.   Also, in an industrial society gasoline is a necessity for which there are no acceptable substitutes. Furthermore, purchases of gasoline are nonpostponable.   And finally, the cost of gasolines to the ultimate consumer represents only a small percentage of the total cost of operating a motor vehicle.

Why is the demand curve for gasoline shaped in an inelastic form? Why is the price (or income) sensitivity of gasoline so slight? Answers to these questions can best be explained in this manner. If gasoline is compared with, say an airplane holiday flight to Beirut, the difference can best be explained in terms of necessity and luxury. Gasoline is a necessity to most people, particularly automobile drivers, while an airplane flight is a luxury to travelers. We don't rush to buy gasoline when prices fall, or at least we don't if we have already purchased a tankful before the fall in price. We are not tempted to buy more gasoline unless we have additional tanks, and that is highly unlikely. But we might be tempted to take an airplane tour to Beirut if the price of the tour dropped suddenly. There are many substitutes for such a trip to Beirut, while there are no substitutes for the gasoline used in motor vehicles. Thus demand for gasoline is inelastic, but the demand for airplane flights is elastic. It is, therefore, the ease or difficulty of substitution that always lies behind the various elasticities of demand.

Time also plays an important role in shaping the demand curve of gasoline. For instance, suppose that the price of gasoline suddenly increased sharply owing to a cutback in crude oil production which left refiners short of crude supplies. The demand for gasoline would still be inelastic in the short run as long as there were ample supplies or reserves of gasoline on hand. As time went on, however, too high a price would stop motorists from driving and car manufacturers from manufacturing. If such an economic condition continued into the long run, car manufacturers would, no doubt, look for power substitutes to run automobiles.

The "relationship of complementarity," which means that some commodities are technically linked so that one cannot use one commodity without the other even though each commodity is sold separately, also influences the determination of the elasticity of a product such as gasoline. For example, if prices of gasoline go up too far, the demand for cars, or at least for large, powerful cars, could drop.

Actually, when reference is made to the inelasticity of demand for gasoline, it means only that the amount taken of the entire product appears to be unresponsive to price changes within a responsible price range. Actually, if prices were to go much past $1.00 (U.S.) a gallon, the effect could be substantial. The degree of price elasticity at this high price level might be considerably greater than that found at lower levels.

The consumption of gasoline, in fact, does not depend so much on price as on the number of cars, trucks, busses, etc., on the highways, the number of miles of highway and improved highway, and the average annual operating mileage of these vehicles.

Determining Elasticity of Demand for Other Refined Products

The demand for lubricating oils is inelastic like that for gasoline, only more so. Lubricating oils have an inelastic demand since they have no substitutes and they are a necessity item, particularly industrially. Furthermore, purchases of lubricating oils are nonpostponable, which also makes demand for them inelastic.

With reference to demand for kerosene, price elasticity of this product is regarded as inelastic. Demand for kerosene is inelastic whether it is used as a principal fuel and illuminant, as it is in developing countries, or for jet fuel, as in industrialized countries. There are few substitutes for kerosene for either use because of its relatively low cost.

On the other hand, the demand for distillate fuel oil seems to be somewhat elastic, largely because distillate fuel oil may be replaced by replacing distillate-fuel-oil-using equipment. The chief short-run factor in consumption of fuel oil is not price so much as temperature in consuming areas. As temperatures become more severe, demand for fuel oil, especially heating oil, tends to rise regardless of price. In the long run, however, it is price that affects demand.

Demand for residual oil is more elastic than that for distillate fuel oil, since there are available substitutes for residual oil.

CONCLUSION ON SUPPLY AND DEMAND

Normally, degree of elasticity of supply is high in the petroleum industry, at least in terms of expansion of product as opposed to contraction of supplies. It takes very little, if any, increase in price to bring additional quantities of crude into the market. However, elasticity of supply is conditioned to some extent by the limitation of time.

On refining, elasticity of supply is conditioned by the joint nature of petroleum product refining. It may take a substantial price increase to bring increasing quantities of a low-value oil product such as kerosene, which is produced jointly with a main product as gasoline whose demand is constant. However, in most cases, especially in the case of large-size refineries, commitments to customers, rather than price advantage, dictate what products are processed from crude at the expense of other products.

Like any other manufactured product, given facilities, raw materials of crude, and time, oil products can be produced in the quantities needed by the market. Actually, adequate supplies of raw materials may not be immediately available, and time may be insufficient to allow the oil industry to process required supplies. Therefore, in the oil industry, as in other industries, one commodity, such as heating oil, may be momentarily in short supply during a particular period of time, especially when weather conditions are more severe than have been expected.

Normally, supply is elastic in both oil products and crude operations. But production of petroleum products is a joint-product operation, so that absolute control cannot be exercised over the production of any one product, even by an individual firm, despite careful forecasting and planning by competent company economists. The supply of one oil product is at least in part a result of the demand for one or more of its other, jointly produced products. Because of this, it is possible to have a surplus in some products due to increased demand for other products. And the elastic supply of refined oil products causes an elastic supply in crude production, since operations of crude depend solely on operations at the refinery.

In summary, demand for crude and its oil products taken as a group is quite unresponsive to changes in price in the short run. But in the long run, if prices continue to rise, consumers will look to substitutes. Actually, the dominant determinants of sales of crude and so oil products are (1) the general level of industrial activity, (2) the standard of living of people in the economy, (3) the level of employment and income, and (4) the weather. In the longer run, the price of petroleum products is important in growth of sales. In the long run, price could affect design and installation of equipment and machinery to use liquid rather than nonliquid fuels, or vice versa.

Finally, to repeat and summarize, there are a number of influences, or clues, to the determination of elasticity of demand. They are as follows:

1. If there are substitutes for the oil product in question, the existence of substitutes renders the demand for that item more elastic. People respond quickly to changes in price of an item if there are available substitutes at a price in competition with a product.

2. If the item constitutes a habit in buying, such as gasoline or fuel oil, or if purchases are made without choice of other items being available, demand is then less elastic than is the case for items where the choice of purchases is not so restricted.

3. When there is inequality of incomes in a market, there is an influence on elasticity of demand for an item. If incomes were equally distributed, which they are not, diminishing utility for the item would then be the only cause affecting elasticity of demand, showing declining satisfactions of buying for successive units of the item in question. But inequality of incomes enables consumers with high incomes to pay higher prices for the first increments (units produced) while those with small incomes cannot purchase the item until the price drops appreciably.

4. Whenever diminishing utility, or diminishing usage of an item, sets in during consumption of that item, an inelastic demand for that item appears.

5. An item that is of the nonpostponable and repetitive buying type always has an inelastic demand.

6. Any item that represents only a small percentage of the total of a product usually has an inelastic demand, or may follow the demand for the whole in the form of a derived demand. For example, gasoline represents only a small percentage of the total operating costs of a car, so gasoline has a price-inelastic demand.

Elasticity of demand can differ considerably among different segments of the demand schedule of an item. That is, response to a change in prices when the price is high can be considerably different from the response to a change in prices when the price is low. So, to be practical, suppliers must give attention to consumer responses to price changes but only within a reasonable range of the existing prices.

Elasticity of demand does not affect the fundamental principle of value. But elasticity of demand recognizes the principle of value, and has an influence on the application of principle of value.

Value on the demand side tends to be determined by the marginal utility (use) of the item to marginal buyers. What marginal buyers are willing to pay at any time determines the price of the item at that time. The elasticity of demand affects this price by causing demand for other items to be affected when the demand for one particular item is either elastic or inelastic.

For instance, the demand for gasoline being inelastic, buyers will usually buy as much gasoline when the price is high as when the price for gasoline is lower. But a higher price for gasoline will tend to lessen the amount of other goods which marginal buyers will buy. In this way, then, the demand for gasoline affects the demand for all other commodities, and the competitive forces that tend to fix the price for gasoline are affected somewhat by the demand for all other items. Thus, differences in elasticities modify values of different products.

NOTES

1. Sullivan, Robert E., Handbook of Oil and Gas Law (Englewood Cliffs, N.J.: Prentice-Hall, 1959), p. 15.

2. Ibid., p. 16.

3. Cassidy, Ralph, Jr., Price Making and Price Behavior in the Petroleum Industry (New Haven, Conn.: Yale University Press, 1954), p. 115.

4. Sullivan, Robert E., Handbook of Oil and Gas Law, p. 20.

5. Nelson, W. L., "Crude Evaluation," Oil and Gas Journal, September 30, 1970, p. 84 (1970).

6. Cassidy, Ralph, Jr. , Price Making and Price Behavior in the Petroleum Industry, p. 115.

7. Russell, Clifford S. , Residuals Management in Industry: A Case Study of Petroleum Refining (Baltimore: Johns Hopkins University Press, 1973), p. 131.

8. Nelson, W. L. , "Refinery Operating Costs," Oil and Gas Journal, vol. 68, no. 9, p. 83 (1970).

9. Bryant, H. S. , "Environment Needs Guide Refinery Sulfur Recovery," Oil and Gas Journal, vol. 71, no. 13, pp. 70-76 (1973).

10. Russell, Clifford S. , Residuals Managements in Industry: A Case Study of Petroleum Refining, p. 121.

Chapter 4

RETURN ON INVESTMENTS:
TIME VALUES IN CAPITAL EXPENDITURES

INTRODUCTION

This chapter is devoted to an examination of some basic forms for application and economic analysis in oil production and in oil marketing which lead to a return of capital investment to investors. The objective of this chapter is to point out some of the "financial tools" that are available to oil management relative to decision making on capital expenditures, that is, some of the time values of money.

The subject of return on investments is continued in Chapter 5 with stress on amortizations—the equivalences of capital sums being returned to investors over a period of time, such as depreciation allowances relative to physical oil properties and depletion allowances relative to the natural oil resources in the ground. Both a study of time values of money and an examination of depreciation and depletion are important in analyzing the financial aspects of oil field and refinery operations.

Time value of money, depreciation, and depletion are often regarded as being almost entirely within the province of the accountant. But the decisions an oil company has to make involving money are influenced largely by economic considerations. Thus, a knowledge of the time value of money, depreciation, and depletion is regarded as necessary background information for the oil engineering economist.

After discussing some basic financial terms and concepts, we will look first at time values of money and the equivalence of time values of money. Compound interest will be discussed here and elsewhere in this text as it affects the conclusions of many economic analyses.

BASIC TERMS AND CONCEPTS

Capital investment: Amounts invested in projects are used, basically, for one of two purposes. These purposes are broadly categorized as follows:

1. Investment of depreciable capital, that is, fixed capital or investment in oil machinery, equipment, buildings, etc.

2. Investment of nondepreciable capital, that is, investment used as working capital. Working capital represents the money necessary to operate the refinery or the oil field. Principally, it is capital tied up in any raw material inventories, in storage, in process inventories, finished product inventories, credit, and cash for wages, utilities, etc.

Working capital must not be ignored in preliminary estimates of needs for capital investment because (1) it is usually a sizable amount of any total investment, and (2) no economic picture of oil processing is complete without inclusion of working capital. Investment in working capital is no different than investment in fixed capital except that the former does not depreciate.

Working capital can theoretically be recovered in full when any refinery, or oil field for that matter, shuts down. Yet capital is "tied up" when the refinery is operating and must be considered part of the total investment. Working capital, however, is normally replaced as it is used up by sales dollars the oil company receives for crudes or refined oil products. Therefore, we can safely say that this capital is always available for return to owners. This is not the immediate case with depreciable capital.

Capital recovery is the replacement of the original cost of an asset and interest. A "sinking fund," explained below, is often used with capital recovery.

Capitalized cost is annual cost capitalized by interest; basically, it is operating costs and compound interest divided by the compound interest rate and the number of years involved, minus one.

Capitalized earning rate is best described by the ratio of depreciation and operating costs to capital investment. It is independent of time and is based on net profit after straight-line depreciation taken together with other operating expenditures. But investment from a time-money standpoint is not included.

Net cash flow represents the difference between cash inflows and cash outflows. Usually net cash flow is net inflows and represents the aftertax profits plus depreciation charges.

Payout, or payback, or years-to-payback is a method for evaluating a proposed investment. Its formula is simply initial investment divided by

annual cash inflow or annual savings, giving the number of years required
to recover the initial investment.

Present value or present worth is a good example of the time value of
money. Basically, its concept is that sums received, or paid, in the future
are worth less than equivalent dollar amounts received, or paid, in the
present. Discount factors, giving the present value of $1 to be received, or
paid, n years hence, are used as multipliers with cash flows to give a discount value, or present worth value.

Capitalization is a variant of the present-value approach. This process
involves dividing net income by an interest rate, giving a dollar amount return on investment to that interest rate.

Rate of return is a measure of the amount of funds invested in a capital
project. It indicates the "pace" at which the investment is being recovered
through earnings made.

A sinking fund is a separate fund into which a periodic deposit is made.
The sum of the deposits plus the accumulated interest on the deposits will,
at the end of a given time period, be equal to a certain sum of money (future
value).

Time value of money can best be explained by use of an example. For
instance, $10,000 cash in hand today is more valuable than the promise of
the receipt of $10,000 at any future time, such as one year from today, even
if the promise is 100% certain. The $10,000 cash in hand today can be in-
vested at interest for one year, and the final sum will exceed $10,000 by the
amount of the interest earned. Therefore, understanding the time value of
money is essentially a matter of understanding compound interest. Oil industry personnel regularly use the concept of compound interest in calculations involving money, since most calculations are made to determine minimum costs incurred in order to obtain a return on invested capital.

PRINCIPLE OF COMPOUND INTEREST

The principle of compound interest is that any interest earned by the original capital investment is added and becomes part of the capital at the end of
the interest period and that interest is earned on all previous interest payments as well as on the original capital. Most oil economic studies are
based on compound interest.

A simple example of compound interest follows.

EXAMPLE 4.1: Compound Interest

A sum of $1,000 is deposited into an account where the interest rate is 6%
compounded annually. What is the future value of the deposit if it is left in
the account for 2 years?

Solution

For the first year, using the formula

Principal + principal × interest = amount

$1,000 + $1,000 (0.06) = $1,060

For the second year, interest is due on both the initial investment and on the first year's return, or $1,000 × $(1 + 0.06)^2$, which equals $1,000 × 1.124 = $1,124 (value of $1,000 after 2 years with compound interest). (Compound interest tables may be found in Appendix B at the back of the book.)

From Example 4.1 it is apparent that compound interest increases future value (here $1,124) faster than simple interest, which in 2 years would be $1,120 (0.06 × $1,000 or $60 for the first year, and 0.06 × $1,000 or $60 for the second year; $60 + $60 = $120 + $1,000 original sum = $1,120).

EFFECTIVE INTEREST RATES

The way interest rates are quoted affects return on investment. For instance, the future value (or return) after 1 year of $1,000 compounded annually at 6% is $1,060. But at 6% compounded quarterly, the return on the same $1,000 after 1 year is $1,000 × $(1 × 0.015)^4$ = 1,061. Both quoted rates, 6% compounded annually and 6% compounded quarterly, are nominal rates. These rates may be compared to each other by calculating the effective interest rate, which is the ratio of return to the amount invested.

EXAMPLE 4.2: Effective Interest Rates

For $1,000 invested at 6% compounded annually, the effective interest rate is

$$\frac{\text{Future value - principal}}{\text{principal}}$$

or

$$\frac{\$1,060 - \$1,000}{\$1,000} = \frac{60}{1,000} = 0.06 \ (6\%)$$

for the $1,000 invested.

At 6% compounded quarterly, the effective interest rate is

$$\frac{\text{Future value - principal}}{\text{principal}}$$

or

$$\frac{\$1,061 - \$1,000}{\$1,000} = \frac{61}{1,000} = 0.061 \ (6.1\%)$$

for the $1,000 invested.

Proof

(6% becomes 1.5% when compounded quarterly.)

F (future value)   3 months = $1,000 + 1,000 \times 0.015 = \$1,015.00$

F (future value)   6 months = $1,015 + 1,015 \times 0.015 = \$1,030.23$

F (future value)   9 months = $1,030.23 + 1,030.23 \times 0.015 = \$1,045.68$

F (future value) 12 months = $1,045.68 + 1,045.68 \times 0.015 = \$1,061.00$
(rounded off)

Actually, effective (stipulated) interest rate can be derived without reference to the amount invested by using the formula

$$\left(1 + \frac{\text{nominal interest rate}}{\text{no. of payments}}\right)^{\text{no. of payments}} - 1$$

or, substituting the figures from Example 4.2,

$$\left(1 + \frac{0.06}{4}\right)^{4} - 1 = (1.015)^{4} - 1 = 0.061$$

EXAMPLE 4.3: Value of Effective Interest Rate

To illustrate the value of knowledge of the effective interest rate to oil management, assume that a short-term loan for 1 year only could be arranged for an oil company in temporary distress. The company needs $100,000 for immediate working capital at either a nominal rate of 12% compounded monthly or a nominal rate of 15% compounded semiannually. The oil company wants to know which arrangement would provide the oil company with the lower debt at the end of the short-term loan period. The use of the effective interest rate formula gives the answer.

Solution

On nominal 12% rate, compounded monthly:

$$\text{Effective interest rate} = \left(1 + \frac{0.12}{12}\right)^{12} = 1$$

$$= (1.01)^{12} - 1$$

$$= 1.127 - 1, \text{ or } 0.127\%$$

On nominal 15% rate, compounded semiannually:

$$\text{Effective interest rate} = \left(1 + \frac{0.15}{2}\right)^{2} - 1$$

$$= (1.075)^{2} - 1$$

$$= 1.156 - 1, \text{ or } 0.156\%$$

The loan at 12% compounded monthly has the lower effective interest rate, or 12.7% to 15.6% for the loan arrangement using a nominal rate of 15% compounded semiannually. Thus the oil company will borrow $100,000 for 1 year at 12% interest compounded monthly, paying back the loan at the end of 1 year with $112,700, which includes $12,700 in interest, instead of borrowing at 15% compounded semiannually which would cost $15,600 in interest, and a total of $115,600. Thus the oil company saves $2,900 by borrowing at 12% compounded monthly.

COMPOUND INTEREST FACTORS

In studying the development of compound interest factors, the following six conversion factors, covering both single payments and uniform payments, are given for the oil engineering economist to study, with simple examples to illustrate their usefulness to petroleum companies. In these conversion factors,

i   represents interest

n   represents the number of periods of interest payments

A   represents annual payments or receipts

P   represents value

F   represents future value

Single payment:

1. Compound amount factor, "to find F, given P" formula

$$(1 + i)^n,$$

     or use the compound interest table in Appendix B and obtain the compound interest factor.

2. Present value factor, "to find P, given F" formula,

$$\frac{1}{(1 + i)^n}$$

Uniform annual series of payments:

3. Sinking fund factor, "to find A, given F" formula,

$$(1 + i)^n - 1$$

4. Capital recovery factor, "to find A, given F" formula,

$$\frac{i(1 + i)^n}{(1 + i)^n - 1}$$

5. Compound amount factor, "to find F, given A" formula,

$$\frac{(1 + i)^n - 1}{i}$$

6. Present value factor, "to find P, given A" formula,

$$\frac{(1 + i)^n - 1}{i(1 + i)^n}$$

Examples illustrating the practical use of each of these compound interest factors follow.

EXAMPLE 4.4: Compound Amount Factor, Single Payment

Given: A study is being made on the possibility of investing $2,500 in a fund for 10 years at 5% compounded annually. At the end of 10 years, the fund is expected to provide enough money for the replacement and addition of 20 typewriters and 5 office calculators.

Wanted: An accumulated value in 10 years, that is, the future value of the present amount of $2,500.

Solution

Using the compound interest table in Appendix B and the factor or formula "to find F, given P," we have $2,500 × 1.629 = $4,062.50 as the future value of $2,500 at the end of 10 years when $2,500 is compounded annually. This amounts to total interest earned on the investment of $1,562.50 ($4,062.50 - $2,500).

If $2,500 had been invested at 5% compounded semiannually, the future value of $2,500 would have been $2,500 × 2.653, or $6,632.50, based on a 2.5% compound amount factor. The $6,632.50 total brings $4,132.50 in interest.

EXAMPLE 4.5: Present Value Factor, Single Payment

Given: $25,584 will be needed in 5 years to replace a furnace currently in use.

Wanted: Management wishes to know what lump sum of money must be set aside now to provide for the needed capital at the end of 5 years. Five percent interest compounded annually is the cost of capital (cost of borrowing).

Solution

Using the compound interest table in Appendix B and the formula "to find P, given F" (the present value factor), we have $25,584 × 0.7835 = $20,045; in other words, $20,045 is the amount that must be put aside in one lump sum so that 5% compounded interest at the end of the fifth year will yield $25,584.

EXAMPLE 4.6: Sinking Fund Factor, Uniform Annual Series of Payments

Given: In 10 years, it is estimated that $144,860 (future value) will be required to purchase several cooling towers; interest available at the bank is 8% compounded annually.

Wanted: Since we wish to create a sinking fund at the bank and deposit each year to provide the required capital of $50,000, we want to know the annuity (A) which will amount to the given fund, $144,860, after 10 years of deposits.

Solution:

Using the compound interest table in Appendix B and the formula "to find A, given F," we have $144,860 × 0.06903 (8% for 10 years) = $10,000 for each annual annuity payment.

Thus each year a payment or deposit of $10,000 should be made into the sinking fund at 8% compounded annually. After 10 years, the fund will contain $144,860 with which the oil company can purchase cooling towers as provided for by the fund. Table 4.1 illustrates the future value at the end of 10 years of $144,860, with total deposits of $100,000. At the end of the second year the fund shows a total of $38,500, and at the end of the fifth year a total of $86,200. Amounts into the fund, including interest, decrease as each year progresses, with no interest being included in the tenth payment.

EXAMPLE 4.7: Sinking Fund Factors, Capitalized Costs

A sinking fund is to be established to cover the capitalized cost of temperature recorders. The recorders cost $2,000 and must be replaced every 5 years. Maintenance and repairs comes to $200 a year. At the end of 5 years the accumulated sinking fund deposits are expected to cover the capitalized cost of continuous expense for these recorders. How much money must be deposited each year, at an interest rate of say 5%, to cover the capitalized costs at the end of 5 years?

Solution

Using the formula "to find A, given F," we have: $2,000 × 0.18097 = $361.94 per year (annual cost of capital recovery). These annual deposits of $361.94 put out at 5% interest will be worth exactly $2,000 at the end of 5 years.

Proof: Using the formula "to find F, given A," $361.94 × 5.526 (compounded interest factor, 5% at 5th year) = $2,000. (This proves that the amount of $361.94 is correct.)

The capitalized cost for the continuous expense of service requires a capitalization of the annual costs for the periodic renewal of the installation, using the money put aside in the sinking fund. In addition, the other uniform annual expenditures must be capitalized. So if we assume maintenance and repairs to be $200 a year, the annual charge to provide $2,000 every 5 years is:

$$\text{Sinking fund deposit} = \$361.94 \ \frac{0.05}{(1.05)^5 - 1} = \$398$$

$$\text{Maintenance and repair expense} = \underline{\ \ 200}$$

$$\text{Total annual expenditures} = \$598$$

The present capitalized value of these annual expenditures is then computed as

TABLE 4.1

Annual Payment into Sinking Fund

Col. 1 Year	Col. 2 Payment into fund	×	Col. 3 Compound interest factor	Col. 4 Compound interest	Col. 5 Payment with interest into fund (Col. 2 × Col. 3)	Col. 6 Amount in sinking fund (Col. 5 × Col. 6)
1	$10,000		1.999	$(0.08)^9$	$19,990	$19,990
2	10,000		1.851	$(0.08)^8$	18,510	38,500
3	10,000		1.714	$(0.08)^7$	17,140	55,640
4	10,000		1.587	$(0.08)^6$	15,870	71,510
5	10,000		1.469	$(0.08)^5$	14,690	86,200
6	10,000		1.360	$(0.08)^4$	13,600	99,800
7	10,000		1.260	$(0.08)^3$	12,600	112,400
8	10,000		1.166	$(0.08)^2$	11,660	124,060
9	10,000		1.080	$(0.08)^1$	10,800	134,860
10	10,000		0	(0)	10,000	144,860
Totals:	$100,000				$144,860	

$$\text{Present value} = \frac{\$598}{0.05} = \$11,960 = \begin{array}{l}\text{capitalized value of sinking} \\ \text{fund deposits and maintenance} \\ \text{and repair expense}\end{array}$$

+ Initial investment in the
 automatic recorders = 2,000

  Total capitalized cost = $13,960

EXAMPLE 4.8: Capital Recovery Factor, Uniform Annual Series of
      Payments

Capital recovery is desired for the sinking fund in Example 4.7. That is,
given a sinking fund of $2,000 invested at 5% compound interest, we wish to
know the sum that can be withdrawn over a 5-year period.

Solution

The formula used is "to find A, given P," or "to find annuity value, given
present value."

 Using the capital recovery factor derived from Appendix B, A = $2,000
× 0.23098 = $461.94. The $461.94 collected each year for 5 years will total
to $2,000, or the amount of the sinking fund.

 Capital recovery, or recovery of investment capital with interest (the
profit to the investor on the investment), is a matter of vital concern to the
oil investor. He usually looks for assurance that any risk he takes with his
investment is proportional to the interest earned.

 Capital recovery is thus important in any study of oil economics,
since it is repayment to the oil investor of his investment plus interest.
Capital recovery is the reward to the oil investor for the use of his money,
and for the risk he was willing to take.

 To illustrate how valuable interest is to investors, consider this example. A company pays $1,000 for an electric motor and pump that has an
assumed life of 10 years. At the end of 10 years the motor and pump are
worn out and the company has nothing to show for its $1,000 investment.
The capital has disappeared. Thus, in this case, the oil investor who put
up the original $1,000 has lost his capital invested plus any interest that
would have been earned by him by investing in another venture that was
profitable.

 A means must be found to protect investors from such losses. The
following equation is one method for estimating the cost of such protection
for the investor. This is the establishment of an annual fund set aside for

a number of years, a sinking fund, with the original investment being recovered plus interest on the investment. The formula used is "to find A,
given P."

$$A = \frac{Xi(1 + i)^n}{(1 + i)^n - 1}$$

where X is the amount invested, A is the annual payment, i is the interest
rate, and n is the number of years for recovery of investment. (See the
compound interest tables in Appendix B.)

If we assume a cost of $1,000 for an electric motor and pump with a
life of 10 years, with a 6% rate of interest added, we obtain an A of $135.87:

$$A = \frac{1,000 \, (0.06) \, (1 + 0.06)^{10}}{1 + 0.06} = \frac{60 \, (1.06)^{10}}{1.06} = \$135.87$$

$135.87 is the amount of capital recovered or earned each year at 6%
interest, the capital recovery charge on the investment of the pump and motor. After 10 years, the total return on investment will be $1,359 (rounded
off), which includes the original investment of $1,000 plus total interest of
$359 earned.

If we wait until after 10 years for return of capital plus interest, the
future worth, or value, of a $1,000 investment would be $1,791. The $1,791
is derived from the compound interest formula as follows:

Future value = present value $(1 + i)^n$

where i is the interest and n the number of years.

Future value = $1,000 \, (1 + 0.06)^{10}$ = $1,791

Actually, $432 is "lost" in interest by receiving annual capital recov-
ery payments, as the following analysis indicates:

$1,791   is the future value of $1,000 if annual interest of 6% is re-
          ceived in one lump sum at the end of 10 years, along with the
          principal of $1,000.

$1,359   (rounded off) is the future value if interest is received along
          with part payment of principal in annual payments for each
          year of 10 years.
_____

  $432   difference
_____

In determining what the value is at the present time for a perpetual series of uniform annual payments in the future, the equivalent "capitalized cost" of a series of annual operating costs, such as repairs and maintenance, that must be paid in an indefinite number of periods in the future in order to continue the given series should be considered. Thus, if repairs and maintenance in a refinery cost $30,000 each year on the average, the capitalized cost of such continuous expense at an interest rate of, say 8%, is

$$\frac{\$30,000}{0.08} = \$375,000$$

which is the equivalent cost of a series of annual operating costs.

The sum of $375,000 must be added to the initial cost of the refinery, say $2 million, to obtain the total capitalized cost for the refinery. For example,

Cost of refinery	$2,000,000
+ Capitalized cost of continuous expenses	375,000
Total capitalized cost of the refinery	$2,375,000

EXAMPLE 4.9: Capital Recovery Factor

Given: $20,000 (present value) is invested at 10% compounded annually.

Wanted: What is the annual sum that can be withdrawn over a 20-year period (capital recovery); or what is the annuity provided by a given capital of $20,000?

Solution

Using the compound interest table in Appendix B and the capital recovery factor or formula "to find A, given P," we have

$$A = \$20,000 \times 0.11746 = \$2,349.20$$

In other words, $2,349.20 is the annual sum that can be withdrawn, or the amount of capital recovery.

This, $2,349.20 is also regarded as the annuity provided by the $20,000 already invested for 20 years (sinking fund), which after 20 years has swelled to $46,984 (2,349 × 20).

If we apply the sinking fund factor which is derived from the formula "to find A, given F," we would have $46,984 as future value times 0.01746, the sinking fund factor from the compound interest table at 20 years, and $46,984 \times 0.01746 = $830.34.

EXAMPLE 4.10: Capital Recovery Factor

Given: An oil company wishes to repay in 10 installments a sum of $400,000 borrowed at 8% annual interest.

Wanted: To determine the amount of each future annuity payment (A) required to accumulate the given present value (debt) of $100,000 when the interest rate, 8%, is known, and the number of payments, 10, is known.

Solution:

Using the compound interest table for 8%, under the formula "to find A, given P," we get

$A = $100,000 \times 0.14903 = $14,903$ annual payment

$149,030 will thus have been paid: $100,000 as principal and $49,030 in interest.

The $100,000 is the present value of the 10-year annuity; $14,903 are the annual payments, or annual capital recovery by the creditor. A capital recovery factor, or $(A/P)_n^i$, was used.

The relationship among time-value annuity factors is apparent from the way the capital recovery of $(A/P)_n^i$, or "to find A, given P" formula, can be converted to the sinking fund factor, $(A/P)_n^i$, or "to find A, given F" formula.

EXAMPLE 4.11: Compound Amount Factor, Uniform Annual Series of
              Payments

Given: $500 invested each year at 8% interest.

Wanted: What is the accumulated value in 10 years, that is, what is the amount of an annuity at the end of 10 years?

Solution

Using the compound interest table, and the formula "to find F, given A," we have

$F = 500 \times 14.487$ (compound amount factor for 10 years) $= $7,243.50.

If we wished to know what F (future value) is after 15 years, we have:
F = 500 × 27.152 (compound amount factor for 15 years) = $13,576.00.

EXAMPLE 4.12: Present Value Factor, Uniform Annual Series of Payments

Wanted: What sum of money must be deposited in the bank today at 6% interest in order to receive $10,000 annually for 20 years?

Given: $10,000 is required annually for 20 years.

Solution

Using the compound interest table, and the formula "to find P, given A," we have: P = $10,000 × 11.470 (present value factor) = $114,700. So $114,700 is the amount that must be deposited today at 6% interest in order to receive $10,000 annually for 20 years.

Present Value of a Future Amount

Present values can be of great use to oil management since many accounting valuations and oil business decisions are based on present-value computations. Basically, the problem is to determine the value now of an amount of money to be received or paid in the future. To do this, the interest rate and the compounding period must be given. Then questions occur, such as "how much must be deposited in a fund now paying 4% compounded annually in order to accumulate $100 in 2 years? The formula "to find P, given F" is used together with the interest factor derived from a compound interest table of 4% for 2 years to give

$$P = \$100 \times 0.9246 = \$92.46$$

$92.46 must be deposited in a fund paying 4% compounded annually to accumulate $100 in 2 years. Thus the present value calculation converts the $100 to a present-day value, which in this case is $92.46.

So a present value calculation converts a single future sum to an equivalent amount at any earlier time, which date does not necessarily have to be the present. We can also use the formula involving compound interest "to find P, given F," which is $(P/F)_n^i$ or the ratio of present value to future value. Thus, if the future value is $215,900, and interest at 8% is compounded for 10 years, present value today of the future value of $215,900 is $100,000 ($215,900 × compound factor of 0.4632 or 8% at 10 years).

Present value calculations also include the converting of a series of future values, with interest rate and number of payments known, to present values. For instance, in an annuity, the capital recovery factor, or $(A/P)_n^i$ (the formula "to find A, given P") is used.

EXAMPLE 4.13: Present Worth Factor, Uniform Annual Series of Payments

Assume 8% interest compounded annually, number of payments as 10, or 10 years, and the debt of the oil company as $67,090, and determine the present value of a 10-year annuity. '

Solution

We calculate as follows:

First, determine the uniform series of payments to be made. This is done by using the capital recovery factor, $(A/P)_n^i$, and substituting:

$$\frac{A}{\$67,090} \ .08/10 \ = \ \$67,090 \times 0.14903 \quad \text{(8\% compound factor for 10 years from the table in Appendix B)}$$

$$= \ \$10,000.$$

Thus, each year for 10 years, uniform payments to be made to repay the $67,090 debt owed by the oil company will be $10,000.

Second, determine the present worth of the uniform series of $10,000 payments by dividing $10,000 by $(A/P)_n^i$ ("to find P, given A" formula) for each year; then add all the years to get the total present worth of $67,090. Table 4.2 gives these figures as illustrated and Figure 4.1 is a summary of Table 4.2. Thus, it will take ten $10,000 payments by the oil company to repay the $67,090 loan made to them.

Total payment is $100,000 for 10 years

Total interest paid is <u>32,910</u> for 10 years

Principal amount of debt $ 67,090

Capital recovery to the creditors is $100,000 for a loan of $67,090 made to the oil company. Capital recovery includes the investment of $67,090 + 8% interest compounded annually, or $32,910, for a total return of $100,000 to the creditors.

Summary

Values involved in compound interest factors are always future (F), present (P), or annuity values (A) tied in with future values or present values.

TABLE 4.2

10-Year Annuity and its Present Value

End of year or time of payment	Amount of payment A	Present value $(F/P)_n^i$		
1	$10,000	$-1.08 \ (1.08)^1$	=	$9,260
2	10,000	$-1.166 \ (1.09)^2$	=	8,570
3	10,000	$-1.260 \ (1.08)^3$	=	7,940
4	10,000	$-1.360 \ (1.08)^4$	=	7,350
5	10,000	$-1.469 \ (1.08)^5$	=	6,810
6	10,000	$-1.587 \ (1.07)^6$	=	6,300
7	10,000	$-1.714 \ (1.08)^7$	=	5,835
8	10,000	$-1.851 \ (1.08)^8$	=	5,400
9	10,000	$-1.999 \ (1.08)^9$	=	5,000
10	10,000	$-2.159 \ (1.08)^{10}$	=	4,630
Total of annuity payment made to pay off the debt.	$100,000	Present value (P) = of 10 years' annuity		$67,090

Steps in the use of compound interest factors or formulas involving F, P, and A for measurement and determination of time values of money for expansion or replacement of older assets are as follows.

1. Determine what is wanted—F, P, or A.

2. Determine what is given—as to F, P, or A.

3. Then apply the formula as to what is given and what is desired, or, use the appropriate compound factor for the formula (found in Appendix B) with the desired rate of interest (i) and the period of time or number of years (n).

EQUIVALENCE

A knowledge of equivalent values can be of importance to oil companies. The concept of equivalence is the cornerstone for comparisons of time values of money comparisons. Incomes and expenditures are identified with time as well as with amounts. Alternatives with receipts and disbursements can be compared by use of equivalent results at a given date, thus aiding in decision making.

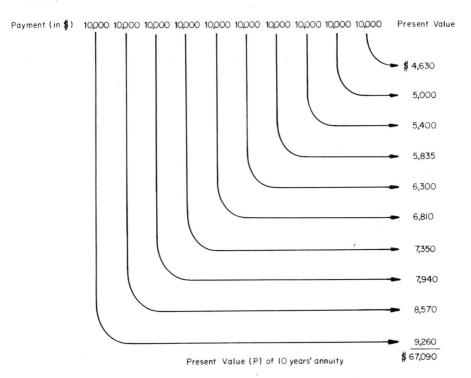

FIG. 4.1. Ten year annuity—8% interest compounded annually.

EXAMPLE 4.14: Equivalence

For example, how can $100,000 to an oil company today be translated into
equivalent alternatives of cash flow?

Solution

$100,000 today is equivalent to $215,900 ten years from now, if 8% is used
as a figure for value or future money worth (using the formula "to find F,
given A," the compound amount factor), and $100,000 today is equivalent to
$25,046 received at the end of each year for the next 5 years (using the for-
mula "to find A, given P," the present value factor).

Cash flow is translated to a given point in time by determining the
present value or the future value of the cash flow.

DETERMINING THE COST OF CAPITAL INVESTMENT

Present Value Method

When speaking of the cost of capital, the implication is actually the "marginal cost" of capital. This can best be understood by use of an example.

EXAMPLE 4.15

Assume that capital investment in a bubble tower is $10,000, and replacement—in this case $10,000 original cost—is recouped in 5 years through annual depreciation = an allowance of $2,000. The cost of capital (what money is worth to borrow) is 10%. Assume also that net income after taxes is $2,600.

Then, after-tax cash flow = after-tax profit + depreciation, or

$4,600 = $2,600 + $2,000

Using the formula "to find F, given P," or use of the compound amount factor

	1st year	2nd year	3rd year	4th year	5th year
$10,000 =	$\dfrac{\$4,600}{(1+i)^1}$	$\dfrac{\$4,600}{(1+i)^2}$	$\dfrac{\$4,600}{(1+i)^3}$	$\dfrac{\$4,600}{(1+i)^4}$	$\dfrac{\$4,600}{(1+i)^5}$
=	$\dfrac{\$4,600}{1+0.10}$	$\dfrac{\$4,600}{1.21}$	$\dfrac{\$4,600}{1.331}$	$\dfrac{\$4,600}{1.4641}$	$\dfrac{\$4,600}{1.61051}$

= $4,181,818 + $3,801,653 + $3,456,048 + $3,141,862 + $2,856,238

= $17,437.619   (present value of after-tax cash flow)

or:

$$\$10,000 = \$4,600 \,(0.10)^5$$

To find P, given A,

$10,000 = 3.79078 × $4,600   or $17,437.619

The present value method is valuable because of its interpretation of the inequality that results. If the present value, $17,437.619, of after-tax cash flows of $4,600 discounted at the cost of capital, which is 10% in this case, is greater than the replacement cost of the investment outlay, which in this case is $10,000, the project is profitable. And the difference between the present value of after-tax cash flow and replacement cost is the profit on the investment.

Summary:

Present value of after-tax cash flows discounted by cost of capital (at 10%) cash flows	$17,437.59
Replacement (original investment) cost	10,000.00
Net present value, or profit on the project	$7,437.59

$7,437.59 is the profit—because the return on investment is $10,000 subtracted from the present value of cash flows, or $7,437.59, and the cost of using the funds each year is accounted for when the cash flows are discounted by the cost of the capital of $4,600 of 10%.

The value of the present value method is in making comparisons of substitute methods of production or alternative machines to use. One chooses the process or machine which results in the highest net present value.

Limit to Use of Present Value Method

Investment under the present value method is carried out until the present value of cash flows equals the replacement cost of the original investment. So the last increment of investment funds yields cash flows which, when discounted by the cost of capital, return the replacement cost of the investment and cover the cost of using the funds.

TIME VALUE OF MONEY

Time value conversions may be explored in more detail by use of several practical examples. Particular emphasis is placed on the economics of selecting alternative equipment in refineries, where a comparison of relative costs on services obtained may be made.

The basic relation for economic selection of alternatives is

$$P = A \frac{(1 + i)^n - 1}{i(1 + i)^n}$$

or the formula "to find P, given A," where

    P = present value (investment cost) of money, or original investment

    A = end-of-period payments, periodically, as annual costs

    i = earned rate of interest on present value, or rate of return method

    n = number of periods for which end-of-period payments are made

Any one of these four variables can be used to compare costs of services offered by oil equipment and to compare earnings obtained when this oil equipment is used in production. (The formula can be substituted by using the compound interest factors discussed earlier.)

Annual Cost Method

When A is used as a basis to compare costs on two or more units of oil equipment being considered for purchase, either as a replacement or as additional equipment, annual direct costs plus capital recovery costs are used.

    The annuity equation is used here where interest rate and time period are known. It features capital recovery with the equation.

EXAMPLE 4.16: Annual Cost Method

Given: Consider two possibilities relative to the purchase of a heat exchanger for an oil refinery to replace an older model for which annual costs are running around $20,950.

    Purchase Possibility A is a heat exchanger constructed with materials of steel and copper. Its investment cost is $15,000. Its economic service life is estimated to be 10 years, and salvage value at the end of the tenth year is estimated at $500. Annual labor, maintenance, repairs, and operational expenses are estimated at $11,500; other annual direct costs are 4% of the investment cost of $15,000, or $600, when operating under optimum conditions.

    Purchase Possibility B is a stainless steel heat exchanger with an investment value of $40,000. Its economic life is also regarded as 10 years, with a scrap value of $1,000 at end of the tenth year. Annual labor, maintenance, repairs, and operational expenses are estimated at $4,000; other annual direct costs are 7% of the investment cost of $40,000, or $2,800, when operating under optimum conditions.

    Capital recovery of the original investment is at 8% or the current cost of capital.

Wanted: Using the annual cost method, determine which purchase possibility would be more economical with respect to annual costs.

Solution

Purchase Possibility A with capital recovery, formula "to find A, given P"

Purchase Possibility B with capital recovery, formula "to find A, given P"

(original cost - salvage value) (recovery factor) + (salvage value) (interest rate)

(original cost - salvage value) (recovery factor) + (salvage value) (interest rate)

($15,000 - $500) (0.1490) + ($500) (0.08) = $2,201 capital recovery of original cost and salvage value

($40,000 - $1,000) (0.1490) + ($1,000) (0.08) = $5,891, capital recovery of original cost and salvage value

Summary of annual costs with capital recovery

Recovery of capital	$2,201
Annual costs, Maint., Rep.	11,500
Annual costs, optimum conditions	600
Total annual costs	$14,301

Summary of annual costs with capital recovery

Recovery of capital	$5,891
Annual costs, Maint., Rep.	4,000
Annual costs, optimum conditions	2,800
Total annual costs	$12,691

With a potential savings in annual costs of $1,610 ($14,301 - $12,691) in favor of the stainless steel heat exchanger, Purchase Possibility B appears to be the more feasible "buy" according to the annual cost method. (Only differences in costs, with cost items common to both purchase possibilities, were used.) Furthermore, the salvage value of the stainless steel exchanger ($1,000) is $500 more than for the steel-copper exchanger.

The annual cost method is used where the same costs for each alternative recur annually almost in the same manner. For a series of costs which are nonuniform, an average annual cost equivalent might be calculated. For alternatives with different lifetimes, the time period for comparison might be that of the alternative with the shortest life.

Whereas the annual cost method does not give the relative amounts of capital, the present value method does. The present value method reduces all costs to equivalent capital at a given date.

Present Value Method

Using the present value method, a series of known uniform annual costs are reduced to an equivalent present value. This allows one to estimate the dollar value at the present time that is equivalent to the amount of annual costs for some fixed years of service by two alternatives. But uniform annual costs must first be determined, and this is what the present value method does. The annual cost method does not determine uniform annual costs.)

EXAMPLE 4.17: Present Value Method

Given: Assume the same two heat exchangers as in Example 4.16, with the same annual costs, economic lives, salvage values, and investments, and with the cost of capital once again 8%.

Wanted: Present worth values for each of the possibilities, as well as total equivalent capital at the "present" time of consideration of purchase of heat exchangers.

Solution

For each of the possibilities, present values of installations and a salvage for present value must be added and deducted, respectively, to current value of annual costs for 10 years in order to get total equivalent capital requirements.

So, in finding equivalent capital at 8%,

	Purchase Possibility A	Purchase Possibility B
1. Present value of original (initial) costs	$15,000	$40,000
2. Present value of salvage value: formula "to find P, given F," or factor	$500 × 0.4632 = -232	$1,000 × 0.4632 = -463
3. Present value of annual costs: (total costs) × factor of formula "to find P, given A"	$12,100 × 6.710 = $81,191	$6,800 × 6.710 = $45,628
4. Total "present" equivalent capital at 8%: (1) - (2) + (3)	$95,959	$85,165

A comparison of the calculations for equivalent capital involved, $95,959 for Possibility A and $85,165 for Possibility B for the present time on an economic basis, indicates $10,794 less favoring Possibility B. In other words, a savings of $1,610 in annual costs, as given by the annual cost method, favoring Possibility B is reflected in a $10,794 reduction in equivalent present value of Possibility B when annual costs are uniform and determined with the use of the present value method.

Under conditions of low interest rates, the present or current value of Possibility B, which is $40,000, can still be less than the current worth of $15,000 for Possibility A. Thus, the interest rate is important in order to determine the present value. Lower interest rates, such as say 5% instead of 8%, favor even more the use of higher initial investments, in this case the $40,000 stainless steel heat exchanger rather than the $15,000 steel-copper exchanger, because the relative cost for the use of money is lower. Example 4.18 will show that this is so.

EXAMPLE 4.18: Annual Cost and Present Value Methods

Compare the relative annual costs and current present values of the two alternatives in Examples 4.16 and 4.17 for 10 years of service if money is worth 5% instead of 8%.

Solution

For the annual cost method at 5%:

	Purchase Possibility A	Purchase Possibility B
Annual Costs:		
Capital Recovery—		
$14,500 × 0.1295		
("to find A,		
given P") +		$39,000 × 0.1295 +
(0.05) ($500) =	$1,903	(0.05) ($1,000) = $5,101
Labor, maintenance, etc.	11,500	4,000
Other direct costs	600	2,800
Total annual costs	$13,903	$11,901

Compared to the costs when the interest rate is 8% (see Example 4.16), total annual costs for each possibility are lower when the interest rate is 5%. But the difference in annual costs is greater when the interest rate is lower. At 8% the difference was $1,610 less in favor of Possibility B, whereas at 5% it is $2,002 in favor of Possibility B. Thus, lower costs of borrowing favor alternatives with large investment amounts more than alternatives with lower investment amounts.

	Purchase Possibility A	Purchase Possibility B
Present worth of original (initial) costs	$15,000	$40,000
Present worth of salvage value ("to find P, given F"):	$500 × 0.6139 = -307	$1,000 × 0.6139 = -614
Present worth of annual costs ("to find P, given A"):	$12,100 × 7.722 = 93,436	$6,800 × 7.722 = 52,509
Total "present" equivalent capital at 5%	= $108,129	$91,895

At 5%, equivalent capital for Purchase Possibility B is $16,234 ($108,129 - $91,895) less than Purchase Possibility A. With lower interest rates, differences in equivalent capital are greater: $16,234 between alternatives at 5% and $10,794 between alternates at 8%. However, total present equivalent capital amounts are greater with lower interest rates: Totals at 5% are $108,129 and $91,895 for Possibilities A and B, respectively, and at 8% are $95,959 and $85,165.

A comparison of these results to those obtained when money is worth 8% shows that (1) the time-money series are equivalent to larger capital requirements, and (2) the difference in equivalent present value is greater in favor of Purchase Possibility B than it is for A when money is worth only 5%.

Capitalization

Comparable capitalized costs are for 5 and 6%. The principal difference in computing capital requirements through capitalization compared to the present value method is that the initial cost, less salvage, is recovered during the period with interest on initial capital omitted.

Use of a sinking fund, with an annual sinking fund deposit, is used. Total capitalized cost thus becomes the sum of the initial cost plus the capitalized value of the annual costs.

EXAMPLE 4.19: Capitalization

This example is actually a conclusion of determining capital requirements for the examples discussed previously through capitalization with interest at 5 and 8%.

Capital requirements through capitalization, with interest at 8%, are as follows:

	Purchase Possibility A	Purchase Possibility B
Total annual costs:		
Net capital inves- ted factor of formula ("to find A, given F"):	$14,500 × 0.06903 = $1,001	$39,000 × 0.06903 = $2,692
Annual labor, maintenance, operational costs, etc.	11,500	4,000
Other direct annual costs	600	2,800
Total annual costs to be capitalized	$13,101	$9,492
Capitalization of annual costs:	$13,101/0.08 = 163,763	$9,492/0.08 = 118,650
Initial costs of investment	15,000	40,000
Total capitalized cost when money is worth 8%	$178,763	$158,650

Capital requirements through capitalization, with interest at 5%, are as follows:

	Purchase Possibility A	Purchase Possibility B
Total annual costs:		
Net capital inves- ted factor of formula ("to find A, given F"):	$14,500 × 0.0795 = $1,153	$39,000 × 0.0795 = $3,100
Annual labor, maintenance, operational costs, etc.	11,500	4,000
Other direct annual costs	600	2,800
Total annual cost to be capitalized	$13,253	$9,900
Capitalization of annual costs:	$13,253/0.05 = 265,060	$9,900/0.05 = 198,000
Initial costs of investment	15,000	40,000
Total capitalized cost when money is worth 8%	$280,060	$238,000

At the lower interest rate of 5%, the capitalized cost is $42,060 less for Possibility B ($280,060 - $238,000). The results illustrate the peculiar effect of the interest rate and emphasizes the potential difficulties in comparing alternates on either a present value or a capitalized cost basis. When cost of capital is high, total capitalized costs become lower, but differences between capitalized costs of higher and lower investment amounts favor higher investments more when cost of capital (interest rate) is lower.

The interest rate is the determining factor, although the relative size of such individual items as initial costs, annual labor costs, annual material, repairs, maintenance, and other costs, when compared to capital recovery costs, can affect total equivalent capital involved.

The important point is that the interest based on the going value of money is always lower than the rate for a venture involving a risk. The engineer using the going rate for interest will bias his comparisons in favor of the alternative equivalent to oil capital requirements. Because of this, the annual cost method is preferred, but the service lives of the alternatives should be equal and annual costs of alternatives should be uniform. When different service lives are involved, or where nonuniform annual expenditures must be compared for alternatives, it is better to use the present value method and put all costs on a comparable basis in order to get accurate results and avoid "distortions" of costs.

Rate of Return Method

The rate of return method is used, to a large extent, for replacement of equipment. The method uses the same equation, the annuity equation, as the present value and annual cost methods as a basis for comparison where the interest rate and time periods are known.

So costs involved for each alternative are reduced to their respective time-money series by determining the interest that makes each time-money series equal, thus obtaining an equivalent interest. All monies involved are earning at the equivalent interest rate. Then if one alternative requires a greater investment than another alternative, this equivalent interest is also the earning rate for the extra investment.

By reducing various costs to an equivalent series, the rate of return method uses a basis of either (1) present value for actual lifetimes, or (2) capitalized costs, or annual costs, for a limited time.

EXAMPLE 4.20: Rate of Return Method)

The rate of return method is used with annual costs. Again the data are the same as were used in preceding examples of the annual cost and present

value methods. The interest rate is 8%. With the same investments in the
two heat exchangers and the same annual costs, calculations are as follows.

Annual costs = (original cost - salvage
value)(capital recovery
factor, "to find A, given
P") + (salvage value)
(interest rate) + annual
maintenance, operating
costs

Annual costs for steel-copper heat exchanger = ($15,000 - $500)(0.14903)
+ ($500)(0.08)
+ $12,100 = $14,301

Annual costs for stainless steel exchanger = ($40,000 - $1,000)(0.14903)
+ ($1,000)(0.08)
+ $6,800 = $12,692

Conclusion: The difference between $14,301 for Purchase Possibility
A and $12,692 for Purchase Possibility B is found to be $1,608, or 11.2%.

Total annual costs for either installation are the same. This means
that the increased annual capital recovery costs for the stainless steel exchanger, since its original cost is more, are just about equal to the annual
savings in operating costs when compared to the steel-copper exchanger.

The 11.2% is known as the "economic rate of return on investment."
Sometimes this percentage is also referred to as the "economic earning
rate." Thus, capital can be invested in either exchanger and recovered at
an interest rate of 11.2% if the annual costs, as given in the previous example, remain the same at $12,100 for the steel-copper exchanger and
$6,800 for the stainless steel exchanger.

Since the stainless steel exchanger requires an additional $25,000 in-
vestment, this additional investment must also earn at the 11.2% rate. The
following is thus the result.

	Purchase Possibility A (steel-copper)	Purchase Possibility B (stainless steel)
Annual depreciation (10 year life, straight-line method) Cost - salvage + total life, in years	$ 1,450	$ 3,900
Annual operating expenses	12,100	6,800
Total annual expenses	$13,550	$10,700

The capitalized earning rate for the extra investment, neglecting differences in salvage, is the difference in total annual expenses of the two purchase possibilities divided by the differences in investments between the two purchase possibilities. The capitalized earning rate is

$$\frac{\$13,550 - \$10,700}{\$40,000 - \$15,000} \times 100\% = 11.12\%$$

The capitalized earning rate of 11.12% is the same as the economic earning rate. Investment, therefore, can be made in either installation if annual costs remain the same. If the capitalized earning rate had favored Possibility B, for instance, because of lower annual costs as given, then Possibility B would have been the more desirable one to purchase.

This example was oversimplified because annual costs were regarded as constant and the physical lifetimes of the two alternatives were the same at 10 years. Thus, the solution came readily. If the physical lifetimes of the two alternatives had been different, the cost of renewals would have had to be included in the annual costs in determining rates of return. For example, if the steel-copper heat exchanger had a life of, say 5 years, with an added 20% for price inflation over the 5 years (20% of $15,000), then cost of renewal would be $18,000; also, a present value basis would be needed to determine the unknown interest rate.

Payout Time (Years-to-Payback) Method

Another method used by engineers and oil management for comparing alternatives is one that determines the number of years (variable n) required for annual earnings, or savings, to pay off an initial investment made in a venture for the purpose of making a profit.

Payout "new" time computations are made on large-scale operations, or replacements, where an oil company has a maximum allowable time (payout time)—which might be 3 to 5 years or even less—for a return on investment. The use of "too low" a payout time may slow down investment for either new or replacement operations.

Payout time can be complicated because of variable annual incomes and depreciation charges for alternatives. But if we assume that annual returns are constant and also that annual returns equal the sum of depreciation charges and net profits minus any financial costs (interest) for borrowed money, we can avoid these complications and confusion about payout time.

Economic payout time (n) is determined from the annuity equation,

$$n = \frac{-\log(1 - iP/R)}{\log(1 + i)} \text{ years}$$

when R is actual annual return available from operations and is constant.

The payout time method can best be explained by use of an example.

EXAMPLE 4.21: Payout Time Method

Given: An oil company plans to expand its refinery and needs $1 million to be raised by selling ownership in the company. Government allowable depreciable life for the addition is 8 years. Expected net profits from the addition are $100,000 per year, plus one-tenth of $1 million for allowable depreciation per year.

 Wanted:

 (a) What is the economic payout time when interest in the annuity equation is 10%?

 (b) What is the economic payout time when interest in the annuity equation is 6%?

 (c) What is the capitalized payout time or payback period?

Solution

(a) Economic payout time n
    when interest is 10%

$$= \frac{-\log(1 - 0.10 \times \$1,000,000/\$200,000)}{\log 1.10}$$

$$= \frac{-\log(0.5)}{0.0414} = \frac{-(1.699)}{0.0414}$$

$$= \frac{0.0414 \times 6,990}{0.0414} = \frac{0.3010}{0.0414}$$

$$= 7.27 \quad \text{years required for annual earnings to pay off the initial investment of } \$1 \text{ million}$$

(b) Economic payout time,
    when interest is 6%

$$= \frac{-\log(1 - 0.06) \times \$1,000,000/\$200,000}{\log 1.06}$$

$$= \frac{1549}{0.0253}$$

$$= 6.12 \quad \text{years time required for return from investment to pay back the amount of investment of } \$1 \text{ million}$$

(c) Capitalized payout time,
    or payback period

$$= \frac{\text{required investment}}{\text{annual receipts} - \text{annual disbursements}}$$

$$= \frac{\$1,000,000}{\$200,000}$$

$$= 5 \text{ years}$$

It appears that the use of a high interest rate (10% vs. 6%) increases the payout time, as it did in this example. With 10% interest, 7.27 years were required for payout, but only 6.12 years were required with 6% interest. If this oil company requires the use of a 10% interest rate in such payout time computations and considers that "about" 6 years is the maximum payout time for which to consider a venture such as this to be feasible, this projected expansion might not be carried out.

However, if an interest rate of 6% had been used with the same maximum allowable payout time of 6 years, this projected expansion would be carried out by the oil company.

The payout time method is one in which depreciation and interest effects are ignored, so calculations are quick and simple. It does give more consideration to cash flows than total earnings and is appropriate for use where alternatives being considered have the same service lives. But the payout method can be deceptive, because it gives no consideration to returns obtainable beyond the payback period.

EXAMPLE 4.22: Payback

Consider two investments, both for $50,000 each. One required investment of $50,000 for 5 years has profits of $40,000 and payments of $20,000. A second required investment of $50,000 for one year has receipts of $60,000 and payments of $10,000. The payback period for the first investment is

$$\frac{\$50,000 \text{ (initial investment)}}{\$40,000 - \$20,000 \text{ (inflows - outflows)}} = 2.5 \text{ years}$$

The second investment yields a payback period equal to

$$\frac{\$50,000}{\$60,000 - \$10,000} = 1 \text{ year}$$

Favoring the alternative with the shortest payback period would rate the second alternative as the best one. But this alternative actually earns nothing, since $50,000 - $50,000 = 0. The first alternative, however, will earn an additional $50,000 by the end of the project in 5 years, with a payout of 2.5 years, and annual net inflows of $20,000 for 5 years ($40,000 inflows - $20,000 outflows).

Summary of Methods Involving Time Value

Generally, where small amounts of capital ($10,000 or less) are involved, the annual cost and present value methods are used.

The rate of return method and the payout time method are used where capital investments are large and profits taxes and other investment costs are to be considered. These two methods are usually applied to complete processes or plants.

Divisions on selecting alternative equipment for new installations are generally easier to make than determining whether a going installation should be replaced or not. In selecting and deciding on a replacement, the present value of the going installation plus the cost of its removal and/or its scrap value must be considered. For example, a special pipestill handling 50×10^{10} Btu/hour located in an oil refinery in a remote location may have a book value of $200,000 in its present location; but if it were replaced, its dollar value to someone else might be less than $10,000 (or 20 times less in value) because of the high cost of dismantling the pipestill and moving it. Also, its salvage value might be no more than $5,000. An economic study for replacement of this pipestill might ignore book value and depreciation that has occurred. Net realizable value of the equipment (that is, the actual sales dollars received for the equipment less the cost of removing it and the cost of lost production, is what is taken into consideration here because (1) of the special nature of the pipestill design for its present location and (2) because of the remoteness of the oil refinery pipestill location to other refineries.

REVIEWING CAPITAL INVESTMENTS: DESIRABLE OR NOT?

Oil companies today are faced with more capital expenditure (investment) opportunities that can be financed with the funds available to them. Because demand is greater than supply, capital funds must be "rationed." In order to apply the "rationing process" more intelligently to available investment opportunities, a measurement of the benefits expected from each proposed capital expenditure needs to be made. Some of the means for measuring prospective benefits from capital expenditures are the use of (1) payback, or payout; (2) average rate of return on investment; (3) time-adjusted rate of return on investment; and (4) present value of future cash flow.

Payback

Payback measures the length of time needed for the investment to be repaid to the investors, through cash flows generated by the capital expenditure. As stated previously, annual cash flow is after-tax profits generated by the capital expenditure added to annual depreciation allowance. For example, a capital expenditure of $2 million for which an annual depreciation allowance of $200,000 is made, and which generates income after taxes of $600,000, has an annual cash flow of $800,000 and a payback of 2.5 years

($2,000,000 ÷ $800,000). So payback is calculated by capital expenditure divided by cash flows.

Usually oil companies seek to recover most of their capital investments in a short payback period, mostly because of uncertainty about the future and the need to have funds available for later investments. This becomes especially important when the company is short of cash—emphasis on rapid recovery of cash invested in capital projects may be a necessity.

The payback period is used by oil companies in ascertaining the desirability of capital expenditures, because it is a means of rating capital proposals. It is particularly good as a "screening" means relative to various capital proposals. For example, expenditures for cracking units may not be made by an oil refinery unless the payback period is no longer than 3 years. On the other hand, the proposed purchase of a subsidiary may not be considered further unless the payback period is 5 years or less.

But payback has its drawbacks. For example, payback ignores the actual useful length of life of a project. Also, no calculation of income beyond the payback period is made. Payback is not a direct measure of earning power, so the payback method can lead to decisions that are really not in the best interests of an oil company.

EXAMPLE 4.23: Payback

For example, observe assumed cash flow for two alternatives of capital expenditures involving an investment in a sulfur-removal plant, as given in Table 4.3. The life of Alternative 1 is 7 years; the life of Alternative 2 is 10 years.

As Table 4.3 indicates, estimated cash flow for Alternative 1 results in a payback of 5 years; the payback period for Alternative 2 from estimated cash flow is 6 years. But Alternative 1 ceases to generate any cash flow after the seventh year, while Alternative 2 continues, through the added cash flow, to generate $400,000 each year after the investment has been paid back at the end of the sixth year.

The payback test alone would recommend the choice of Alternative 1 for capital investment rather than Alternative 2. However, over the period from year 7 to year 10, $1 million in added cash flows would be generated by Alternative 2, and a total cash flow of $1.3 million more by Alternative 2 over Alternative 1 for the 10-year period.

Example 4.24 is an example of the payback method with alternatives having the same payout period but different trends of cash flows.

TABLE 4.3

Cash Flow for Two Alternative Capital Expenditures
(Sulfur-Removal Plant, Investment in Each is $2 Million)

Year	Cash flow	
	Alternative 1	Alternative 2
1	$500,000	$200,000
2	500,000	300,000
3	400,000	300,000
4	350,000	400,000
5	250,000	400,000
6	200,000	400,000
7	100,000	400,000
8	—	400,000
9	—	400,000
10	—	400,000
Total cash flow	$2,300,000	$3,600,000

TABLE 4.4

Comparison of Two Boiler Investments
(Investment in Each Boiler is $50,000)

Year	Cash flow	
	Boiler 1	Boiler 2
1	$20,000	$ 5,000
2	15,000	10,000
3	10,000	15,000
4	5,000	20,000
Total cash flow	$50,000	$50,000

EXAMPLE 4.24: Payback

The payback means of measuring the feasibility of capital investments also
indicates the weaknesses of (1) ignoring the useful life of an asset for which
the expenditure is being considered, and (2) ignoring the cost of money.
Table 4.4 further illustrates.

The estimate of cash flow for both Boilers 1 and 2 indicates a payback period of 4 years. The recovery of investment in Boiler 1 is more rapid than is the recovery of investment in Boiler 2. From the standpoint of the cost of money, investment in Boiler 1 is preferable to investment in Boiler 2.

This example points out that, when using the payout period method, oil management should also observe the rapidity of cash flows between alternatives. The alternatives may have the same number of years-to-payback, as they do here, but one may be more favorable than the other because the largest amount of cash flow comes in the first few years. This could be an excellent point in favor of investment in one alternative over another when both have approximately the same payout periods. It could be a strong factor in selection of one especially if a greater amount of cash "back" is needed early in the investment.

A "refinement" of payback is illustrated in Table 4.5 when cash flow with assumed amounts is divided between return investment and recovery of the principal, with return on investment assumed at 10%. But this "refinement" is not without fault, since it still makes no provision for receipts after payback in the ninth year.

Average Rate of Return on Investment

The rate of return is a measure of the amount of funds invested in a capital project and the anticipated returns from the investment. This method, unlike the payback measure, uses net earnings after allowances for depreciation.

The average rate of return adds all projected net earnings with depreciation allowances deducted during the life of the project, and divides this total by the number of years of the life of the project.

EXAMPLE 4.25: Rate of Return on Investment

Table 4.6 illustrates the method for two projects, each of which requires an initial investment of $1 million. The useful life of Project 1 is 4 years; the useful life of Project 2 is 5 years. The earnings pattern during the life of Project 1 shows a declining trend, while that of Project 2 is a rising one. The average rate of return, when calculated, indicates a return of 16.25% per year on the life of Project 1, and 22.2% per year on the life of Project 2.

The average rate of return on investment method is superior to the payback method because it takes into consideration the earnings generated by each project over their useful lives. The big fault with the average rate

TABLE 4.5

Return Applied to Cash Flow

Year	Investment first of year - cols. 1-5 previous years 1	Annual cash flow 2	Return on investment, 10% of col. 1 3	Annual recovery of investment, col. 2-3 4	Balance of investment unrecovered at end of year - cols. 1-4 5
1	$1,000,000	$100,000	$100,000	—	$1,000,000
2	1,000,000	150,000	100,000	$50,000	950,000
3	950,000	200,000	95,000	100,000	845,000
4	845,000	200,000	84,500	115,500	729,500
5	729,500	200,000	72,950	127,050	602,450
6	602,450	200,000	60,245	139,755	462,695
7	462,695	200,000	46,270	155,730	308,965
8	308,965	200,000	30,897	169,103	139,862
9	139,862	200,000	13,986	186,014	—

TABLE 4.6

Average Return on Investment

	Project 1			Project 2		
Year	Income before depreciation	Depreciation allowance	Net earnings after depreciation	Income before depreciation	Depreciation allowance	Net earnings after depreciation
1	$400,000	$250,000	$150,000	$ 75,000	$200,000	$ -125,000
2	350,000	250,000	100,000	180,000	200,000	- 20,000
3	300,000	250,000	50,000	300,000	200,000	100,000
4	275,000	250,000	25,000	400,000	200,000	200,000
5		—		600,000	200,000	400,000
			$325,000			$ 555,000

Project 1:

Average investment (no scrap value) = $500,000

$$\text{Average earnings } \frac{325,000}{4} = \$81,250$$

$$\text{Average rate of return} = \frac{81,250}{500,000} = 16.25\%$$

Project 2:

Average investment (no scrap value) = $500,000

$$\text{Average earnings } \frac{555,000}{5} = \$111,000$$

$$\text{Average rate of return} = \frac{111,000}{500,000} = 22.2\%$$

of return on investment method is the fact that money received in the future
is treated with equal importance as money at present time. Thus, the time-
adjusted rate of return, which features present values, is often used in place
of the average rate of return on investment method.

Time-Adjusted Rate of Return on Investment and Present Value

In determining the time-adjusted rate of return on investment the formula
used is one that considers the present value of future receipts of money:

$$\text{Present value of future receipts of money} = \frac{\text{amount to be received in future}}{(1 + i)^n}$$

Using this formula, that is, applying a discount factor based on "given future value, find present value," if an oil company expects $800,000 in 10
years from the present date, and 10% is selected as the interest rate, the
cost of capital, the present value of the $800,000 is

$$\frac{\$800,000}{(1 + 0.10)} \, 10 \; = \; \$308,400$$

$$\$800,000 \times 0.3855 \; = \; \$308,400$$

Thus, $800,000 in 10 years is worth $308,400 today at 10% for analysis
purposes.

A table can be prepared which presents expected annual cash expenses,
including taxes but not depreciation, of a particular investment, as well as
the expected annual cash receipts, or savings, from the investment. Then
expenses can be deducted from revenues and the net difference discounted at
the rate of what it would cost to obtain funds necessary for the investment
(or what money is worth when borrowed from a financial source) so as to get
the present value of net income.

EXAMPLE 4.26: Time-Adjusted Rate of Return With Present Value

Assume that a distillation tower with an initial cost of $200,000 is expected
to have a useful life of 10 years, with a salvage value of $10,000 at the end
of its life. Also, it is expected to generate a net cash flow above maintenance and expenses amounting to $50,000 each year. Following are the cal-
culations of the time-adjusted rate of return with a selected discount rate
of 10%.

Initial investment	$200,000
Salvage value at end of 10 years	$10,000
Selected discount rate	10%
Net cash flow annually expected cash receipts	$50,000
Present value of cash flow of $50,000 annually, for 10 years at 10% (6.145)	$307,250
Present value of net receipts of $480,000 for 10 years, minus original investment of $200,000 at 10%	$180,000
Present value of $10,000 salvage value to be received at end of 10 years at 10%, $10,000 X 0.386	$3,860
Difference between initial investment of $200,000 and present value of net cash receipts plus present value of anticipated salvage value (this is the sum of the last two items)	$184,460

If alternative boiler investment proposals are to be considered, the above calculations can be made for each proposal to discover which alternative is the most promising for investment in terms of present value.

The selection of 10% as the discount rate factor is arbitrary. If money is borrowed for investment, the cost of the loan is usually the discount rate, or sometimes the assumed cost of retained income if money used in the investment is from one's own internal funds and is not borrowed.

Adjustment for the time value of money requires the selection of a discount rate. In the above example, at the rate of 10%, the present value of the future stream of cash (cash flow) of $50,000 annually for 10 years is $307,250. On present value of future cash flow, it is obvious that a lower discount rate generates a higher present value; also, a higher discount rate generates a lower present value.

Example 4.27 is another example of present value, using three different discount rates.

EXAMPLE 4.27: Present Value

If two projects each requiring the same initial investment of $1 million are compared on the basis of discounted cash flow (or present value of future cash flows), a low discount rate on one project can favor this project over another project with a higher discount rate. Table 4.7 gives present values of cash flows for the two projects with 8, 10, and 12% discount rates. (Discount factors are taken from published tables of compound interest.)

In summarizing Table 4.7, if the cash flows of Project 1 and Project 2 are discounted at 8%, Project 2 is preferable; if the cash flows are discounted at 10%, Project 2 is preferred to Project 1 because the present value of Project 2 is almost $14,000 more; and if the cash flows are discounted at 12%, Project 1 is slightly preferable to Project 2, and will continue to be more preferable to Project 2 as discount rates go higher than 12%.

Therefore, as the example shows, the choice between the two projects depends on the discount rate used. Usually, the oil company's cost of capital for investing in the project will determine which project is selected.

Figure 4.2 gives the present value curves for both Project 1 and Project 2 resulting from the three discount rates used. The "point of indifference" appears to be between 10 and 12%. Before this point, Project 2 had the more favorable present value; after this point, Project 1 is favored. As discount rates become higher past the "point of indifference," Project 1 will continue to be more desirable for investment purposes.

From the data, it can be seen that at a discount of 12%, the present value of cash flow from Project 2 is $1 million; and at a discount of over 12% the present value of cash flows gives us the discount rated amount of under $1 million. This analysis of the present value of cash flows gives us the discount rate at which anticipated cash flow equals the initial investment. This is known as the rate of return on investment. For Project 2, the rate of return on investment is about 12%; for Project 1, it is about 13%.

SUMMARY ON CAPITAL EXPENDITURE (INVESTMENT) PROPOSALS

Capital expenditure proposals must be sufficiently specific in order to permit their justification (money expenditures) for either expansion purposes or for cost-reduction improvements and/or necessary replacements. Actually, an evaluation of capital expenditure proposals is both technical and economic in nature. First, there are the technical feasibilities and the validities on the assumptions about production volumes, market potentials, and engineering consequences; second, there is an economic phase to the evaluation process.

In the economic phase of evaluation, oil management may find that it has more investment opportunities than capital to invest, or more capital to invest than investment opportunities. Whichever situation exists, oil management needs some economic criteria for selecting or rejecting investment proposals. Management's decision is, in either case, likely to be based

TABLE 4.7

Initial Investment and Anticipated Cash Flow

Year	Project 1	Project 2
1	$ 400,000	$ 100,000
2	320,000	200,000
3	300,000	300,000
4	200,000	400,000
5	100,000	500,000
Total anticipated cash flow for 5 years	$1,320,000	$1,500,000

At 8% discount

		Project 1		Project 2	
Year	Discount factor for 8%	Cash flow	Discounted value (col. 1 × col. 2)	Cash flow	Discounted value (col. 1 × col. 4)
	1	2	3	4	5
1	0.926	$400,000	$370,400	$100,000	$ 92,600
2	0.856	320,000	274,240	200,000	171,400
3	0.794	300,000	238,200	300,000	238,200
4	0.735	200,000	147,000	400,000	294,000
5	0.681	100,000	68,100	500,000	340,500
			$1,097,940		$1,136,700

At 10% discount

		Project 1		Project 2	
Year	Discount factor for 10%	Cash flow	Discounted value (col. 1 × col. 2)	Cash flow	Discounted value (col. 1 × col. 4)
	1	2	3	4	5
1	0.909	$400,000	$363,600	$100,000	$ 90,900
2	0.826	320,000	264,320	200,000	165,200
3	0.751	300,000	225,300	300,000	225,300
4	0.683	200,000	136,600	400,000	273,200
5	0.621	100,000	62,100	500,000	310,500
			$1,051,920		$1,065,100

TABLE 4.7 (continued)

		At 12% discount			
		Project 1		Project 2	
Year	Discount factor for 10%	Cash flow	Discounted value (col. 1 × col. 2)	Cash flow	Discounted value (col. 1 × col. 4)
	1	2	3	4	5
1	0.893	$400,000	$357,200	$100,000	$ 89,300
2	0.797	320,000	255,040	200,000	159,400
3	0.712	300,000	213,600	300,000	213,600
4	0.636	200,000	127,200	400,000	254,400
5	0.567	100,000	56,700	500,000	283,500
			$1,009,740		$1,000,200

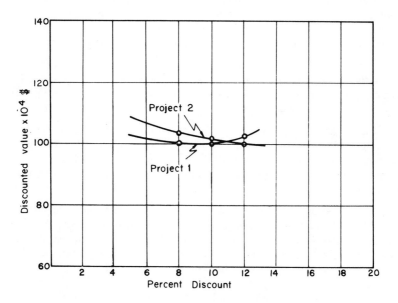

FIG. 4.2. Effect of discount rates on present value of cash flow for each of two projects.

largely on the measures of financial return on investment we have discussed, with the most common measures, or methods, of economically evaluating return being (1) payout period or years-to-payback, (2) average

rate of return, and (3) discounted net cash flow (present value) rate
of return.

All three methods determine in one way or another the return on investment (ROI). To evaluate whether a project, or a proposal on a project for
the future, is yielding, or will yield, a good or bad return, ROI must be
compared to a standard acceptable level of profit which the oil company
wishes to maintain. The internal cutoff rate (or breakeven point for return)
is the cost of capital, which is the rate of borrowing money at the time of
use of these measures for calculating return on investment. There is no
precise agreement on how oil management calculates cost of capital, but it
should include both the cost of borrowed funds and the cost of equity financing.

EXAMPLE 4.28: Debt and Equity Financing

Assume that oil management has determined that the cost of borrowing
(debt financing, as often called) is 9% at current loan rates. Also, the
growth potential of the oil company, according to monthly accounting statements, indicates that after-tax earnings will be $5 per share of stock
to owners of shares in the oil company. This will cover the cost of
equity financing. (If the oil company is not a corporation, the cost of equity
financing should still be estimated and included.)

With a book value of, for example, $25 per share, the earnings (growth
potential) would represent a 20% return on equity ($5 ÷ $25 × 100%), or the
cost of equity financing.

It might also be assumed, then, that the cost of capital is simply the
addition of the cost of borrowing at 9% to the cost of equity financing at 20%,
yielding a total cost of derivation of the "capital base" of 29%. But the
capital base, with which management invests, is proportioned between debt
and equity.

If, for this oil company, debt represents 40% and equity represents
60% of the capital base, or percent of total capital investment, the cost of
deriving of this capital base should also reflect this debt-to-equity ratio,
and should be the weighted average of the costs of each type of capital. The
calculation would be performed in the following manner.

	Col. 1 (cost percent)	Col. 2 (percent of total capital base)	Col. 1 × Col. 2 (in percent)
Debt	9	40	3.6
Equity			
($25 – $5)	20	60	12.0
Total	29	100	15.6

For debt, 3.6% (col. 1 × col. 2) is the cost of deriving the 40% of the total capital base by borrowing; and for equity, 12.0% (col. 1 × col. 2) is the cost of deriving the 60% of the total capital base by selling ownership in the business.

Thus, the total cost of deriving the capital base for this oil company is approximately 15.6%, not 29%.

Now, how does this 15.6% compare with other oil companies? Having determined the cost of deriving this capital base, the management of this oil company can compare the 15.6% cost of derivation of capital base with those of other major oil companies in the oil industry. Such a comparison can act as a guideline in the development of the objectives of this particular oil company. For instance, such an analysis could indicate many things to an oil company, such as a change in company policy relative to the 40:60 ratio of debt to equity in financing the capital base. For instance, if costs of borrowing appear to be going up over 9% consistently, the ratio may be changed to more equity financing, or vice versa.

There is another factor to consider relative to evaluating capital expenditures. Every oil company has to consider that certain investments will not yield a "measurable profit" because some investments may be needed to improve employee or community goodwill, or to meet legal requirements of the government under which the oil operations are located. For example, investments in equipment to reduce air or water pollutants and investments in the social well-being of the community may not contribute dollars to equity of a company. These are examples of those investments which will not yield a measurable profit. And oil companies must face some of these "opportunities," especially when their operations are in countries other than that in which their main administrative offices are located.

An efficient oil company is aware of this type of investment and makes plans in advance for these expenditures. Oil management must, therefore, increase its return on those investments yielding measurable profits accordingly, so that their portfolios of profit and nonprofit investments taken together yield a sufficient overall return. For example, suppose that an oil company has calculated its "needed" return (or cost of capital for owners of the oil company) to be approximately 15%. But 25% of its investments are nonprofit, or "necessity" projects. This means that 75% of its investments are "profitable" ones. To cover the 25% that are nonprofit investments, the returns on the 75% that are profitable will have to be approximately 20% (15% ÷ 75%). Thus, oil companies need not only to appraise all potential investments individually, but also constantly view the position of their portfolios of profit and nonprofit investments taken together.

EXAMPLE 4.29: A Summary Problem

The following example is a problem involving the need for a decision relative to two alternatives involved in installing a complete electrified pump
station for an oil company. To help management decide which alternative
to select, cash recovery, cash flows cumulative, payout period, rate of
return, and a present value factor are all used. Assume that other alternatives have been "eliminated" and the choice is now down to two possibilities.

Before making calculations of cash recovery, cash flow, payout period,
and rate of return, an estimate of annual maintenance repair, and operating
costs (MRO costs) of the project must be made.

First, for direct operating labor, assume that full-shift coverage requires 1.58 men on direct payroll for each 8-hour man-shift. (Foremen
and their clerks are not included as a part of direct operating labor. They
are in supporting manpower ratios.)

Total direct operating manpower assigned to the plant is, then:

Staff: 3 shifts × 3 men/shift × 1.58 × salary/year

Intermediate: 3 shifts × 3 men/shift × 1.58 × salary/year

General: 3 shifts × 10 men/shift × 1.58 × salary/year

Direct maintenance labor includes staff supervision, service organization, intermediate management such as foreman at the pump station, and
workers, including those under contract. Other labor is mostly indirect,
such as cleanup, security, and so on.

Chemicals for operation of the pump station, and materials for maintenance, the latter based on 10% of high-maintenance equipment, valued at,
say $2.25 million and 8% of low-maintenance equipment of $2 million,
follows.

Utilities are divided into two costs: One is power based on billion kilo-
watt hours used, and the other is treated water based on billion gallons used.

Finally, MRO includes equipment expenses for annual costs of operating
and maintaining motor equipment for staff and service personnel.

The estimated MRO cost summary for both Alternative 1 and Alternative 2 is given in Table 4.8. This information then leads to cash flow, payment period, and rate of return for constant annual cash recoveries.

In addition to estimated MRO cost summaries for the two alternatives,
the following figures are assumed.

Direct project costs, or direct capital needed, is assumed to total
$8 million for Alternative 1, and $11.2 million for Alternative 2, broken
down as shown in Table 4.9.

TABLE 4.8

MRO Cost Estimate Summary

Costs		Alternative 1
Direct operating labor:		
Senior staff: 3 shifts × 3 men × 1.58 men		
× $10,000	$142,000	
Intermediate: 3 shifts × 3 men × 1.58 men		
× $3,000	42,600	
Workers: 3 shifts × 20 men × 1.58 men		
× $1,500	142,200	$326,800
Direct maintenance labor:		
Staff: 6 men at $8/hr × 1,800 hr/year	$86,400	
Service organization: 6 men × $6/hr		
× 1,800 hr/year	64,800	
Intermediate: 15 men at $2/hr		
× 1,800 hr/year	54,000	
General workers: 15 men at $1/hr		
× 1,800 hr/year	$27,000	
Arab contract: 50 men × $0.50/hr		
× 1,800 hr/year	45,000	$277,200
Other labor, mostly indirect:		
Chemicals for occupation, materials for		
maintenance:		
Operating chemicals	$266,100	
Maintenance materials:		
12% of high-maintenance equipment of		
$2.25 million	270,000	
80% of low-maintenance equipment of		
$2 million	160,000	$696,100
		$1,518,500
Utilities:		
Power: 250 billion kWh at $0.007	$1,750,000	
Treated water: 100 million gal at $0.0079	790,000	$2,540,000
Equipment expense (annual cost of motor		
equipment):		
40% of senior staff, direct labor		
operating: $142,000 × 0.40	$56,800	
90% of staff men: 86,400 × 0.90	77,760	
Others, misc., mostly service staff	106,940	$241,500
Total MRO cost estimate for		
Alternative 1		$4,300,000

TABLE 4.8 (continued)

Costs		Alternative 2
Direct operating labor:		
Senior staff: 3 shifts × 3 men × 1.58 men		
× $10,000	$142,000	
Intermediate: 3 shifts × 3 men × 1.58 men		
× $3,000	42,600	
Workers: 3 shifts × 13 men × 1.58 men		
× $1,500	92,200	$276,800
Direct maintenance labor:		
Staff: 6 men at $8/hr × 1,800 hr/year	$86,400	
Service organization: 4 men × $6/hr		
× 1,800 hr/year	43,200	
Intermediate: 12 men at $2/hr		
× 1,800 hr/year	43,200	
General workers: 10 men at $1/hr		
× 1,800 hr/year	18,000	
Arab contract: 40 men at $0.50		
× 1,800 hr/year	36,000	$226,800
Other labor, mostly indirect:		$326,100
Chemicals for operation, materials for		
maintenance:		
Operating chemicals	136,000	
Maintenance materials:		
12% of high-maintenance equipment of		
$1.5 million	180,000	
80% of low-maintenance equipment of		
$1 million	80,000	$396,000
		$1,225,700
Utilities:		
Power: 150 billion kWh at $0.007	$1,100,0000	
Treated water: 30.4 million gal at $0.0079	240,000	
	(rounded off)	$1,340,000
Equipment expense (annual cost of motor		
equipment:		
40% of senior staff, direct labor		
operating: $142,000 × 0.40	$56,800	
90% of staff men: 86,400 × 0.90	77,760	
Others, misc., mostly service staff	99,740	$234,300
Total MRO cost estimate for		
Alternative 2		$2,800,000

TABLE 4.9

An Electrified Pump Station

	Alternative 1	Alternative 2
Direct project costs (direct capital needed)		
Direct construction labor	$1,200,000	$1,800,000
Engineering and technical services	700,000	1,500,000
Materials	6,100,000	7,900,000
Total direct capital needed	$8,000,000	$11,200,000
Indirect project costs (supporting facilities required—estimated costs; see table below)		
Increase in direct capital investment		
Shops	$250,000	$250,000
Motor equipment	22,000	22,000
Garages and related facilities	16,000	16,000
Air conditioning	2,000	3,000
Steam and diesel power plants	2,800,000	3,400,000
Gas turbine power plants	650,000	1,150,000
Water systems	500,000	750,000
Sewer systems	160,000	300,000
Electrical distribution facilities	1,600,000	2,500,000
Others	500,000	509,000
Total indirect costs	$6,500,000	$8,900,000
Total direct and indirect project costs	$14,500,000	$20,100,000
Financial gains and costs		
Estimated useful life of pump	10 years	10 years
Anticipated gross annual revenues (est.)	$36,000,000	$42,000,000
Annual MRO costs (see Table 4.8)	4,300,000	2,800,000
Annual depreciation (total direct-indirect costs divided by 10)	1,450,000	2,010,000
Total annual costs	$5,750,000	$4,810,000
Net income before taxes (gross annual income minus annual costs)	30,250,000	37,190,000
Less income tax at 55%	13,612,500	16,735,500

Indirect project costs, or supporting facilities required which need an increase in direct capital investment of $8 million and $11.2 million for Alternative 1 and Alternative 2, respectively, are $6.5 million and $8.9 million as shown in Table 4.9. Thus, total direct capital investment required for Alternative 1 is $8 million + $6.5 million = $14.5 million, and for Alternative 2 is $11.2 million + $8.9 million = $20.1 million.

Financial gains and costs are given in Table 4.9. Estimated useful life of the pump and station are given as 10 years for both alternatives. Anticipated gross annual revenues from use of this electrified pump station are estimated at $36 million for Alternative 1 and $42 million for Alternative 2. MRO cost estimates from Table 4.8 are carried into Table 4.9 as given. Annual depreciation charges are calculated by the straight-line method, dividing total capital investment of $14.5 million and $20.1 million for Alternatives 1 and 2, respectively, by 10.

Arriving at net income before tax, the total of MRO costs and annual depreciation charges is subtracted from anticipated gross annual revenues. This amounts to $30.25 million annually for Alternative 1 and $37.19 million for Alternative 2.

Income tax is assumed to be 55%, so net income after tax is calculated to be $13,612,500 for Alternative 1 and $16,735,500 for Alternative 2.

The summary (Table 4.10) involves four factors: cash recovery, net cash flow, payout period, and rate of return on incremental cash flow.

Cash recovery is calculated by adding depreciation and net income after taxes. Cash recovery for Alternative 1 is $15,062,500 and for Alternative 2 is $18,745,500.

Net cash flow, cumulative, is calculated by subtracting direct capital investment from the cash recovery. Tables 4.11a and 4.11b give the net cash flow, cumulative, for 10 years with each alternative.

The payout period, in years, is calculated by dividing total direct capital investment by cash recovery. For Alternative 1 it is 0.96 years, or 11.5 months. For Alternative 2 it is 1.07 years or 12.8 months. This means that in less than one year, total direct capital investment of $14.5 million will be received back with Alternative 1, and $20.1 million will be received back with Alternative 2 in almost 13 months.

Finally, the rate of return on incremental cash flow is determined by dividing capital recovery by direct capital investment. For Alternative 1 the rate of return is 104%, and for Alternative 2 it is 93.3%.

This example assumes a constant annual cash recovery after startup. Thus Tables 4.11a and 4.11b illustrate annual cash flow calculations for 10 years, indicating no charge after the first year for Alternative 1, since the payout period is 0.96 years, and no charge after the second year for Alternative 2, which has a payout period of 12.8 months (1.07 years).

TABLE 4.10

Summary

	Alternative 1	Alternative 2
1. Cash recovery (annual cash returned by depreciation + net income after taxes) Since this example has a constant annual recovery after startup, it is only necessary to show annual cash flow calculations for the first year after startup.	($1,450,000 + $13,612,500) $15,062,500	($2,010,000 + $16,735,500) $18,745,500
2. Net cash flow (cumulative) Cash recovery: direct capital investment needed – total direct and indirect cost	($15,062,500 in – $14,500,000 out) $562,500 in For 10 years—cumulative $150,625,000 – $14,500,000 or $136,125,000	($18,745,500 in – $20,100,000 out) $1,354,500 out For 10 years—cumulative $187,455,000 – $20,100,000 or $167,355,000
3. Payout period (in years): total direct and indirect costs ÷ cash recovery	($14,500,000 ÷ $15,062,500) 0.96 years (11.5 months)	($20,100,000 ÷ $18,745,500) 1.07 years (12.8 months)
4. Rate of return on incremental cash flow Cash recovery: total direct and indirect costs ÷ direct capital investment needed × 100%	($15,062,500 ÷ $14,500,000 × 100%) 104.0%	($18,745,500 ÷ $20,100,000 × 100%) 93.3%

TABLE 4.11a

Net Cash Flow Summary for Alternative 1

Year	Cash flow in	Cash flow out	Balance on investment of $14,500,000	Accumulated net cash flow in	
1	$15,062,500	$14,500,000	—	562,500	
2	15,062,500	—	—	15,625,000	
3	15,062,500	—	—	30,687,500	
4	15,062,500	—	—	45,750,000	
5	15,062,500	—	—	60,812,500	
6	15,062,500	—	—	75,875,000	
7	15,062,500	—	—	90,937,500	
8	15,062,500	—	—	106,000,000	
9	15,062,500	—	—	121,062,500	
10	15,062,500	—	—	136,125,000	(total ac-
End of					cumulated
tenth year	$150,625,000	$14,500,000	—	—	cash inflow)

TABLE 4.11b

Net Cash Flow Summary for Alternative 2

Year	Cash flow in	Cash flow out	Balance on investment of $20,100,000	Accumulated net cash flow in
1	$18,745,500	$18,745,500	$1,354,500	—
2	18,745,500	1,345,500	—	$17,391,000
3	18,745,500	—	—	36,136,500
4	18,745,500	—	—	54,882,000
5	18,745,500	—	—	73,627,500
6	18,745,500	—	—	92,373,000
7	18,745,500	—	—	111,118,500
8	18,745,500	—	—	129,864,000
9	18,745,500	—	—	148,609,500
10	18,745,500	—	—	167,355,000
End of tenth year	$187,455,000	$20,100,000	—	—

There are differences between Alternative 1 and Alternative 2. Alternative 2 costs more in capital investment by $5.6 million ($20.1 million as against $14.5 million). But the cash recovery of Alternative 2 of $18,745,500 is $3,638,000 more than Alternative 1 with $15,062,500. The net cash flow of Alternative 2, however, shows a minus $1,354,500, or $1,354,500 less cash returned to direct capital investment made ($18,745,000 coming in as against $20,100,000 put out in investment). Even at money worth 5% interest, if Alternative 1 is used and the $1,354,500 is "saved," this money would return $67,725 in simple interest in 1 year or $837,980 in compounded interest in 10 years. Alternative 1 has a slight advantage of 1.3 months in terms of payout period, or 11.5 months to 12.8 months.

Thus, there does not appear to be any problem of selection of the less expensive pump station. Alternative 1 has the shorter payout period even though the more expensive pump station, Alternative 2, has the higher ultimate cash recovery. Furthermore, the additional capital cost of Alternative 2 does not appear to be justified by Alternative 2's higher annual cash recovery as the rate of return on incremental cash flow so indicates, 104% for Alternative 1 and 93.3% for Alternative 2, or 10.7% greater rate of return for Alternative 1. But if the refinery is interested in moving more oil at a cheaper cost of operation, Alternative 2 is better, since its anticipated gross annual revenues are $6 million more per year and its MRO operating costs are estimated at $1.5 million annually.

Rate of return can be referred to Figure 4.3, a graph which is valid only because the two projects, Alternative 1 and Alternative 2, are assumed to have a constant annual cash recovery.

It is important also, before making a final selection as to which electrified pump station to take, to determine whether the additional capital cost of Alternative 2 is justified by that Alternative's higher annual cost recovery. Figure 4.3 illustrates the payout period and rate of return for the incremental cost of Alternative 2 over Alternative 1. The graphs of cumulative net cash flows and present value curves, given in Figure 4.3, immediately follows Tables 4.10, 4.11a, 4.11b, 4.12a, and 4.12b, which give cash flow, payout period, and rate of return calculations.

The point of intersection of the cumulative net cash flow curves and the Y axis shows the payout period on total investment for each alternative. The point of intersection of the curve for Alternative 1 and the curve for Alternative 2 shows the payout period for the additional cost of Alternative 2. The graph indicates an incremental payout period of 1.07 years for additional investment of Alternative 2.

Present worth values are given in Tables 4.12a and 4.12b for Alternatives 1 and 2. Present value factors are given. These tables demonstrates the use of present value factors to determine the present value of

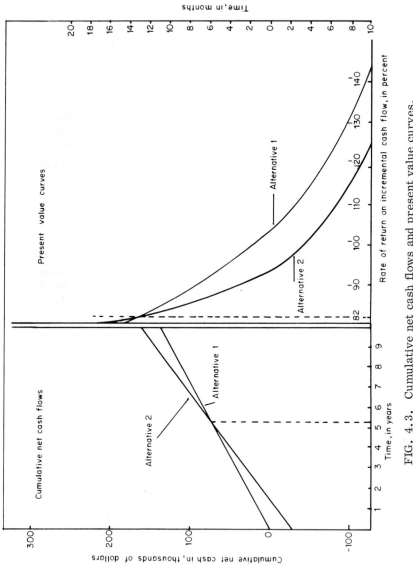

FIG. 4.3.   Cumulative net cash flows and present value curves.

TABLE 4.12a

Present Values for Alternative 1

Year	Net cash flow	Present value factor	Present value
1	- $14,500,000	0.7453560	- $10,807,662
	+ 562,500	0.7453560	+ 419,263
2	+ 15,062,500	0.4140867	+ 6,237,181
3	+ 15,062,500	0.2300482	+ 3,465,101
4	+ 15,062,500	0.1278046	+ 1,925,057
5	+ 15,062,500	0.0710026	+ 1,069,477
6	+ 15,062,500	0.0394459	+ 594,154
7	+ 15,062,500	0.0219144	+ 330,086
8	+ 15,062,500	0.0121747	+ 183,381
9	+ 15,062,500	0.0067637	+ 101,878
10	+ 15,062,500 1	0.0037576	+ 56,599
			$25,189,839
			- 10,807,662
			$14,382,177

TABLE 4.12b

Present Value for Alternative 2

Year	Net cash flow	Present value factor	Present value
1	- $18,745,500	0.7453560	$13,972,071
	- 13,354,500	0.4140867	560,880
2	+ 17,391,000	0.4140867	+ 7,201,382
3	+ 18,745,500	0.2300482	+ 14,312,369
4	+ 18,745,500	0.1278046	+ 2,395,756
5	+ 18,745,500	0.0710026	+ 1,330,979
6	+ 18,745,500	0.0394459	+ 739,433
7	+ 18,745,500	0.0219144	+ 410,796
8	+ 18,745,500	0.0121747	+ 228,221
9	+ 18,745,500	0.0067637	+ 126,789
10	+ 18,745,500	0.0037576	+ 70,438
			$31,349,114
			- 14,532,951
			$16,816,163

Annual cash recovery	Annual cash recovery
$15,062,500 × 10 = $150,625,000 for Alternative 1	$18,745,500 × 10 = $187,455,000 for Alternative 2

the cumulative net cash flow for Alternatives 1 and 2 at an 80% interest rate basis.

The intersection of the present value curve with the Y axis of the graph in Figure 4.3 indicates the rate of return on total investment for each alternative. The graph indicates that the rate of return for Alternative 1 is 104% and the rate of return for Alternative 2 is 93.3%. The intersection of the two present value curves shows the rate of return for the additional cost of Alternative 2. In this case, the graph shows an "incremental" rate of return of 82% on the additional cost of Alternative 2 or $5.6 million ($20.9 million - $14.5 million).

With the above information available, management can decide which course of action is financially most attractive. If they consider that a payout period of 1.07 years and a rate of return of 82% is attractive, the additional direct capital investment of $5.6 million could be justified. If, however, the $5.6 million could be invested elsewhere to yield a shorter payout period than 1.07 years (or return of the $5.6 million back to its owners) and a higher rate of return than 82%, then Alternative 1 is the most attractive financial alternative.

Chapter 5

RETURN ON INVESTMENTS, CONTINUED:
AMORTIZATION OF FIXED CAPITAL

VALUATION OF ASSETS USING DEPRECIATION AND DEPLETION

Petroleum company management frequently must determine the value of oil
engineering properties. An adequate discussion of the methods used to arrive at the correct value of any property would require at least a good-
sized volume, so only a few of the principles involved will be considered
here—those intimately connected with the subjects of depreciation and
depletion.

There are many reasons for determining the value of oil field and refinery assets after some usage. For instance, these values may be needed
to serve as a tax base or to establish current value for company statement
purposes. Taking depreciation first, the primary purpose of depreciation
is to provide for recovery of capital that has been invested in the "physical"
oil property. Depreciation is a cost of production; therefore, wherever
this production causes the property to decline in value, depreciation must
be calculated. Indirectly, depreciation gives a method of providing capital
for replacement of depreciated oil equipment. In short, depreciation can
be considered as a cost for the protection of the depreciating capital, without interest, over the given period (minimum set by government) during
which the capital is used. Finally, the process of valuation is usually an
attempt either to make an estimate of present value of future oil profits
which will be obtained through ownership of a property, or to determine
what would have to be spent to obtain oil property capable of rendering the
same service in the future at least as efficiently as the property being
valued.

Investment of depreciable capital is used for one of two purposes in
the oil fields. It is used as (1) working capital for everyday operating

expenses such as wages, materials, and supplies, and (2) to buy oil drilling machinery, rigs, etc., used in development and production of oil wells.

Normally, working capital is replaced by sales revenue as it is used up. Thus, this part of investment capital is always available for return to investors.

Investment used for oil drilling machinery, well casings, etc.—that is, fixed capital—cannot be converted directly to original capital invested in oil equipment and machinery, because these physical properties decrease in value as time progresses. They decrease in value because they depreciate, wear out, or become obsolete. Recovery of this investment of fixed capital, with interest for the risks involved in making the investment, must be assured to the investor. The concept of capital recovery thus becomes very important.

The valuation of oil resources in the ground is something else. Oil resources cannot be renewed over a period of years like some other natural resources, such as timber or fish. Also, oil resources cannot be replaced by repurchase as such depreciable physical properties as machinery and equipment can be. Some provision is thus needed to recover the initial investment, or value, of oil reserves and reservoirs, sometimes referred to as an oil lease if purchased by others who are not owners of the land.

One way for investors to recover capital investment in an oil lease, known as depletion capital, is to provide a depletion allowance with annual payments made to the owners of the oil lease. Payments are based on the estimated life of the resource where such an estimate can be made with some degree of accuracy.

Another way to recover capital investment in an oil lease or other depletable capital is to set up a sinking fund with annual deposits based on one interest rate for the depletable capital plus another interest rate or profit on the investment.

In the case of exploration costs and development costs, or money spent for exploration and operations preliminary to actual recovery (production) of oil, such costs are usually recovered by "write-offs" (an accounting term) against other revenues in the year they occur or through a depletion allowance. In the case of foreign oil companies, that is, foreign investors with other outside revenues, these costs can be subtracted from their other revenues along with other expenses in arriving at net income for tax purposes in their own countries. For example, in exploring and developing new leases in the Arabian Gulf area, which could involve millions of dollars before production or perhaps even with little chance of production success, Shell Oil or British Petroleum could write off these costs against their overall revenues. This would reduce their taxable income, and thereby reduce income taxes they would be liable to pay in their home countries.

If a depletion allowance is to be used, there are two possible methods of calculating its value. One method is computation of depletion as a fixed percentage. In the United States, for instance, a depletion allowance of 27.5% that had been used in the oil and natural gas industries, now is cancelled by the government.

Another method for calculating depletion is the cost per unit. A depletion cost per unit rate is computed by dividing intangible development cost and other depletable costs by estimated total units, or barrels, potentially recoverable.

To illustrate how both depreciation and depletion are calculated, several methods of determining depreciation and depletion are given, with examples of each.

METHODS FOR DETERMINING DEPRECIATION

There are several ways of determining depreciation for a given period. The following are some of the more popular methods used in most industries. Some are more applicable to the oil industry than others.

Pattern of Use or Production Method

The length of useful life of an asset is expressed in terms of units produced rather than in terms of time.

For instance, the petroleum industry uses complicated, high-speed automatic machinery in the cracking process. Most oil companies purchase this equipment with the thought that it will satisfactorily process an estimated number of barrels, or tons, of distilled oil products during its lifetime. Thus, the life of the cracking equipment can be estimated and stated in terms of millions of barrels or tons processed.

Using this method, depreciation is computed by dividing the capital depreciable cost by the number of barrels to determine the "unit cost of depreciation." Then the total amount of depreciation in any given time period during the lifetime of this equipment is found by multiplying the unit cost by the number of units produced in that time period.

EXAMPLE 5.1: Pattern of Use or Production Method

An example of where this method might be used in the oil industry is a heat exchanger. Suppose that the heat exchanger has a depreciable cost of $50,000 and will last for, say, 20 million bbl. The depreciation factor per barrels is then

$$\frac{\$50,000}{20,000,000} = \$0.0025 \text{ per barrel}$$

Then if the heat exchanger processes 600,000 barrels in any one year, its annual depreciation for that year will be $1,500 (600,000 × $0.0025 per barrel).

Obviously, the amount of depreciation per time period can vary greatly, depending on the activity level achieved by the oil company in that period. As demand for oil increases, there is an increase in depreciation expense resulting from the increased use of the equipment.

Declining Balance Methods

The declining balance method assumes that the equipment in question will contribute more to the earning of revenues in the early stage of useful life than it will as the equipment gets older.

A valid use of a declining pattern of depreciation occurs when it is felt that obsolescence will exert a strong influence on the life of the equipment but there is no way of predicting when it will occur.

Two ways of calculating a declining pattern are in general use today. They are the double declining balance method and the sum-of-the-digits method.

Double Declining Balance Method. The double declining balance method is based on the assumption that the amount of depreciation to be charged off each year is a function of the remaining asset value at the beginning of that year. The calculation is made by applying a designated percentage to the remaining original cost (original cost minus depreciation to date) of the equipment. The allowance can be no greater than twice any "straight-line percentage."

Therefore any equipment lasting 5 years would have a 20% straight-line percentage and thus an allowable 40% for purposes of making the double declining balance calculation.

EXAMPLE 5.2: Double Declining Balance Method

An example as to how the double declining balance method is calculated is given here. If we assume that an evaporator had an original cost of $17,000 and a 5-year life, allowable new depreciation deduction would be 40% by the double declining balance method. This 40% is applied each year to the remaining original cost. The depreciation schedule would then be as shown in Table 5.1, with end-of-the-5-year salvage value of the equipment estimated to be $2,000.

TABLE 5.1

Depreciation Schedule for Example 5.2

Year	Depreciation expense	Book value	Remaining depreciable cost
Start		$17,000	$15,000 (with $2,000 salvage value off)
After 1st year	$6,800 (40% of $17,000)	$10,200	$8,200 ($15,000 - $6,800)
After 2nd year	$4,080 (40% of $10,200)	$6,120	$4,120 ($8,200 - $4,080)
After 3rd year	$2,448 (40% of $6,120)	$3,672	$1,672 ($4,120 - $2,448)
After 4th year	$1,469 (40% of $3,672)	$2,203	$203 ($1,672 - $1,469)
After 5th year	$203 (depreciation before salvage value)	$2,000	0

Sum-of-the-Digits Method. There is another way of calculating a de-clining pattern of depreciation. The calculation is made by adding the digits for the years of estimated useful life of the asset. Thus, for example, a check valve with an estimated useful life of only 5 years would have a sum-of-the-digits of 15 (1 + 2 + 3 + 4 + 5).

The depreciable cost and resale value can also be considered. It would then be allocated to the 5 years of the life of the valve as follows:

1st year   5/15
2nd year   4/15
3rd year   3/15
4th year   2/15
5th year   1/15

Note: With this method, the fractions are applied each year to the original cost (minus) the salvage value, and not to the declining balance after each year, as is the case with the declining balance method.

EXAMPLE 5.3: Sum-of-the-Digits Method

If we take the same amount of $17,000 on a new evaporator as was used in Example 5.2, it would appear as in Table 5.2.

TABLE 5.2

Depreciation Schedule for Example 5.3

Year	Depreciation expense + $2,000 for salvage	Book value	Remaining depreciable cost
Start	0	$17,000	$15,000
After 1st year	$5,000 (5/15 of $15,000)	$12,000	$10,000
After 2nd year	$4,000 (4/15 of $15,000)	$8,000	$6,000
After 3rd year	$3,000 (3/15 of $15,000)	$5,000	$3,000
After 4th year	$2,000 (2/15 of $15,000)	$3,000	$1,000
After 5th year	$1,000 (1/15 of	$2,000	0

Constant Percentage Method

Still another method for calculating depreciation is the constant percentage method. This method assumes that the annual cost of depreciation is a fixed percentage of the salvage value at the beginning of the year. The ratio of the depreciation in any one year to the salvage value at the beginning of that year is constant throughout the life of the structure and is shown in Example 5.4.

EXAMPLE 5.4: Constant Percentage Method

Assume that an office calculator costs $120 when new, and has an estimated life of 10 years and a salvage value of $20 at the end of 10 years.

Then, to determine the rate of depreciation and salvage value of the calculator at the end of, for example, the sixth year, the calculation is as follows:

Ratio of depreciation in any one year to the salvage value at the beginning of that year

$$= 1 - \sqrt[10]{\frac{20}{120}} = 0.1641$$

Then the depreciation during the sixth year is

$$\$120 (1 - 0.1641)^6 = \$40.94$$

Thus, the heaviest depreciation costs occur during the first years of the life of the calculator. Proponents of this method feel that the results more nearly parallel the actual second-hand sales value than do those obtained by other methods. The important thing is to obtain a measure of future profits which can be expected.

Straight-Line Method

The straight-line method of computing depreciation assumes that the loss in value is directly proportional to the age of the asset.

EXAMPLE 5.5: Straight-Line Method

If an electric typewriter has an original cost of $120, an estimated salvage value of $20, and an estimated life of 10 years, the uniform annual depreciation charge is

$$\frac{\$120 - \$20}{0.10} = \$10 \text{ per year}$$

The shortcoming of the straight-line method is that it does not take into account such things as interest, operation, maintenance costs, or profits. Its proponents hold that these factors tend to balance each other and that, since operation and maintenance—as well as useful life—must be estimated, there is little reason to use a more complex formula that indicates maintenance, interest, and operation profit.

Sinking Fund Method

With the sinking fund method, the annual cost for depreciation is uniform. Also, interest is taken into account.

EXAMPLE 5.6: Sinking Fund Method

Assume a petroleum company investment of $1 million for an expansion to a current refinery, allocated $100,000 for land and $700,000 for fixed and other physical properties subject to depreciation. Additional capital of

$200,000 is available for operation purposes, but this sum is not subject to depreciation. Investors want a 15% interest rate (or earning rate to investors) on their money for a 10-year period. The sinking fund method will be used, with depreciation figured at 15% per year. No income taxes are involved in order to simplify the example.

First-year profit before deducting the sinking fund depreciation charge made at the earning rate of 15% interest, and assuming no salvage value for the physical properties, is $0.15 \times \$1,000,000$, or $150,000 per year.

But the oil company must earn enough additional money annually to pay for the depreciation occurring on the depreciable capital of $700,000.

Using sinking fund depreciation and a 15% interest rate for the sinking fund, the annual deposit in the fund is

$$\frac{\$700,000 \times 0.15}{(1.15)^{10} - 1} = \$34,440$$

Thus, company profits before depreciation must total $184,440 ($150,000 + $34,440) and not merely $150,000 in the first year. Actually, the $184,440 in the first year represents:

$34,440   = the sum of annual depreciation charge

$105,000 = the 15% interest on the undepreciated part of the depreciable capital which is in the first year or before any deductions, $0.15 \times \$700,000$

$45,000   = the 15% interest on the nondepreciable capital, or $0.15 \times \$300,000$

$184,440 = the total for the first year

Thus $139,440 ($105,000 + $34,440) is needed to cover (1) the depreciation deposit in the sinking fund, and (2) the interest on the depreciable capital for that year. This is also calculated as:

$$\$700,000 \frac{0.15(1.15)^{10}}{(1.15)^{10} - 1} = \$139,440$$

In each succeeding year, the book value of the depreciable capital decreases, but the depreciation reserve increases in such a manner that the sum of the two always equals $700,000 and the total annual interest remains constant at $105,000 even though the interest charges on each component vary.

The biggest drawback to the actual use of the sinking fund method in business is the fact that businesses rarely maintain an actual depreciation sinking fund. The interest rate which could be obtained on such deposits would be small, probably not over 6% in the petroleum business, according to financial experts in the oil industry. An active business, such as an oil company operation, is constantly in need of working capital. This capital will usually earn much more than 6%.

A reasonable rule is that all values should be kept invested in the oil business and not remain idle. As a result, a fictitious depreciation fund is often used: The amounts which have been charged to depreciation are actually left in the business in the form of assets, and a "reserve for depreciation" account is used to record these funds.

Where such a "depreciation reserve" is used, the company is actually borrowing its own depreciation funds. Therefore, there is no place from which interest on these values could be obtained except from the business itself. This would create a situation in which a business pays itself interest for the use of its own money. To accomplish this, the cost of depreciation equal to the sinking fund deposit has to be charged as an operating expense, and then interest on the accumulated sinking fund has to be charged as a financial expense. Such a procedure accurately accounts for all expenses, but might require considerable explanation to government income tax authorities. Hence interest is not used when sinking fund deposits are not made to an outside source.

Summary of Methods for Determining Depreciation

Our discussion of depreciation concludes with an example comparing the sinking fund method to the straight-line method.

EXAMPLE 5.7: Comparison of Sinking Fund and Straight-Line Methods

Table 5.3 illustrates depreciation over 10 years for the investment in Example 5.6 as calculated by both the sinking fund and straight-line methods. Figure 5.1 compares the book values obtained by the two methods as a line graph. As Table 5.3 and Figure 5.1 show, at the end of the second year the depreciation deposit into the sinking fund is $34,440, but interest on the previous deposit is $0.15 \times \$34,440$ (deposit for the first year) or $5,160. This is repeated for the third year with 15% interest on $39,600 ($34,440 + $5,160), and so on for each year. Before the petroleum company can earn interest of $150,000 for the second year it must deposit $34,440 in the sinking fund and pay $5,160 interest on a total of $39,600 to the sinking fund depreciation reserve.

Figure 5.1 shows how the straight-line and sinking fund methods differ. The curve of the sinking fund bulges from the straight-line method curve, yet both eventually meet at the end of the 10th year.

TABLE 5.3

Examples of Straight–Line and Sinking Fund Depreciation Methods

End of year	Total in sinking fund depreciation reserve	Annual interest, 15% on column 2	Annual deposit	Annual charge	Book value at end of year	Annual charge	Book value
1	2	3	4	5	6	7	8
Start	0	0	0	0	$700,000	0	$700,000
1	$34,440	0	0	$34,440	665,560	$70,000	630,000
2	74,040	$5,160	$34,440	39,600	625,960	70,000	560,000
3	119,580	11,100 (15% of 74,040)	34,440	45,540	580,420	70,000	490,000
4	171,960	17,940	34,440	52,380	528,040	70,000	420,000
5	232,200	25,800	34,440	60,240	467,800	70,000	350,000
6	301,540	34,900	34,440	69,340	398,460	70,000	280,000
7	381,280	45,300	34,440	79,740	318,720	70,000	210,000
8	472,920	57,200	34,440	91,640	227,080	70,000	140,000
9	577,960	70,600	34,440	105,040	122,040	70,000	70,000
10	700,000	87,600	34,440	122,040	0	70,000	0
		$355,600	$344,400	$700,000		$700,000	

$700,000 is a constant deduction.

Conclusions: The sinking fund method requires a lesser profit before depreciation in the first year; the straight–line method requires a higher profit, or $220,000 in the first year.

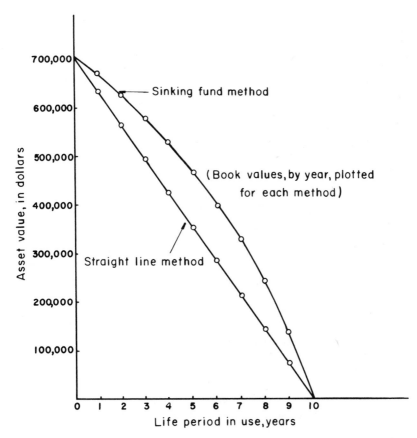

FIG. 5.1. Comparison of straight-line and sinking fund methods of calculating depreciation.

Depreciation and Obsolescence

Before turning to the subject of depletion, it is well to look at another aspect of depreciation: the effect of obsolescence. Obsolescence means that that are certain "sunk costs" in equipment that cannot be recovered. Depreciation of equipment that becomes obsolescent before the end of its useful life must take into account not only book value (original cost - total depreciation to date), but also salvage value (trade-in or resale value) and the cost of alternative equipment.

EXAMPLE 5.8: Effect of Obsolescence on Depreciation

Consider a control valve that becomes obsolete 3 years before it has been fully depreciated. When fully depreciated, the valve will have a salvage value of $400, but at this time (3 years before), it has a trade-in (or resale) value of $1,000. If the book value (original cost - total depreciation to date) is $760, there is a favorable "bonus" to management of $240 in trading-in.

But the bonus of $240 is irrelevant as a sunk cost. If a minimum rate of return is assumed as 10% before taxes, the question is whether the obsolete control valve with 3 years to go before being fully depreciated should be replaced now by a new valve. Calculations are needed to compare the old valve with a new valve, which would cost $5,000 and have an eventual salvage value of $500.

Annual cost of old valve:

Capital recovery costs (760)(0.40211)
  10% for 3 years + 0.10 × $400   = $346.00
Operating and maintenance costs   = $1,820.00
Total annual cost of old valve   = $2,166.00

Then, tentative annual cost of new valve:

Capital recovery costs ($4,500)(0.16275)
  10% for 10 years + 0.10 × $500   = $782.00
Operating and maintenance cost
  (estimated)   = $1,000.00
Total annual cost of new valve   = $1,782.00

By comparing the old control valve with the new valve, we can see that purchasing the new valve now would mean an annual savings of $384 ($2,166 - $1,782). If the old valve is depreciated out, only the salvage value of $400 could be allowed on capital recovery.

METHODS FOR DETERMINING DEPLETION

When limited natural resources, such as crude oil and natural gas, are consumed, the term "depletion" is used to indicate the decrease in value which has occurred. As some of the oil is pumped up and sold, the reserve of oil shrinks and the value of the oil property normally diminishes. Unless some provision such as depletion charges is made to recover the invested capital as the crude oil is pumped and sold, the net result will be loss of capital. This is prevented by charging each barrel, or ton, of crude with the depletion it has caused.

The depletion allowance may be computed on either (1) a fixed percentage basis, or (2) a cost-per-unit basis.

For the fixed percentage method, the percent depletion is usually set by government ruling (in the United States it has been 22% of net sales), but in no case can the fixed percentage exceed 50% of net income before deduction of depletion.

In the cost-per-unit method, the amount of depletion charged to each barrel, or ton, of crude produced is determined by the ratio of intangible development cost plus the depletable costs divided by the estimated total units potentially recoverable. This then gives a cost per unit, which is in either barrels or tons depending on how the estimated total units potentially recoverable are given.

The total units recoverable may be estimated if the number of years of production and the production rates can be estimated. For oil and gas wells the calculations vary with the nature of the production curve and the allowable flow permitted by conservation authorities of the government of the oil-producing country. A mathematical analysis follows for estimating the total barrels of oil potentially recoverable under certain assumed conditions.

EXAMPLE 5.9: Fixed Percentage Method

An oil company has an oil lease on a field with an estimated proven reserve of 1.2 million bbl. The engineers consider the reserve to be 60% recoverable in 3 years; in other words, 0.72 million bbl (720,000 bbl) can be recovered in 3 years, or 240,000 bbl in 1 year.

Intangible development costs, excluding a $20,000 bonus to the landowner, all occur in the first year and are estimated at $160,000. Depreciable capital casing, machinery, derricks, rigs, etc., are $900,000, with an estimated economic life of 9 years.

Now, if 300,000 bbl of crude are sold the first year at $2.00/bbl, and operating and other expenses are $50,000 for the year, by the fixed percentage method of calculating depletion, the depletion charge on a 3-year basis is as follows:

Cost items for the year:

Net sales 300,000 bbl × $2   = $600,000
Annual depreciation, $900,000 ÷ 9 =   100,000
First-year operating expense   =   50,000

If we assume a fixed percentage rate of 27.5% of net sales, the depletion is 0.275 × $600,000 or $165,000, and if the $165,000 does not exceed

50% of net income before allowing for depletion, the depletion allowance is $165,000. But net income must be computed to determine if the $165,000 does not exceed 50% of net income or profit.

The $20,000 bonus in this problem is "recovered" as part of the depletion charge.

Net sales		$600,000
First-year expense	$50,000	
Development expense	160,000	
Annual depreciation expense	100,000	310,000
Total net income (profit)		$290,000
And 50% of net income is		$145,000

Thus, the maximum allowable depletion in this problem by use of the percentage depletion method is $145,000, since $165,000, or the computed depreciation at the fixed percentage of 27.5%, is more than 50% of the net income of $145,000.

EXAMPLE 5.10: Cost-per-Unit Method

With the cost-per-unit method of depletion, using the same figures as in the previous example, the depreciation rate is $0.25/bbl of crude or refined product:

$$\frac{\$160,000 \text{ (development expense)} + \$20,000 \text{ (bonus for the lease}}{0.60 \text{ (recoverable)} \times 1,200,000 \text{ (total estimated proven reserves)}}$$

$$= \$0.25/\text{bbl}$$

Then the allowable depletion based on the cost-per-unit method for the first year is $0.25 \times 300,000$ bbl (amount of crude sold in the first year) = $75,000 in depletion. The $75,000 amount would be allowed even if it exceeded the amount permitted by the fixed percentage method; but the cost-per-unit method must be used each year once it has been adapted.

Summarizing the two methods given for determining depletion allowances, total allowable first-year charges for capital recovery may be summarized as follows:

Basis	Percent depletion changed to dollars	Unit cost depletion
Annual depreciation (equipment)	$100,000	$100,000
Development expense	160,000	not included, but here in formula
First-year depletion	145,000	75,000
Total first-year charges for capital recovery	$405,000	$175,000

Net income (before taxes) is net sales minus the allowable deductible costs; for the first year, these are as follows:

	Percentage depletion method		Cost-per-unit depletion method	
Net sales		$600,000		$600,000
First-year operating expense	$50,000		$50,000	
Development expense (100% incurred in first year)	160,000		(included in formula)	
Depreciation expense on equipment	100,000		100,000	
Depletion on oil reserve	145,000	455,000	75,000	225,000
Net income		$145,000		$375,000

Then, in the second year (assuming the same sales figures):

	Percentage depletion method		Cost-per-unit depletion method	
Net sales		$600,000		$600,000
Second-year operating expense	$40,000		$40,000	
Depreciation expense	100,000		100,000	
Depletion on oil reserve	165,000	305,000	75,000	215,000
Net income		$295,000		$385,000

In this case, 50% of net income before depletion with use of the percentage depletion method is $230,000 (600,000 - 140,000 ÷ 2), so the full 27.5% of net sales is allowed since $165,000 is less than $230,000.

Summary on Depletion

On the calculations above, the percentage depletion method promotes recovery of a greater amount of oil reserve depleted value, or $145,000 to $75,000 for the cost-per-unit method. But there are no more development costs incurred after the first year's $160,000. The $160,000 was a large factor in determining the amount of depletion. Of course, new development charges could be incurred in other years, and would then be included in determining the amount of depletion.

Also, any additional development costs or any changes in estimated recoverable oil will require a recalculation of the cost-per-unit depletion rate which is then used to determine depletion in subsequent years.

Although the net income by the cost-per-unit depletion method is greater in both the first and second years for the example given, the total net income plus capital recovery for 2 years added together by the percentage depletion method ($440,000 + $160,000 + $200,000 + $310,000 = $1,110,000) is equal to the cost-per-unit depletion method total ($760,000 + $200,000 + $150,000 = $1,110,000).

However, the percentage method has an advantage in a lower profits tax over the cost-per-unit method with reported lower net incomes for each year (in the first year, $145,000 to $375,000, and in the second year, $295,000 to $385,000). But the cost-per-unit method does have an economic advantage where rights to oil resources are purchased outright or leased at a relatively higher price to the seller because the net income figures are greater with this method.

Accounting for depletion can be complicated because of the uncertainties of future development costs, uncertainties about actual recoveries of oil from proven reserves, uncertainties about future value of oil reserves as selling prices go down or up, and uncertainties about the scale of operations, that is, the magnitude of production of oil. Variations in all or any of these factors may result in changes in the depletable value, necessitating separate calculations each year for the depletion charge.

When there is an increase in value of oil reserves as opposed to an increase in amount of proven oil reserves, or a big increase in selling prices, accretion rather than depletion is practiced to show the increase or "growth" in the oil reserve. When such increase in value results, an allowance for it must be made in the accounts of the oil company.

Chapter 6

OIL FIELDS

BASIC TERMS AND CONCEPTS

In discussing oil fields, terms such as "reservoir" or "resource base," "reserve," and "resource" are often used interchangeably, but in fact mean quite different things.

The "reservoir," or "resource base," is the sum total of crude oil, natural gas, and natural gas liquids in the ground within an identified geographic area. The reservoir thus includes all stocks, including some stocks which are unrecoverable and therefore not included in "proven reserves."

"Reserves" include the proven reserve stocks of immediate or short-term economic feasibility of extraction; therefore stocks which are known to exist but cannot profitably be extracted are excluded from reserves. The cost limits, or as far as one can go on profitably employing these reserves, are those costs consistent with the taking of "normal" risk and commercial production.

Reserves can be either on-shore and off-shore types. All on-shore and off-shore reserves are regarded as "ultimate reserves" based on the proven reserves concept that includes only that oil which is recoverable with current technology. To convert an ultimate reserve estimate into an inferred resource base (or reservoir), the estimates of all reserves (both primary and secondary estimates, where primary are the original reserve estimates with current technology applied, and secondary are the reserves with advanced technology applied) are divided by the average recovery factor of both the primary and secondary recoveries. So if total reserve estimates are 50 million bbl and the average recovery factor of both primary and secondary recoveries is 50%, the reservoir is reputed to have 100 million bbl.

"Resources" are that portion of the reservoir which will possibly become available, given certain technological and economic conditions. Resources are often known as "secondary reserves."

The term, "finding oil," is used for purposes of simplicity to include exploration (search) for oil, development of successful exploration discoveries, including the drilling of wells, and finally, the drilling and preparing of oil for commercial production, including the laying of gathering pipelines and pump installation for the movement of oil to central points for gas separation.

"Proration" is a common term used today to denote "deliberate conservation," or a "holding back" of oil supplies by oil companies and oil-producing countries to "slow down" fast production of dwindling oil reserves. There are several reasons for proration, based mostly on government policy of oil-producing countries relative to conservation of productivity and marketing of their principal natural resources, oil.

"Recovery" represents the actual number of barrels or tons of crude oil which will be obtained from an oil reservoir. The "recovery factor" is expressed as a percent, or ratio, of the amount to be obtained to the total amount available in the reservoir.

EXPLORATION AND DEVELOPMENT

The first prerequisite to satisfying man's requirements for refined petroleum products is to find crude oil. Oil searchers, like farmers and fishermen, are actually in a contest with nature to provide the products to meet human needs. They are all trying to harvest a crop.

But the oil searcher has one problem that the farmer does not have. Before the oil man can harvest his crop, he has to find it. Even the fisherman's problem is not as difficult, since locating a school of fish is simple compared to finding an oil field.

The oil searcher is really a kind of detective. His hunt for new fields is a search that never ends; the needle in the haystack could not be harder to find than oil in previously untested territories. But new petroleum must constantly be found to replace that which has been used. Only in this way can oil companies maintain their crude reserves and keep sending raw material to the refineries and oil products to consuming markets.

Finding Oil

The search for oil begins with a preliminary examination of evidence found on the surface in the area concerned and involves mapping and geophysical surveys. Geological surveys are usually made by the seismic method or creation of artificial earthquake (seismic) waves, in which high explosives are fired into holes and the rates of travel of these waves are analyzed by echo sounding techniques. Other means of exploring for oil include detailed ground geological surveys aided by preliminary results of aerial photography and photogeological work.

All this activity leads only to an evaluation of the probability that oil is in the particular location. Once it seems probable that there really is oil, test wells must be drilled. Costs rise rapidly at this stage. The cost of drilling a test well may be several thousand dollars a day, including a hiring accommodation for a drilling crew and engineers. Actually, the drilling of test wells is the most costly single operation in oil exploration.

In oil well drilling, the rotary drilling method is the one most popularly used. This involves drilling by use of a bladed (tooth) cutter rotating into the ground while pumping a fluid through the ground to clear away sand and rock debris. Some of the drilling equipment required (fixed costs) which adds to the cost of drilling are the oil derrick itself (which can stand several stories high), and the drill pipe equipment and fittings and tools for the drill pipe, including equipment for lowering and raising the drill pipe in the hole, equipment for rotating and increasing length of drill pipe as the hole goes down, and pumping equipment.

Furthermore, one exploratory well alone does not indicate extensive oil accumulation. Other wells, carefully located near the well where oil has been discovered, are drilled to discover if there is a reservoir in the area and how much approximately is available and can be recovered. Thus, it is desirable (1) to obtain reliable information as to the quantity of oil (and gas) which is recoverable, so an economic and proper size and type of surface crude oil production plant can be set up, and (2) to determine from the samples of the reservoir the characteristics of the oil itself as to the nature and amount of oil in the reservoir. The raising of oil to the ground surface and then the handling of the oil at ground surface will depend to a great extent on the nature of the oil itself and its associated gas.

Crude oil can range from very heavy viscous oil, almost a tar, with little or no gas dissolved in it and under very low pressure, down to an extremely light, straw-colored oil with a considerable volume of gas, known as a condensate-type crude. The condensate-type crude is more likely to be found at great depths. Under conditions of high pressure and temperature which exist at deep levels, the crude is usually in the gaseous stage. Between the extremes of a heavy viscous oil and a very light oil there is an infinite variety of crude oil. The manner of producing these crudes is decided only after examining samples which show their characteristics and physical attributes.

When a large oil reservoir that is commercially feasible is known to exist, such questions as "how many wells should we drill in the reservoir," "how many wells do we need before we can lay pipe economically," "what spacing is necessary between well heads," etc., will need to be answered.

Proven reserves can tie up large sums of capital. As a result, the oil industry usually desires high production wherever possible. The ratio of proven reserves to production, therefore, tends to be small. Furthermore, the oil industry attempts to hold technology and costs constant as long as they can. But when technology is limited to a current status, there is often a strong downward bias in determining future productive capacity as such.

Once the oil has been explored, developed, and produced, all costs involved in getting the oil to the surface, where it becomes a commodity as it is piped in gathering lines to central points for gas separation, are called the cost of oil field operation. The basic question of "what does oil cost to find, to develop, and to ready for commercial production" would be comparably simple to answer if, during a short period of time—say 1 to 3 years—an oil company could start in the oil producing business, discover say 10 million bbl of oil, develop that 10 million bbl, and finally produce the 10 million bbl of crude. The cost of finding, developing, and producing could then simply be found by dividing the total amount spent for exploratory, developing, and producing effort by 10 million bbl, which would give a cost per barrel of crude.

But this is just "grocery store accounting." Actual accounting for costs in the oil-producing industry is not that simple. When a company searches for oil, it may spend several years and millions of dollars on exploration and development before any substantial, and commercially feasible, amount of oil is located. In development alone, a company may work for several years and spend many dollars developing the oil reservoir which is to produce over an even greater number of years; and also, all this time, the process is constantly repeating itself as more oil is being discovered, more oil is being developed, and more oil is being produced.

Determining true costs of exploration, development, and production can thus be difficult. For example, a company might claim that it had a finding cost of $1.00/bbl last year. While spending $2 million on exploration, development, and production, the company brought up 1 million bbl of oil to ground surface, and also increased its estimated proven reserves by 1 million bbl. Only under unusual conditions would these figures give a true picture of the company's position. The figures indicate that the company spent $2 million a year and no money on exploratory or development work last year, yet the company produced 1 million bbl of oil last year and adjusted its reserve estimates upward by 2 million bbl on the basis of oil well performance. This, of course, is obviously erroneous. Thus, determining the true cost of finding oil is a problem in which the various major oil companies differ in their calculations.

Starting with costs of exploration dependent on drilling depths, before test drilling, there are numerous costs not associated with drilling and depth, of which only some costs can be identified. For instance, crew costs may vary according to complex shooting patterns. Other costs are due to increasing material and personnel costs, common depth-point procedures, increases in right-of-way permit costs, lease bonus payments to governments, seismic crews, trucks, digital equipment, etc. There are also average land costs, at times, in the form of rentals.

Budget expenditures before production, as given above, may either be capitalized, in which case they are depreciable assets, or may be written off each year as revenue expenditures. For example, salaries, supplies, and contract services of geophysical crews are regarded as revenue expenditures, along with rentals, taxes paid, and costs of geophysical core drilling tests; but transportation equipment, digital equipment of seismic crews, as well as cash bonus payments to acquire a lease on property are capitalized. Thus, the latter are regarded as depreciable assets. When test drilling, and later development drilling stages, are reached, most costs are then associated with "depth" drilling.

Finally, an oil company's success is measured by its ability to discover reserves. In its search for oil, it spends substantial amounts of money in many different ventures in widely scattered areas. The oil company does this knowing that many of these ventures will be nonproductive and will eventually be abandoned.

On the other hand, the oil company recognizes that successes in other areas must be large enough to recoup all money spent in order to break even or to provide a profit. Thus, the true assets are the oil reserves, and these costs are capitalized. But the costs of nonproductive exploration activities and of dry holes are also a necessary part of the full cost of finding and developing these oil reserves.

The Cost of Finding Oil. There are several costly stages involving several months before crude oil can be ready for production for refineries. For instance, as stated before, costs of exploration, which can continue for many months if oil is discovered at all, will also include the costs of crews for exploration, seismic surveys, geophysical surveys, test drilling, etc. After several months at different drilling sites which may involve, finally, a number of dry holes, the stage of development of wells begins. This involves a number of costly items such as oil rigs with derricks, well casings, etc. To be successful, the drilling of a particular well must reveal enough crude oil for it to be commercially feasible to bring in pumps and to lay pipe to transport the crude to tank farms.

Knowledge of costs of searching for oil in terms of average finding cost per barrel of oil and the average cost per exploratory well is important.

Knowledge of these figures makes one able to more favorably compare them with revenues to be received per barrel of crude and revenues received per well. If cost data reveal, for instance, that the average cost of finding a barrel of oil is $2.70/bbl and the cost of drilling the average exploratory well, with additional costs of rentals, geological, geophysical, and overhead is $115,275, then revenues must at least exceed $2.70/bbl or $115,275, and by an amount that favors a "reasonable" return on investment.

Suppose, however, that for each net barrel of crude oil produced, the oil industry spends $3.19 as compared with an average wellhead value of $2.92/bbl. Relative to production, an approximate breakdown of this cost of $3.19 might be: exploration, $1.07; development, $1.21; lifting and overhead cost, $0.91. This indicates that the oil producer is confronted with a serious cost-price squeeze. A situation such as this cannot continue for too long before exploration and development suffers.

One aspect that might be observed, which may have been increasing costs, is the dry hole ratio and the discovery ratio to the exploratory dollar. When the number of dry holes is increasing while the quantity of new reserves found per exploratory dollar is decreasing, it is evident some positive action is needed in cutting costs of exploration. Exploration expenditures are estimated to represent approximately one-third (See Table 6.1 below) of the oil industry's combined expenditures in finding, developing, and producing petroleum liquid, according to Middle East experts in oil fields. Furthermore, like other business pursuits, the search for new oil reserves involves higher expenditures as the years progress.

Replacement of "used reserve" costs have increased tremendously, which means that any margin of profit on current production represents the difference between present high-cost reserves and "higher-than-high-cost" reserve price plus high depletion allowances for decreasing reserves. The margin of profit on current production must be high enough that there is no possibility that in the future sufficient risk capital will not be available for continued explorations. Thus, an unrelenting search for oil is necessary as world demand for oil begins to exceed supplies.

The management of any oil company should know how their exploration costs compare with crude values and reserves found per well. A table such as Table 6.1 might be constructed to give an up-to-date view of costs to values. The column headings in Table 6.1 represent some hypothetical data related to exploration costs and crude values for 6 years, using 1968, 1969, and 1970 as 100 for index purposes.

As indicated in Table 6.1, both value index in column 5 and cost index in column 7 are rising, but since 1969 or beginning with 1970, costs have gone ahead of values, as the percent of exploration costs to value also increase.

TABLE 6.1

Combined Expenditures of Exploration and Development for Finding Oil

Year	Gross crude production in barrels	Average price per barrel	Gross value of production	Value index 1968–1970 = 100	Exploration and development cost	Exploration cost index 1968–1970	Exploration cost, percent of value
1	2	3	4	5	6	7	8
1968	$1,857,000	$1.93	$3,578,000	80	$904,000,000	82	29.5
1969	2,020,000	2.60	5,245,000	106	1,135,000,000	103	25.3
1970	1,842,000	2.54	4,675,000	104	1,255,000,000	114	31.4
1971	1,974,000	2.51	4,963,000	110	1,439,000,000	131	33.9
1972	2,248,000	2.53	5,690,000	126	1,740,000,000	157	35.7
1973	2,290,000	2.53	5,785,000	129	1,987,000,000	181	40.2

This indicates a precarious situation, and means that this particular oil company needs higher prices for its crude or a cut in costs of exploration if the oil company is to remain in business. The only reason there is any profit for this oil company in producing crude oil, as illustrated, is due to the fact that much of the oil produced currently was possibly discovered in years earlier than 1968 when costs of finding and development per barrel were much lower.

As stated in Chapter 5, every new oil field discovered is a declining asset from its inception. This should emphasize the need for undiminishing efforts by oil men to offset the natural decline of productive wells with new sources of supply. Obviously, costs do not remain stationary, and experience has shown that oil becomes harder to find each year. But margins of profit must be adequate for oil companies to continue in business. Thus, any value index must always exceed the cost index. This means higher prices for crude oil as costs of finding and developing rise. Prices and price theory relative to crudes, as well as refined products, are discussed in Chapter 8.

After oil is discovered and it has been established that there are sufficient quantities for extraction to be economically feasible, if the oil property is not in the Middle East or North Africa under a concession agreement, an evaluation of the property may be made preparatory to leasing or buying of the property. This is a necessary step before investment in capital equipment and in pipelines begins.

Evaluation of Oil Property

For various financial purposes, it is often desirable to know the present worth of property being evaluated (see Chapter 4). This can usually be obtained by applying an appropriate discount factor (or formula), usually computed at a rate of from 4 to 8% per year compounded monthly.

EXAMPLE 6.1: Evaluation of Oil Property

Table 6.2 gives data for a hypothetical oil property. A discount factor of 8% could be applied to column 13 to give present value for each of the 11 years given in the table. Table 6.3 gives present values for these years.

Fair market value of any oil property should be based on the principle that in a transaction between a buyer willing to buy and a seller not forced to sell, the buyer should recover his investment and make a profit commensurate with the number and character of risks assumed with the property and the present value of money.

TABLE 6.2

Evaluation of 100% Working Interest in a
Hypothetical Oil Field as of January 1, 1973

Future years	Prod. oil wells	Estimated future oil reserves, bbl			Value of future net oil reserves	Operating expenses less taxes			Ad valorem prod. taxes	Net operating income (Col. 6 − Col. 8 + Col. 10)	Net investment and salvage	Est. undiscounted future net cash realized	
		Daily ave. oil prod. (365)	Gross prod. (365)	Net production		Per well month	Net total	Per net barrel				Col. 11 + Col. 12 total	Col. 13 accumulated
1	2	3	4	5	6	7	8	9	10	11	12	13	14
1	30	598	218,270	190,980	$540,470	$300	$108,000	$0.57	$19,100	$413,370	($16,000)	$429,370	$429,370
2	28	480	175,200	153,300	433,840	300	100,800	0.66	15,330	317,710	(8,000)	317,710	747,080
3	28	384	140,160	122,640	347,070	300	100,800	0.82	12,270	234,000	(16,000)	242,000	989,080
4	27	305	111,320	97,410	275,670	300	97,200	0.98	9,740	168,730	(16,000)	184,730	1,173,810
5	25	244	89,060	77,930	220,540	300	90,000	1.15	7,790	122,750	(8,000)	138,750	1,312,560
6	23	195	71,170	62,270	176,220	300	82,800	1.33	6,230	87,190	(8,000)	95,190	1,407,750
7	22	155	56,570	49,500	140,080	300	79,200	1.60	4,950	55,930	(16,000)	71,930	1,479,680
8	20	124	45,260	39,600	112,070	300	72,000	1.81	3,960	36,110	(16,000)	52,110	1,531,790
9	18	98	35,770	31,300	88,580	300	64,800	2.07	3,130	20,650	(16,000)	36,650	1,568,440
10	16	78	28,470	24,910	70,500	300	57,600	2.31	2,490	10,410	(8,000)	18,410	1,586,850
11	15	68	24,820	21,720	61,470	300	54,000	2.49	2,170	5,300	(180,000)	185,300	1,772,150
Total future, or average, values			996,070	871,560	$2,466,510	$300 (ave.)	$907,200	$1.04 (ave.)	$87,160	$1,472,105	($300,000)	$1,872,150	

A fully developed property having an established rate of decline naturally presents the least risk to a potential buyer. Recognizing this fact,
the buyer in this case may be willing to purchase on the basis of receiving
$1.00 profit per $1.00 invested, which would make the purchase price
$906,400 (net production of 871,560 bbl × $1.04 average profit).

The buyer would also examine his investment on the basis of the dollar
value of net reserve he is purchasing. The $906,400 purchase price would
be equivalent to $1.04 per net barrel of oil reserve, which, let us assume,
is consistent with standards existing in the market at this time.

Another consideration is the rate of payout. According to column 11
the $906,400 investment would pay out in somewhat less than 3 years
($413,370 + $317,710 + $234,000) for income tax considerations. There-
fore, from the standpoint of payout, the $906,400 appears to be somewhat
low, since oil management usually wants a payback of less than 2 years on
most projects. A look at Table 6.3 on present values, discounted net cash
flow realized, for future years, reveals that total present values for 11
years are $1,328,773, in line with total net cash flows of $1,772,150 when
discounted at 8%. The foregoing reasoning suggests that a fair market
value for the example property might be in the range of $900,000 to $1 million disregarding secondary recovery possibilities.

TABLE 6.3

Present Value of Net Cash Flows at 8%

Future years		Net cash flow realized	Discount factor	Present value
1		$429,370	0.926	$397,597
2		317,710	0.857	272,277
3		242,000	0.794	192,148
4		184,730	0.735	135,776
5		138,750	0.681	94,489
6		95,190	0.630	59,970
7		71,930	0.583	41,934
8		52,110	0.540	28,139
9		36,650	0.500	18,325
10		18,410	0.463	8,524
11		185,300	0.429	79,404
	Total	$1,772,150		$1,328,773

Concession Agreements

Foreign oil corporations, especially in the Middle East, seek oil reserves on the basis of special concession agreements. These agreements give them exclusive rights to exploit petroleum resources in a defined area for a long period of time. In return for this privilege, these foreign oil corporations undertake to explore for and to produce oil, and to pay the host country royalties and certain other sums in connection with various phases of their operations, such as an income tax, a production tax based on a barrel- or ton-produced basis, or an ad valorem tax on their production operations.

Although each concession contains special stipulations that distinguish it, a number of provisions are common to practically all concessions granted by the producing countries. They are as follows.

1. The exclusive right to explore, prospect, extract, refine, and export crude oil and related items, such as natural gas, within the area of the concession, but not operation subject to any separate agreement.

2. A time limit within which exploration must begin. If exploration does not begin on time, the concession can be revoked by the host country.

3. Extensive area in the concession. (For example, Aramco originally began their concession with 93,000 square miles, which was later reduced by the Saudi Arabian government.)

4. The right to establish their own system of transportation and communication for efficient conduct of operations, such as radio, telegraph, telex, telephone, railroads, vessels, and even airplanes. But the host country usually reserves the right to requisition these facilities in case of war or of a national emergency on payment of "fair" compensation by the government involved.

5. Acquire title to lands either from the host country or individuals, if necessary, in conduct of operation of the company. Prices of purchase from individuals should conform to local prices at time of purchase. Holy places, monuments, and historical sites are generally exempted.

6. The host country must be supplied by the companies with specified amounts of oil and oil products for local consumption.

7. Subsidiary companies can be established by the oil companies to conduct specific phases of their operations, such as a subsidiary company for marketing or a subsidiary for refining. Profits of such subsidiaries are included in royalty calculations.

8. Companies accept the right of host countries to be represented in dealing with the companies by specially appointed delegates, whose salaries and expenses are borne by the companies. These delegates also have the right to inspect company books of accounts as well as operations in the field and in the refinery.

9. Unskilled laborers employed by the companies must all be nationals of the host country. Engineers, managers, skilled technicians, and other skilled workers and clerks can be aliens if qualified personnel cannot be found locally. Furthermore, agreements usually provide for the companies to train local nationals in skills required for these positions.

10. Annual operating reports, including data on discovery of new oil deposits and geological plans and records, must be presented to the host country by the companies. This is treated as confidential information by the host country.

11. Any new issuance of capital stock must be open to subscription to the public of the host country. Sometimes, concessions specify the percentage of new issues which are to be made available. For example, in the case of Aramco, 20% of any new stock issue must also be made available to the Saudi public.

12. Companies are either exempted from all direct and indirect taxation (Aramco excluded), or are assured that such taxes as may be applied shall not be different or higher than those taxes imposed upon other industrial undertakings in the country. However, for any year, income taxes are not to exceed 50% of the net operating profits of the company, which profits to be calculated after payment of foreign income taxes.

13. Concessions exempt companies from customs duties and licenses for the importing of all machinery, equipment, and materials necessary to conduct operations of the oil company.

14. Companies must provide for (pay cost of) arbitration involving any disputes between the host country and the companies. Usually, this involves an arbitration for each side with the arbitrators selecting the "umpire." Sometimes, the World Court supplies the "umpire" if arbitrators cannot agree.

15. In major concessions the financial arrangements usually include: (a) a "bonus" payment of a lump sum to the host country at the time the concession is granted; (b) payment of a "dead rent" during the period of exploration; (c) payment of royalties (plus income taxes on profits at times) calculated either on the basis of volume of oil produced, sold, or exported or on the basis of profits, or on the basis of both volume and profits.

16. Finally, the existing major concessions today do not provide for a partnership between the host government and the company, but for a foreign-owned and foreign-managed enterprise. But changes have been made in concessions relative to this point, as witness those of I.P.C., Aramco, and the Anglo-Iranian Oil Company.

Some departures from standard patterns of concession agreements may be classified into two basic categories: (a) revisions and changes affecting still-existing older concessions; (b) departures from standard procedure in new agreements as in the case of the development of offshore areas by foreign companies with the Iranian government, and also agreement of Japanese interests with Saudi Arabia and Kuwait relative to offshore territory in the Neutral Zone.

In some countries, particularly in the Middle East, the sovereign governments open to international bidding specified areas and regions for petroleum exploration. Usually, the procedure is for governments to send notices to all interested oil companies, inviting them to participate in an international tender for oil permits in the specified areas.

When an oil concession is granted in the Middle East, an oil agreement is drawn up between the country and the oil companies involved.

EXAMPLE 6.2: Concession Agreement

The following is a summary of an oil agreement between the Government of Abu Dhabi and three oil companies for a concession of two offshore areas, totaling 3,150 km^2 [1]. The details of the summary gives an idea of what transpires between a producing country's government and the oil companies in a concession.

I. Ownership of the agreement is as follows:

Pan American Oil Corporation	60%
Syracuse Oils, Ltd.	20%
Wingate Enterprises, Ltd.	20%
	100%

II. The term of the agreement is 35 years, from June 7, 1970.

III. Work obligations:

1. The company must commence exploration within 6 months of June 7, 1970, and complete initial geophysical operations within 18 months.

2. The company must commence drilling a test well within 24 months. Having started drilling, the company is expected to proceed with due diligence to completion of other test wells drilled, and in no event shall the aggregate depth of all test wells be less than 30,000 ft, the minimum amount for drilling.

3. The company must spend at least $19 million (minimum investment) on prospecting, exploration, drilling, and development of wells during the first 8 years prorated as follows:

In the first year	$500,000
In the second year	1,500,000
In the third year	1,500,000
In the fourth year	1,500,000
In the fifth year	2,000,000
In the sixth year	3,000,000
In the seventh year	4,000,000
In the eighth year	5,000,000
	$19,000,000

4. Penalty: If the company gives notification of surrender of concession area and the company has not fulfilled the above minimum investment obligation of $19 million prorated over the period of 8 years from the effective date, June 7, 1970, to the date of notification of surrender, the company shall pay an amount equal to one-half of the under expenditure. (Thus, if only $9 million has been spent by date of notification of surrender, as of June 7, 1970 (or 6 years), the company is obligated to pay $500,000 to the government of Abu Dhabi, which is one-half of the underexpenditure of $1 million at the end of the sixth year. By the end of the sixth year, $10 million should have been invested by the joint companies in exploring and developing the concession.

IV. Bonus payments: The company agrees to pay the government a nonamortizable bonus of $11.5 million as follows: (a) $2.5 million within 30 days of the bonus date; (b) $2 million within 30 days after the date of discovery of crude oil in commercial quantities; (c) $3 million within 30 days after the date on which regular exports of crude oil have first reached and maintained an average rate of 100,000 bbl/day; (d) $4 million within 30 days after the date on which regular exports of crude oil have first reached and maintained an average rate of 200,000 bbl/day for 30 consecutive days.

V. Annual rentals:

1. $100,000 rental to be paid by the company within 30 days of effective date, June 7, 1970.

2. $100,000 within 30 days after each anniversary of the effective date, including the anniversary preceding the date of discovery of crude oil in commercial quantities.

3. $100,000 within 30 days after the discovery of crude oil in commercial quantities and each anniversary afterward, up to and including the anniversary preceding the export of oil commencement date.

VI. Relinquishment:

1. The company must relinquish not less than 25% of the concession area within 3 years from the effective date, June 7, 1970.

2. Also, the company is expected to relinquish 25% of the original size of the concession area within 5 years from the effective date, June 7, 1970.

3. The company shall relinquish another 25% of the original size of the concession area within 8 years from the effective date, June 7, 1970.

VII. Royalties: Royalties are based on posted prices, as explained in detail in Chapter 8.

1. The company will pay cash royalties or equivalent to cash at the option of the government at the rate of 12.5% of the posted prices of crude oil on production up to 50,000 bbl/day.

2. The company will pay cash royalties at a rate rising to 14% when production reaches an average rate of 50,000 bbl/day in any calendar year.

3. Upon reaching 100,000 bbl/day production in any calendar year, royalties will be 15% of posted prices.

4. Upon reaching 150,000 bbl/day production in any calendar year, royalties of 16% will be paid on posted prices.

5. Royalties can be expensed, but in accordance with the OPEC formula.

6. An additional royalty of 12.5% of the value of all natural gas sold or exported or used in Abu Dhabi to manufacture products for sale is also payable.

VIII. Natural gas: As a result of oil operations, the natural gas produced shall be conserved to the maximum extent, and in the best manner consistent with accepted standards of the petroleum industry.

Any associated gas not used by the company in its oil operations shall be the property of the Abu Dhabi government.

IX.   Taxation (based on posted prices): The company is liable to income tax, according to government decree of September 19, 1965, at a
fixed rate of 50%.

An annual allowance of 5% per annum is made in respect of intangible
assets and 10% per year in respect of physical assets. (The annual 5%
allowances means that the company is allowed to expense or take off their
profit, thus allowing some income tax relief to the company, at the rate of
5% of the original cost value of intangible assets and 10% each year of the
original cost of purchase price of physical assets. Intangible assets, thus,
have a "life" of 20 years, and physical assets have a life of 10 years.)

When crude oil is exported by the company, the aggregate value must
not be less than the amount which equals the number of barrels of crude
exported times the applicable posted prices per barrel, and reduced by the
following amounts (if any):

1. A marketing allowance of 0.5¢ (U.S.)/bbl of crude oil exported;

2. A percentage allowance equal to a percentage of the posted price of
    each barrel of crude oil exported in years of 1970 to 1971 inclusive
    as follows:

Year	Percentage allowance
1970	3.5%
1971	2.0%

and for 1971, and later years, no percentage allowance.

3. A gravity allowance per barrel of crude oil exported in years 1970
    to 1974 is made, inclusive for each full degree by which the gravity
    of such crude oil exceeds 27°API gravity, provided that crude oil
    which has an API gravity in excess of 37° is treated so as to calcu-
    late this allowance as if it had an API gravity of 37°. The amount
    of such allowance for each year is to be the sum of the figures of
    columns A and B for such year in the following table:

Year	Gravity allowance (U.S. ¢/bbl)	
	(A)	(B)
1970	0.264700	0.178677
1971	0.264700	0.238236
1972	0.264700	0.297795
1973	0.176467	0.198530
1974	0.088233	0.099265

For 1975, and later years, no allowance.

In determining taxable income, none of the following is to be permitted as a deduction from taxable income, neither as an expense for the year, nor by way of depreciation or amortization.

1. Foreign taxation paid on income derived from sources inside Abu Dhabi.

2. Interest, or other consideration, paid or suffered by the company in respect of the financing of their operations in Abu Dhabi.

3. Expenditures related to organization and commencing of petroleum operations in Abu Dhabi.

4. Bonuses paid to the government.

5. Annual rentals paid to the government.

X. Participation: Within 6 months from the date of discovery of crude oil in commercial quantities, the Ruler of Abu Dhabi may elect himself, or an entity nominated by him, to acquire a participating interest of 50% in all the rights and obligations under the agreement.

The government will pay a sum of 50% of total accumulated costs and expenses for such 50% participating interest. Payment will be in 10 equal installments, together with interest at the rate of discount of the Federal Reserve Bank of the United States plus 1.5%, but not exceeding 7%.

After date of discovery of oil in commercial quantities, the company is the entity that provides money advance for expenditures required for exploration, drilling, and development of production structures. This advance must cover both the company's share and the Ruler's share in the undertaking.

XI. Reimbursement: Reimbursement by the government to the company shall be made in 10 equal annual installments, together with interest at the rate of discount of the Federal Reserve Bank of the United States plus 1.5%, but not exceeding 7%. The first installment shall be paid within 6 months after the export commencement date.

The company is expected to market the Ruler's share of the crude oil production of 50%, at current realized prices. Realized prices shall mean the total consideration received for crude oil sold to nonaffiliates. (Sales to any affiliate of the company shall not be considered in determining realized prices.)

XII. Investment: When production of crude oil from the concession area reaches and maintains an average rate exceeding 100,000 bbl/day for 90 consecutive days, the company shall make, or cause to be made, studies as to the feasibility of carrying out one or more of the following hydrocarbon processing activities:

1. Methanol—1,000 to 2,000 tons/day shipment to Japan.

2. LPG recovery—10,000 to 30,000 bbl/day shipment to Japan.

3. Gas reinjection—100 to 300 MMCFD.

4. Nitrogen fertilizer—1,000 to 2,000 tons/day.

5. Crude oil desulfurization.

6. Liquefied natural gas—300 to 500 MMCFD.

Where any of the above proves feasible, the company is expected to prosecute to completion the construction of such a suitable plant to carry out the activity as reasonably practical, or at least within 3 years of the study.

If the feasibility studies indicate an alternative project requiring a comparable investment as better serving the interests of the company or the government of Abu Dhabi, or that a comparable investment in a joint venture with another concessionaire might better serve interests involved, the company will proceed with the alternative project or joint venture, after obtaining approval from the government. The company is expected to invest at least 10% of its net profits in one or more of the above feasible economic projects.

XIII. Educational, medical, and other services: Within 1 year of the export commencement date, the company is expected to contribute $150,000/ year for the establishment of various facilities pertaining to educational, medical, hygiene, and other services to be agreed with the Ruler of Abu Dhabi.

Summary: Concessions or Leases. In the Middle East and other countries with natural resources of oil, excepting the United States and Canada, the "negotiated" concession system—negotiation between the concessionaire oil company and the oil-producing country—has advantages as well as disadvantages for both government and the oil company. For one thing, concessions are normally large in a system where the government determines administratively who is to get what. Frequently these concessions range up to several thousand square miles or kilometers.

Also, there is no duplication of effort on the part of other oil producers to obtain information, since a single concessionaire controls a given area. Consequently, less money is spent, even though exploration and development expenditures are high, because several prospective oil companies are not required to seek the same information.

The concession system also has the advantage to the oil company of permitting the company to select and retain the best productive area. Thus, major developmental expenditures are made in the production area of the concession only, with a resulting reduction in the risk factor. For example, if an oil company is granted a concession for an identified 50,000 mi^2 area and if oil is discovered, through geophysical surveys and test drillings, in an area of approximately 500 mi^2, the risks of "development investment" will be been minimized. The development expenditures that follow exploration costs will, of course, be concentrated in the 500 mi^2 area.

The large initial size of the concession also minimizes the possibility of expending a large bonus payment for a barren property, which can occur with the U.S. "system" of leasing land for oil prospecting.

In the United States, through the nominating system, the government (or private owners if nongovernment property is involved) determines where, when, and how much land to offer. The lessee (or concessionaire) is selected through the medium of sealed bids. This system has the economic advantage of substituting market forces for administrative judgment. Because a "bonus" must be paid before the lease is issued, the system tends to assure the selection of an efficient oil producer. Presumably, the more efficient the crude oil producer, the lower his costs, and the higher the bid. Consequently, the more efficient oil producers will usually tend to win the bids. Bids, of course, will also reflect the bidder's appraisal of the oil property's potential and of the risks involved.

Virtually all countries, except the United States, have a relinquishment requirement for nonproductive acreage. The term of the U.S. lease is generally shorter than that in most countries, so a relinquishment requirement may not be needed.

The United States is also one of the few countries that does not have an exploration commitment. In most other countries either an approved exploration and development plan must be filed, or the minimum number of feet to be drilled or the minimum amount of exploration money to be expended must be specified. Such a commitment may not be important in the United States since the bonus bid, with its large initial cash investment at the start of the lease, tends to assure exploration of the deposit. Bonuses in other countries often cover large blocks of real estate, and on a per-acre basis would probably be much lower than the bonusus paid in the United States. For instance, a Continental Oil Company agreement with Iran involved a $10 million bonus for a 5,000 mi^2 area. Considering the required relinquishments, this averages $12.50/acre. The United States in 1968 collected an average bonus of $1,441.00/acre sold.

Ultimate Recovery

When ultimate recovery is known, it is possible to determine, within narrow limits, the future production rate. Once a significant proportion of the

ultimate reserves is produced, the shape of the future production curve
can be determined.

EXAMPLE 6.3: Oil Recovery from Reservoir

For example, let us assume a minimum ultimate recovery of 250 billion
bbl. From this assumption, a curve can be drawn as illustrated in Figure
6.1.

Indicated by this technique is a peak production rate which approaches
11 million bbl/day. If the ultimate recovery should prove to be 300 billion
bbl instead of 250 billion, peak production would prove to be no more than
10% higher, and would be delayed at most by 10 years.

On a reserve-production basis, in the face of rising production, annual
reserve increments need to increase constantly. In other words, the oil
industry needs to find at least as much oil as it produces and consumes. As
a result, the reserve-production ratio must be favorable each year. For
example, if proven reserves of crude oil total 40 billion bbl and the reserve-
production ratio is 10:1, production and consumption have been 4 billion bbl.

Actually, a ratio of 10:1 is widely thought to be near the lower limit of
efficient, sustainable production. When related to known current reserves,
this ratio can indicate what a nation's maximum producibility can be. Then
if a nation's crude oil producibility is to grow, there must be a sharp increase in annual additions to reserves.

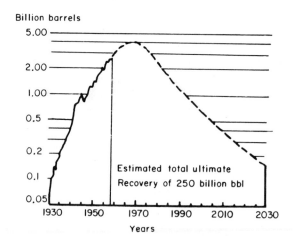

FIG. 6.1. Computed pattern of crude oil production.

So oil companies are constantly searching for new sources of oil in order to improve, or at least keep up with, current reserve-production ratio in the face of a rapid increase in rate of demand.

As reservoirs are discovered, the recoverable parts of the reservoirs become exceedingly important, since these constitute the proven reserves on which companies and countries base their production plans. Naturally, the limits of these reserves become important. Figure 6.2 illustrates one graphical approach used by oil engineers to define an oil reservoir relative to the amount of recoverable oil (reserves) available. This is, of course, in reference to primary recovery. The results indicated in Figure 6.2 can be interpreted as follows:

Well 1 penetrated the water zone.

Well 2 found the oil-water contact.

Well 3 shows little, thinning oil.

Well 4 shows little, thinning oil.

Well 5 shows nothing.

A study of all wells in the field in a similar manner defines the vertical and lateral limits of the oil-productive portion of the reservoir. In actual practice this procedure is difficult because of incomplete and conflicting information. Concurrent with this type of analysis, the engineer prepares maps that aid in establishing the volume of oil originally in place in the reservoir. Figure 6.3 shows what one of these maps might look like.

In addition to volume of oil available, it is also important to know quality or type of crude in the reserve. Figure 6.4 illustrates gas solubility, or gas-oil ratio relationship with pressure. The pressure increases as the gas-to-oil ratio becomes greater, until a saturation pressure point is reached, at which point natural pressure tapers off. Figure 6.5 illustrates the oil viscosity in the reservoir, with pressure diminishing as the crude becomes lighter; and Figure 6.6 illustrates shrinkage of the oil discovered in a reservoir, where natural pressure increases up to the saturation pressure as the ratio of reservoir oil to stock tank oil increases.

The above fluid characteristics of gas solubility, oil viscosity, and oil shrinkage are important in calculating oil reserves from production data. The relationships set forth in gas solubility (gas-oil ratio to pressure in the reservoir), reservoir oil viscosity (ratio of viscosity of oil to the pressure in the reservoir), and oil shrinkage (ratio of reservoir oil to stock tank oil and the pressure of the reservoir) were derived from an analysis of a sample of a typical reservoir oil.

When reservoir oil is lifted to ground surface and placed in storage tanks, pressure is decreased and dissolved gases separate from the oil,

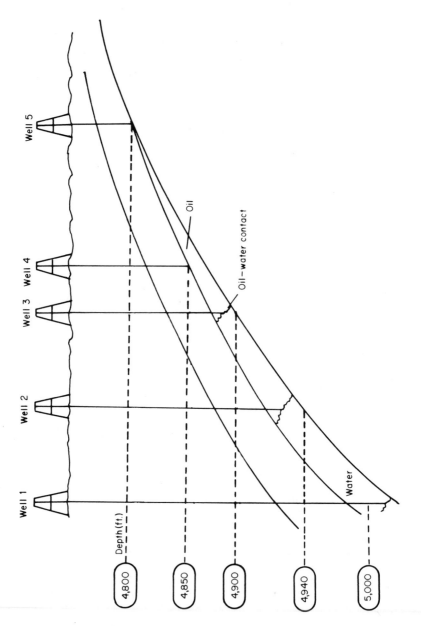

FIG. 6.2. Schematic cross section illustrating the defining of an oil reservoir.

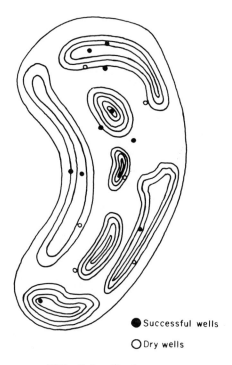

Successful wells

Dry wells

FIG. 6.3. Contour map.

thus shrinking the volume of the original liquids. The characteristics of
shrinkage (reservoir volume factor, as it is often referred to) are essential
in volumetric calculations of oil reserves.

In most reservoirs, natural drive, or energy, is available to move the
oil in the reservoir to the well bore, because the oil is trapped under conditions of relatively high pressures and temperatures. There are five natural energy sources. These are (1) reservoir rock and fluid expansion,
(2) free (external) gas cap, (3) solution (internal) gas injection, (4) water
injection, and (5) gravity drive. These five energy sources may act independently in any particular reservoir, or they may act in combination.

The importance of these recovery mechanisms may be seen as follows:
A solution gas drive may get oil recoveries from 5 to 35% of the original
stock-tank oil in place, leaving a large percentage of oil in the reservoir
unrecovered at depletion. A gas cap drive can bring oil recoveries in excess of 25%, and in combination with gravity drainage get as high as 60% in
oil recoveries. A water drive, or injection, is considered an efficient natural recovery mechanism, and recoveries range from 35 to 65% of the stock

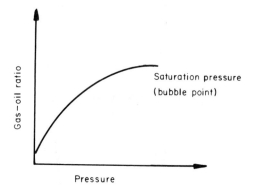

FIG. 6.4. Gas solubility.

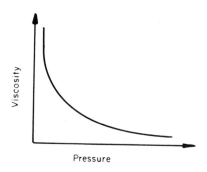

FIG. 6.5. Reservoir oil viscosity.

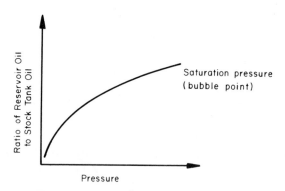

FIG. 6.6. Oil shrinkage.

tank oil in place. Gravity drainage and reservoir rock and liquid expansion are supplemental to other forms of energy, and seldom act independently.

The nature of the rock in which the oil is found must be studied by reservoir engineers to determine porosity, permeability, and amount of oil and water in the pore space (see Figure 6.7). When the rock nature or formation of the reservoir is known, studies of this type can indicate the type of recovery mechanism to use, or which mechanisms might later prove to be the most effective sources of energy.

Usually early development will show whether or not a gas cap is present. Early development might also present evidence of the continuity of the reservoir rock. If the reservoir rock is lenticular and there is no evidence of a gas cap, the engineer might conclude that the source of energy is solution gas. On the other hand, if the evidence indicates a continuous reservoir rock over a broad area and that the area is water-bearing below the oil reservoir, the engineer would recognize the possibility of a water drive.

Methods of Determining Oil Reserves

There are three basic methods of determining oil reserves. These are (1) volumetric calculation, (2) material balance reserve estimate, and (3) decline curve reserve estimate. Each method will be presented by an example, and then the three methods will be compared.

Volumetric Calculation. An example of the use of the volumetric calculation in reserve analysis is to determine (1) the number of barrels of stock tank oil in 1 acre-ft of reservoir, (2) the number of barrels of stock tank oil in the reservoir, and (3) the number of barrels of stock tank oil recoverable.

EXAMPLE 6.4: Volumetric Calculation

The following formula is only as good as the information that goes into it. Therefore the given data are assumed to be highly reliable. The reservoir rock characteristics assumed are as follows:

Given:

1. Porosity: 21.2%; permeability 500 ml; connate-water saturation, 22% of pore space.

2. Oil reserve limits, defined by the reservoir engineer: 1,200 acres in area, based on structure contour maps. Thickness 10 ft based on an "isopach thickness" map.

Sand grain

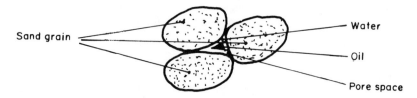

Water

Oil

Pore space

FIG. 6.7. The nature of rock.

3. Reserve fluid characteristics: Solution gas-oil ratio, 600 ft^3/bbl of oil. Viscosity, 1.0 c.p., formation volume (a shrinkage factor). Reservoir oil per barrel of tank oil, 1.3 bbl, by a bottom-hole-sample analysis.

4. Energy mechanism and efficiency of recovery: Energy mechanism is a solution-gas drive; recovery efficiency is estimated at 25% of stock tank oil in place, based on permeability, porosity, solution gas-oil ratio, fluid viscosity, and comparative analysis of similar oil accumulations.

Solution

1. Number of barrels of stock tank oil in 1 acre-ft of reservoir $= \dfrac{7,758 \times 0.212\,(\text{porosity}) \times (1 - 0.22)\,(\text{pore space})}{1.3\,(\text{reservoir oil per barrel of tank oil})} = 987$ bbl

Thus, 987 bbl is the number of barrels of stock tank oil in 1 acre-ft of reservoir.

2. Number of barrels of stock tank oil in reservoir $= 987 \times 10$ ft (thickness) $\times 1,200$ acres

$= 11,844,000$ bbl

Thus, 11,844,000 barrels of stock tank oil is reputed to be in the reservoir of 1,200 acres.

3. Number of barrels of stock tank oil which are recoverable $= 11,844,000 \times 0.25 = 2,961,000$ bbl

Therefore 2,961,000 bbl is the ultimate reserve, on the estimated basis of 25%. This is the number of gross barrels of oil which should be recoverable on primary recovery. Secondary recovery, using water or gas injection, should improve this amount.

Actually, there are 15,397,200 bbl of reservoir oil, but not all of stock tank standards. This is based on the given ratio of 1.3 bbl of reservoir oil to 1 bbl of stock tank oil and 11,844,000 bbl total stock tank oil in the reservoir.

It is often not advisable to use volumetric reserve estimates in new reservoirs as the "last word" on reserves. Estimates can be subject to a considerable margin of error. Volumetric calculations are often revised as more production and pressure history become available through years of operations, and as company operating policies become known.

Material Balance Reserve Estimate. In this example, it is assumed that 3 years have elapsed since the oil reservoir was first discovered. The reservoir is presumed to be fully developed, and gas, oil, and water production records have been carefully kept. Gas-oil ratio data have been accumulated, and reservoir pressures have been measured.

Sufficient data are at hand to try to estimate future reserves by a series of solution-gas drive, material-balance equations. Such a formula might be: Free gas + gas oil in solution = stock tank oil produced + original space occupied by gas - present liquid (oil) space ÷ a conversion factor + number of barrels of stock tank oil remaining × gas per barrel at any given pressure. Actually, this equation is nothing more than an expression of the fundamental laws of conservation of matter and energy as applied to an oil reservoir.

Use of an equation is complicated and cumbersome. For simplicity, an equation is not used, but data is given in tabular form as in Table 6.4. The technique of applying an equation on material balance to predict future reserves and determine recovery efficiency involves assumptions of incremental oil production for specific decreases in reservoir pressure. When the assumption of oil produced is correct, a group of computations show that certain variables in the equation are simultaneously taken care of and the equation will "balance." The procedure is repeated in successive steps until the prediction reaches depletion conditions.

EXAMPLE 6.5: Material Balance Reserve Estimate

A series of material balances assumed to be run on a hypothetical field are given in Table 6.4. Reserve estimate and an ultimate recovery of 2,778,000 bbl of oil results.

TABLE 6.4

Material Balance Reserve Estimate

	Average daily rate of production, bbl	Annual rate of crude production, bbl
Actual assumed for first 3 years		608,080
4th year	1,700	620,500
5th year	1,030	375,950
6th year	710	259,150
7th year	525	191,630
8th year	410	149,650
9th year	340	124,100
10th year	285	104,030
11th year	240	87,600
12th year	210	76,650
13th year	185	67,520
14th year	165	60,220
15th year	145	52,920
Total future reserve, sum of 4th year through 15th year		2,169,920
Total ultimate reserve (608,080 + 2,169,920)		2,778,000

The data in Table 6.4, giving a total estimated ultimate reserve of 2,778,000 bbl, indicate that the solution-gas drive will not be as efficient as originally estimated in the volumetric calculations, which claimed 2,961,000 bbl as the ultimate reserve.

Decline Curve Reserve Estimate

EXAMPLE 6.6: Decline Curve Reserve Estimate

Now, if we assume that another 3 years have passed, continuing analyses of the decline curve, as given in Table 6.5, shows that a decline trend is beginning to be established. An extrapolation of the decline curve, which is based on production rates of all wells in the reservoir, indicates the ultimate reserve by this method to be 2,841,000 bbl of oil.

Table 6.5 indicates that 2,626,380 bbl is the total ultimate reserve. The reserve estimate is lower than that of the original volumetric estimate.

TABLE 6.5

Decline Curve Reserve Estimate

	Average daily rate of production, bbl	Annual rate of crude production, bbl
Actual assumed for first 6 years		1,844,930
7th year	521	180,000
8th year	383	140,000
9th year	300	108,000
10th year	225	82,000
11th year	172	63,000
12th year	132	48,000
13th year	102	37,000
14th year	78	28,500
15th year	60	22,000
*16th year	47	17,000
Total future reserve, years 7 through 16		725,500
Recoverable, as per decline curve, years 17 through 48 (uneconomical)		55,950
Total ultimate reserve (725,500 + 55,950 + 1,844,930)		2,626,380

*In 0.3 of the 16th year, or 9.3 years after the first six years' time from discovery to date of estimate, 713,500 bbl

It is possible that the reservoir could be benefiting from the effects of gravity drainage, possibly in later stages of depletion in a solution-gas drive reservoir.

    Comparison of the Three Methods of Determining Oil Reserves. A comparison of reserve estimates by the three methods presented are given in Table 6.6. As Table 6.6 shows, the reserve estimates are quite different; that is, the material balance and decline curves differ appreciably from the volumetric determination. Thus, there is certainly no degree of conformance as we desire. Quite often, and this may be the case here, there is a lack of good basic data, which forces the reservoir engineer to assume critical information, and eventually the assumptions prove wrong in part or in entirety. But, in general, reserve estimates have considerable reliability based on conscientious and intelligent efforts of oil reservoir engineers' records and judgment.

TABLE 6.6.

Comparison of Reserve Estimate Methods

Type of estimate	Time from discovery to date of estimate	Estimate, ultimate oil reserve, in bbl	Percent change from volumetric calculation
Volumetric	6 months	2,961,000	—
Material balance	3 years	2,778,000	-6.2
Decline curve	6 years	2,626,380	-11.0

Use of the Production Decline Curve. The production decline curve
is probably the most useful tool for the oil economist who is analyzing
progress of oil wells. It illustrates decline in production conforming to
several shapes, as linear, parabolic, hyperbolic, etc., and the estimated
amount of the reserve still available for recovery from the reservoir. The
curve includes the decline ratio, expressed as $P/\Delta P$, or current production
in any period ÷ the change in production between that of the given period and
that of the previous period.

P (production) is a variable and ΔP, or the rate of change of P, is
usually (but not always) constant; the decline in production can be other than
exponential.

When the decline ratio is constant, the current production ratio is
decreased by a constant percentage in each time period.

EXAMPLE 6.7: Decline Curve Reserve Estimate

This is known as the constant percentage decline of current production,
and is illustrated in Figure 6.8.

The production decline curve is a straight line, plotted as a result of
3 years of actual operation. At the end of 3 years, the line is extrapolated
(extended) to illustrate the expected production each year, since the decline
ratio is regarded as constant. The shaded part illustrates what is left in
the reserve. The minimum amount, or rate minimum is assumed to be
20,000 bbl/year. This amount proves to be so small that the costs of
getting the oil out of the ground exceed the revenues to be received for the
crude. Thus, at 20,000 bbl, we can say we have reached the economic
limit of the oil lease, since the more oil recovered after this amount, the
less is the amount, as well as costs, in natural oil pressures. To recover

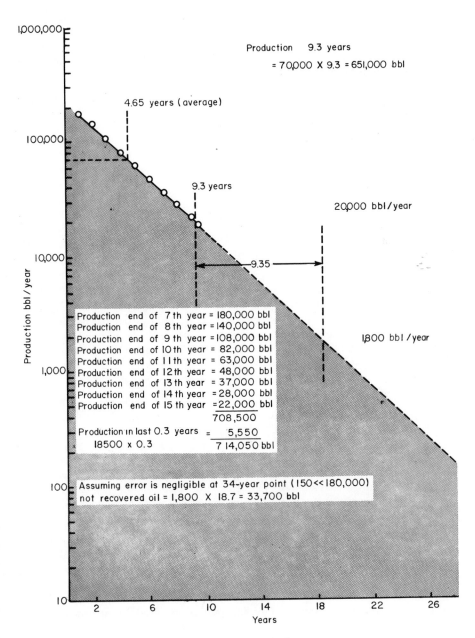

FIG. 6.8. Production decline curve.

the remaining amounts in the reserve, either water or gas injections will need to be applied. This is hardly worthwhile, in this case, since the amount left to be recovered is relatively small. By the time the 713,500 bbl are recovered in 9.3 years plus the 1,844,930 bbl produced in the first 6 years, a total of 2,558,430 bbl will have been recovered, cr 97% of the reserve total of 2,626,380 bbl.

With 20,000 bbl as the final minimum production per year, it would have been economical to recover 713,500 bbl after the original 6 years. By plotting the first 3 years with productions of 180,000, 140,000, and 108,000 bbl (428,000 bbl total), in that order, we have extended the constant percentage decline line to show production for the next 6.3 years and thereafter. At the end of the third year, we can see by the shaded area under the curve what the estimated recoverable reserve of the total 713,500 bbl will be. In this case, it is approximately 285,500 bbl, recoveragle in 6.3 years. Thus, we can say that the wells in the reservoir area have delivered, or recovered, 59% (428,000 bbl ÷ 725,500 barrels × 100%) in 3 years. Similar production decline curves, with different decline ratios, not necessarily constant ones, can be plotted and extrapolated as well.

Table 6.7 shows the data in Figure 6.8 in tabular form, similarly to Table 6.5 but further extrapolating the production line to what the production decline would be in barrels per year for several years up and including the 34th year (after the original 6 years). By the 28th year, production would be only 150 bbl/year. This would leave only 510 bbl for years 35 through 48. Total reservoir before the oil well operation run of 9.3 years for establishing the production decline curve is 725,500 bbl.

After 9.3 years, when production reaches the 20,000 bbl/year level, we have reached the minimum economic life of the oil well lease.

The value of the production decline curve is in its extrapolation of production to come each year, and in pointing out what recoverable reserves are left for future years before the economic life of the oil well is reached. Thus for each well, the economic point in barrels must be determined where it is uneconomical to continue pumping oil up to the surface.

Usually, the higher the production "peak," or the sooner the top or capacity of production is reached, the sooner the decline curve can be expected to begin and the steeper that decline curve may be. A logistic growth curve fitted to a cumulative record of oil production which is just another mathematical way of handling the decline curve concept, can be set up. An analysis of decline curves, generalizing from past production records of exhausted oil fields, gives a curve that indicates possible future behavior of output.

TABLE 6.7

Production of Reserves

Years after "time from discovery to date of estimate	Annual rate of crude production bbl
End of 7th year	180,000
End of 8th year	140,000
End of 9th year	108,000
End of 10th year	82,000
End of 11th year	63,000
End of 12th year	48,000
End of 13th year	37,000
End of 14th year	28,500
End of 15th year	22,000
End of 16th year	17,000
End of 17th year	13,000
End of 18th year	10,000
End of 19th year	7,500
End of 20th year	5,800
End of 21st year	4,500
End of 22nd year	3,500
End of 23rd year	2,700
End of 24th year	2,000
End of 25th year	1,600
End of 26th year	1,250
End of 27th year	950
End of 28th year	730
End of 29th year	570
End of 30th year	425
End of 31st year	325
End of 32nd year	250
End of 33rd year	190
End of 34th year	150

After the 34th year, approximately 510 bbl are left for years
35 through 48

The question usually arises: "What is the producing life of a reserve?" The method of treating production over time can best be described, in economic concepts, in terms of the oil production possibility curve leading up to peak production and in terms of the decline curve analysis.

In records of any oil field production, production level rises to a peak as the oil field produces fully, then falls as the effect of depletion exerts a greater limitation on output. Such a production curve for a single field is called a decline curve because the curve traces the long period of decline subsequent to peak production. Thus, according to the decline curve theory, which includes the proven reserves-to-production curve, production should reach a "peak" of activity and then go into a long period of decline until ultimate reserves are fully depleted.

DRILLING

Economic Balance in the Oil Fields

The recovery of oil from underground, or offshore, reservoirs is a good application of the "principle of economic balance." The problem is one of determining the optimum number of wells to drill, and the accurate spacing of these wells, to get maximum profit.

Actually, the greater the number of wells, the larger will be the ultimate recovery, provided that the recovery rate does not exceed the "most efficient engineering rate." But the most efficient engineering rate (economic balance) does not necessarily mean the optimum rate for maximum profits.

Economic balance, therefore, consists of a balance of (1) greater fixed costs for a larger number of wells drilled plus usually higher operating costs for higher production rates against (2) greater ultimate recovery from the larger number of wells. Thus the principle of economic balance in the oil fields is to drill as many wells as possible and needed within fixed costs and operating cost limits relative to the greatest ultimate recovery in terms of the realizable value (sales value) for the recovery. There is an upper limit to the number of wells that can be drilled, however, because of technical considerations.

In other words, greater fixed costs plus higher operating costs must be considered when increasing the number of wells to be drilled in an attempt to obtain a greater ultimate recovery of oil.

Upon discovery of large enough reserves for commercial drilling, the concept of well spacing becomes important to the oil engineer. The characteristics of reservoirs largely control the well-spacing pattern. For example, reservoirs with thick or multiple zones of oil will usually require more wells, and possibly closer spacing between wells, to take advantage of natural drainage (gravity flow) at its maximum than those reservoirs with thin crude oil composition located in single zones. Furthermore, porous reservoirs will produce more barrels of oil than "tight" reservoirs.

Other factors of a technical nature which should be considered in the spacing of wells, besides thickness vs. thinness of the crude itself and the multiple zones vs. single zones, include depth to the productive zones of the oil, viscosity of the oil, gravity of the oil, reservoir pressures, and reservoir properties.

Therefore, in well spacing, economics of anticipated recoveries based on thickness of oil and saturation of the pay zone become important. Obviously, the greater the number of wells drilled in a single reservoir, the greater the ultimate recovery per surface area of oil and/or gas will result.

There is a practical limit to the number of wells, and hence the spacing of wells, that can be drilled, however, which is controlled by the cost of drilling and operation. This limit to the number of wells to be drilled is based on estimated ultimate recovery, in barrels of oil, from each well. Since depth is the principal factor governing drilling costs, depth has a bearing on the problem of well spacing.

There is no hard and fast rule on spacing of wells; the technical and nontechnical factors relative to the oil reservoir must be considered separately.

Generally, if the anticipated ultimate recovery per well per surface acre is, say 20,000 bbl, and the average drilling cost per well is $150,000, plus the cost of production, it is evident that a spacing of less than 7.5 acres per well (average cost of drilling per well, $150,000 ÷ the anticipated ultimate recovery per well per surface acre, 20,000 bbl) would be uneconomical.

EXAMPLE 6.8: Well Spacing

The following simple example offers two alternatives relative to the number of wells to be drilled and spaced in a reservoir involving approximately 110 acres.

	Alternative 1 Drill 6 wells	Alternative 2 Drill 2 wells
Capital investment (total)	$1,800,000	$750,000
Operating costs, estimated total per year	$360,000	$80,000
Production (total)	100,000 bbl/day	20,000 bbl/day

Required: Determine which alternative to take: Drill 6 wells with closer spacing, at a total capital investment of $1,800,000 and total estimated running costs of $360,000 annually, or drill 2 wells, with wider spacing and at a total capital investment of $750,000, with total yearly operating expenses of $80,000.

Solution

For 6 wells, spacing between wells is estimated to be about every 18 acres, based on $300,000/well ÷ 16,667 bbl/day per well.

For 2 wells, spacing per well is estimated to be 37.5 acres between wells, or $375,000 per well ÷ 10,000 bbl/day per well.

Although operating costs are greater in total and on a per-well basis with 6 wells, total production is greater, and hence total revenues earned including profits, will be greater. Furthermore, the payout period favors the 6 wells alternative over the payout period of the alternative on 2 wells, since more overall production of 6 wells will increase total revenues received, sufficient to return investment more quickly.

Finally, capital investment per barrel produced per day favors Alternative 1. Capital investment per barrel per day with 6 wells drilled is $18.00 ($1,800,000 ÷ 100,000), whereas capital investment per barrel per day with 2 wells drilled is $37.50 ($750,000 ÷ 20,000).

Obviously, Alternative 1, or 6 wells, assuming everything else favors this alternative, including reservoir pressures, no limit on production, permeability and porosity features are favorable, etc., is the selection.

Usually, the greater the depth to reach productive zones of oil, the less the closer spacing of wells. Furthermore, since viscous oils do not possess the mobility of ready passage through reservoirs, as lighter, less viscous oils do, a closer spacing of wells is usually needed with oils of heavy viscosity properties in order to effect maximum efficient drainage. In the case of gravity, the lighter-gravity oils (with the higher API) contain more dissolved gases, have more mobility, and are less viscous than the lower-gravity oils, and so will require fewer wells and less closer spacing to effect maximum efficient drainage. On reservoir pressures, reservoirs with high pressures, particularly if pressures are maintained by some recycling operations such as use of water, gas, or air, offer higher recovery per well. Thus a wider spacing can be employed in reservoirs with high pressures.

Such reservoir properties as porosity, the ability to contain fluids, and permeability have an influence on well spacing. Porous and permeable reservoirs, which allow fluids such as oil to flow through the reservoir to the well bore, means that reservoirs can be effectively drained, so fewer wells with wide spacing is suitable under such conditions. Closer spacing of wells is necessary when "tight" reservoirs, with low porosity and permeability, are involved.

Some nontechnical factors also affect well spacing. These include, for instance, (1) the rate of production desired because of terms of the oil lease, market price of crude, market demand, etc. Also, proration laws of a government can dictate the amount of oil or gas an oil company can produce. When this is the case, the number of wells drilled, and the spacing, may be affected. Where the rate of payout desired is lengthened, and deferment of income over a wide period because of income tax problems is the objective, the number of wells drilled may be cut back. Thus spacing will tend to be wider under such conditions. The opposite of this, where the rate of payout desired is for a short period, dictates more wells drilled with closer spacing.

Probability of Success or Failure in Drilling

Employment of mathematics, particularly the use of binomial expansion probability, can be of definite value to the oil engineer and economist. Mathematics associated with binomial expansion is simple to apply if the number of wells drilled is small. For example, if we assume F (failure) and S (success) to have an equal chance (50% probability) and we drill only 2 wells, then the equation FF + FS + SF + SS tells us that:

We have a 25% chance of drilling 2 dry hole wells, or (FF).

We have a 50% chance of drilling only 1 successful well, or (FS + SF).

We have a 25% chance of drilling 2 successful wells, or (SS).

We have a 75% chance of drilling at least one successful well, or (FS + SF + SS).

When we go to a larger number of wells and other probabilities for F and S, the mathematics becomes more complex. For instance, consider the binomial expansion for a 5-well program, for which the equation is

$$F^5 + 5F^4S + 10F^3S^2 + 10F^2S^3 + 5FS^4 + S^5 = 1.00 \ (100\%)$$

Fortunately, there are tables of binomial probabilities related to number of trials, and for us the number of trials will be the number of wells in a program. The tables show individual probabilities (the odds of exactly X or more successes in an N-well program), or in cumulative terms (the odds of X or more successes in an N-well program). From these tables, graphs such as Figure 6.9 can be constructed to show the odds for various numbers of discoveries given certain "wildcats" (total drilling) with success rates. Figure 6.9 assumes a 10% success rate.

Explorers for crude oil try to determine how often success will be gained from a given program of N wells (wells drilled). "What are the odds of success?" a company might ask. A company drilling, say, 20 or 30 wells per year might want to know the odds of making 1, 2, 3, or 5 discoveries, with discovery meaning simply a producing well and not profitability of the well. How much oil there is, is not part of discovery, but comes under field size distribution.

To find these odds of success to total wells drilling, a mathematical technique called binomial (two numbers) expansion is used.

EXAMPLE 6.9: Binomial Expansion

To simplify, assume that each well in the program has the same chance of success with an assumed 10% success rate. Oil explorers know that some prospects have better "odds" or chances of success than others. For most exploration programs, we can assume an "average success" rate with reasonable safety.

F indicates probability of failure (a dry hole), and S indicates probability of success.

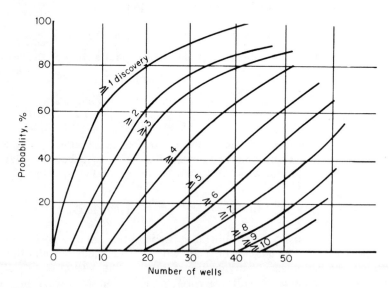

FIG. 6.9. Cumulative binomial probability (assuming 10% success rate).

For 1 well (one outcome) $F + S = 1.00$, or we can write $F + S (F + S)^1$. For 2 wells, there are 4 possible outcomes, $FF + FS + SF + SS = 1.00$; and, of course, $FS + SF$ can be written $2FS$. Then $F^2 + 2FS = S^2 = 1.00$.

Now, if you remember your algebra, $F^2 + 2FS + S^2$ is the product of $(F + S)(F + S)$ and can be written as $(F + S)^2$. So $F^2 + 2FS + S^2 = (F + S)^2$. The left half of this equation is the expansion of the binomial $(F + S)$ to $(F + S)^2$.

But we can set up a table, such as Table 6.8, with an assumed 10% success rate, for any large number of wells to be drilled and we will get some probabilities of success in number of discoveries to total number of wells drilled.

From Table 6.8, a graph can be drawn, similar to Figure 6.9, to illustrate tables of cumulative binomial probabilities. Figure 6.9 tells us that:

1. At least 1 discovery or more is 88% (or 88 chances of success in a total of 100 chances), or with 4.4 changes of S in five chances.

2. At least 2 discoveries is 60% (or 60 chances of success in 100 total chances), or 3 in 5 chances.

3. At least 3 discoveries is 30% (30 chances of success in 100 chances), or about 1.5 in 5 chances.

4. At least four discoveries is 13% (13 chances of success in 100 chances), or about 1 in 8 chances.

The chances of drilling any number of dry holes in succession, like the chance of one dry hole "in succession," is $1.00 - 0.10$, or 0.90 (90%). For additional wells, they are as follows:

2 dry holes in succession $= 81\%$, or 4 in 5 chances

5 dry holes in succession $= 59\%$, or 3 in 5 chances

10 dry holes in succession $= 35\%$, or 1 in 3 chances

20 dry holes in succession $= 12\%$, or 1 in 8 chances

Thus, even with a 10% success rate, even in drilling 20 holes, we still face a 12% chance that all holes will be dry.

The employment of such a table and graph is a possibility for explorers for crude oil in their efforts to predict success and failure, or discoveries to dry holes. It can also be useful to oil engineers in estimating probabilities, or odds of success.

TABLE 6.8

Cumulative Binomial Probability (Using a 10% Success Rate)

No. of wells drilled	No. of discoveries	Probability of success in no. of discoveries (%)	Odds of success
10	1	60	1 in 10
10	2	26	1 in 5
10	3	15	3 in 10
20	1	80	1 in 20
20	2	61	1 in 10
20	3	50	3 in 20
20	4	25	1 in 5
20	5	10	1 in 4
30	1	90	1 in 30
30	2	73	1 in 15
30	3	70	1 in 10
30	4	50	4 in 30
30	5	25	1 in 6
30	6	12	1 in 5
30	7	5	7 in 30
40	1	98	1 in 40
40	2	83	1 in 20
40	3	78	3 in 40
40	4	64	1 in 10
40	5	48	1 in 8
40	6	30	6 in 40
40	7	18	7 in 40
40	8	9	1 in 5
40	9	3	9 in 40
50	1	98	1 in 50
50	2	90	1 in 25
50	3	86	3 in 50
50	4	78	2 in 25
50	5	62	1 in 10
50	6	50	6 in 50
50	7	35	7 in 50
50	8	18	8 in 50
50	9	16	9 in 50
50	10	5	1 in 5

Drilling Operations

There are two methods of drilling a well, the cable tool and the rotary methods. No matter which method is used, a derrick is necessary to support the drilling equipment.

Cable tool drilling is the older method of drilling. In this method a hole is punched into the earth by repeatedly lifting and dropping a heavy cutting tool, a bit, hung from a cable. Today, however, practically all wells are drilled by the rotary method.

Rotary drilling bores a hole into the earth much as a carpenter bores a hole into a piece of wood with a brace and bit. In the middle of the derrick floor there is a horizontal steel turntable which is rotated by machinery. This rotary table grips and turns a pipe extending through it downward into the earth. At the lower end of the pipe, a bit is fastened to it.

As the drill chews its way farther and farther down, more drill pipe is attached to it at the upper end. As section after section of drill is added, the drill pipe becomes almost as flexible as a thin steel rod. Controlling the drill pipe under such conditions, and keeping the hole straight as well, is very difficult and requires great skill in drilling.

During the drilling, as Figure 6.10 illustrates, a mixture of water, special clays, and chemicals, known as drilling mud, is pumped down through the hollow drill pipe and circulated back to the surface in the space between the outside of the pipe and the walls of the pipe. This drilling mud serves several purposes, including lubricating and cooling the bit, and flushing rock cuttings to the surface.

As the drilling hole is deepened, it is lined with successive lengths of steel pipe, called casings. Each string of casing slides down inside the previous one and extends all the way to the surface. Cement is pumped between these successive strings of casing, and seals against any leakage of oil, gas, or water.

To achieve large annual additions to reserves and to output, the rate of drilling must be stepped up sharply. Barrels added per foot drilled are one of the best indicators of the results of drilling effort. This measure should not show a decline. A projection of the trend of barrels added per foot of drilling might be made as in Figure 6.11, where the assumed trend is downward.

In contrast with barrels added per foot drilled, drilling costs per foot should be compared. Figure 6.11 traces trends, indicating how a rising level of drilling cost per foot might appear along with a declining barrels added per foot drilled. When these opposing trends are related to the average well-head price of crude oil, as given in Figure 6.11, a shrinking of anticipated gross revenue per dollar of drilling cost is indicated.

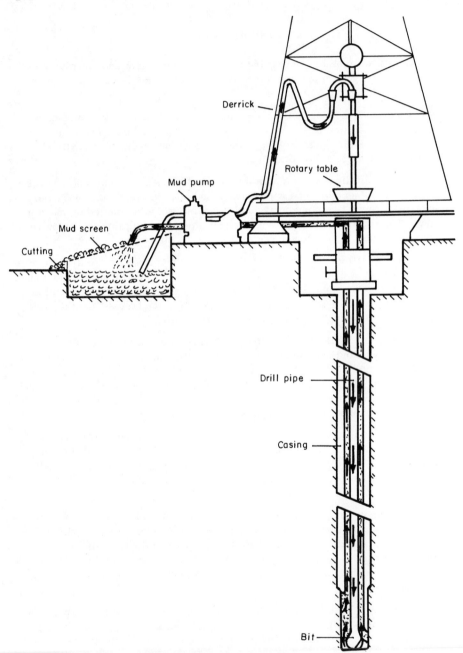

FIG. 6.10. Rotary drilling.

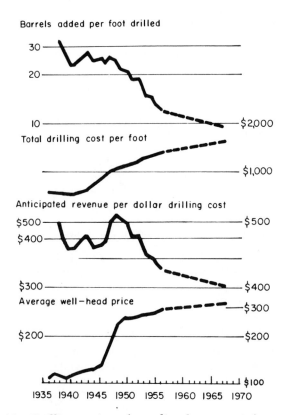

FIG. 6.11. Drilling costs and results: 3 years moving average.
[From "Future Growth of the World Petroleum Industry," Chase Manhattan
Bank, New York, 1966.]

Pointed up also is the fact that the economic limit of drilling costs per
barrel of added reserves may be fast approaching. In this case, exploratory efforts must be improved in terms of success, and drilling costs reduced, if the oil company in this case is to operate within the price framework as given.

Costs of Drilling

An increase in depth increases drilling costs. Actually, costs increase
exponentially with depth, even for a "normal," trouble-free well. Also,
an increase in depth can increase the chances of mechanical problems.
This adds to the cost of drilling.

Increased depth also reduces available information about potential reservoirs, as to quality of crude oil and quantity available (proven reserves). Risks increase with uncertainties as to reservoir quantity and quality available.

Costs of drilling depend on the kind of oil and what potential energy the oil possesses by virtue of its initial pressure in its reservoir, and by the amount of dissolved gas it may contain. In many cases the crude may have enough potential energy to permit a well to flow large quantities of oil to the surface without any artificial assistance, such as use of gas or water injection. (This is quite prevalent in oil wells in the Middle East.) But when oil cannot flow unaided, or when the pressure in the reservoir has decreased to a pressure that is too low to be economical, costly mechanisms which lift oil to the ground surface must be employed. Furthermore, low pressure in the reservoir and low gas content generally go together. This kind of crude, therefore, must be handled in a different manner.

The average cost of drilling oil and gas in the United States reached a peak of $19.03/ft in 1971, based on average depth of wells drilled, according to a study by the American Petroleum Institute, the Independent Petroleum Association, and Mid-Continent Oil and Gas Association. This figure is 1% higher than for the year earlier and 48% higher than for 1961. The average cost per well was $86,718 versus $78,431 in 1970 [2].

Marginal costs, or costs for each additional foot drilled, usually increase as depth increases. Table 6.9 indicates the drilling costs per foot by different levels of depth in 1956 and again in 1967.

TABLE 6.9

U.S. Drilling Costs per Foot in Various Level Depths (at Current Prices)

Depth, ft	Average cost per foot in 1956	Average cost per foot in 1967
0–1,250	$8.34	$11.49
1,251–2,500	7.76	9.92
2,501–3,750	8.58	11.42
3,751–5,000	9.15	13.01
5,001–7,500	10.67	13.40
7,501–10,000	14.41	18.09
10,001–12,500	21.31	30.27
12,501–15,000	29.02	37.82
15,001 and over	47.00	45.32
All depths	$12.35	$16.61

Source: American Petroleum Institute, and Oil and Gas Journal, February 1968.

As Table 6.9 indicates, average costs per foot of drilling increase as depth increases for both 1956 and 1967, at current prices. But in 1967, through improved drilling techniques and drilling bits, average costs per foot actually declined as wells of over 15,000 ft were drilled.

Most oil companies are not concerned with how far down drilling proceeds, but with how high the cost will be to get that deep and what the cost will be to go, say, another 100 ft or more.

Marginal costs are some direct function of depth. If, then, we let Y be those costs which vary with depth, but no overhead costs, and let X be depth itself, a formula can then be written as $dY/dX = C(X)$, the costs per foot. Thus, depth affects marginal costs. For example, the rise of temperature with depth, among other tings, increases the probability that a drilling bit will have to be replaced an additional time in a well drilled an additional 100 ft, because mechanical energy is lost as the drilling process continues. But also, some costs, such as the costs of additional "mud materials," needed to drill a deeper well may actually increase rather slowly in relation to increase in depth, thus giving a decreasing marginal cost in relation to depth. This is indicated in Table 6.9: In 1967 for depths of over 15,000 ft, average cost per foot of wells drilled declined in comparison to 1956.

Possibly the one factor that most affects the costs of drilling is the average footage drilled per hookup. As more information on drilling tendencies in any one oil field become available, the number of changes in drilling hookup is reduced and the speed of the drilling operation is increased. Also, feet per hour at the bottom of the well, combined with the amount of time spent at the bottom, are perhaps the best measures of the relative efficiency and speed of a drilling operation in a particular oil well and for a given amount of controlled footage.

In sum, costs of drilling increase because of the following, usually in some combination: (1) a poorly desinged casing program; (2) an inadequate rig or incompetent personnel on the test drill; (3) poor selection of proper drilling bits for the formations to be penetrated; (4) insufficient drilling bit weight for maximum penetration (economic balance here relative to weight to use and cost of the bits); (5) lack of hydraulic efficiency with good fluid hydraulics or an insufficient mud program (a good mud program is important to lower viscosities and reduce solid contents to a minimum; also, a good mud program will reduce sand content to a minimum, thus increasing drilling bit efficiency.

EXAMPLE 6.10: Costs of Drilling

To illustrate, assume a problem involving 4 wells to be drilled, all with the same casing depth, and with the same rig and same rig crew for each well. Then results might be as given in Table 6.10.

TABLE 6.10

Performance of 4 Wells

Depth, ft	Hole size, in.	Program casing, in.
250	20	16
2,500	13-3/4	10-3/4
10,600	9	7

	Well A	Well B	Well C	Well D
	Rig performance, 0 to 3,900 ft			
Drilling days	7	8	6.5	6.5
Bits, number and size	Each well has 5 bits, 12-1/4" reamer bits, and 9" bits			
Feet per bit	780 ft penetration per bit for all wells			
Average mud circulation rate, g.p.m.	630	675	675	755
Nozzle velocity, ft/sec	Conventional	290	290	390
	Rig performance, 3,900 to 5,300 ft			
Drilling days	6	6	4.5	6
Bits, number and size	9 9"	9 9"	5 9"	5 12-1/4"
Feet per bit	155	205	279	279
Average mud circulation rate, g.p.m.	490	505	580	665
Annulus velocity, ft/min	195	205	235	125
Nozzle velocity, ft/sec	Conventional	145	175	230
	Rig performance, 5,300 to 7,300 ft			
Drilling days	13.5	11	12	11
Bits, number and size	26 9"	19 9"	15 9"	10 8-5/8"
Feet per bit	77	105	135	200
Weight per inch of bit diameter, lb	3,900	4,500	5,500	5,800
Average mud circulation, g.p.m.	425	445	575-6,700	640-6,700 475-7,300
Annulus velocity, ft/min	170	180	180	220
Nozzle velocity, ft/sec	Conventional	160	140-6,700 240-7,300	220-6,700 260-7.300

Summary of Problem

1. Shows effect of nozzle velocity, cuts down on average mud circulation rate, when increased

2. The improved drilling rate on well D in comparison with the other wells was due to (a) reduced sand content, which increased the life of the drilling bit; (b) reduced weight of drilling fluid, which decreased column pressure; (c) an increase in water loss, which allowed penetration of drilling fluid into the well bore; and (d) a low solids percentage, which increased the life of the bit and reduced the abrasive action of the drilling fluid.

3. High bit consumption of well A was due to the fact that the weight per inch of bit diameter was 3,900 lb or 18% less than well B and about 30% less than the bit weight used in drilling well D.

4. On well B, footage was drilled in 91 fewer drilling hours, not including well D, so more feet drilled per hour at well B.

5. Wells B and D penetrated 7,300 ft one day earlier than well C and 2.5 days earlier than well A, although well B used more drilling bits than well D. Thus well D was the most efficient here, with less bits used and more feet per bit; it also used more weight per inch of diameter bit.

Drilling Bits

There are numerous diamond drilling bits of various designs on the market. For the engineer to be economic-minded, it is important to understand the purpose, advantages, and disadvantages of each design. The capabilities, as well as the limitations, of each bit can often be determined by studying past performance of the various bits. Reviews of past performance of bits should be made before purchasing new bits for use in oil well drilling. There are definite economic and operating advantages when the capabilities and limitations of these tools are understood. For instance, the longer life of diamond bits can eliminate one or more round trips of rig through a given interval. In addition, improved bit life also increases rotating time of drilling at the bottom of the well, and so increases footage, thus reducing drilling costs.

Total rig time involved, together with overall bit cost, must be determined in order to evaluate properly the economics involved in initial investment of bits to penetration rates which are affected by reservoir formations on hand.

Hourly rig operating cost has an important influence on the economics of diamond bit drilling. On small, low-cost land rigs, under $50/hr cost of rig operation, the net cost of one diamond bit equals the expense of many round trips. On high-cost, offshore rigs, estimated cost of rig operation up to $1,200/hr, as little as a 2-hr saving in rig time can offset the cost of a diamond bit. In calculating diamond bit cost, an average of 50% of the original bit price, consisting of diamond content plus setting charge, can normally be "recovered" when the bit is returned for salvage cut-out. So if the bit cost originally $7,200, 50% or $3,600 will be "recovered" by the oil operator when he returns the bit to the manufacturer for salvage cut-out.

Bit Performance. Proper bit selection affects performance of that bit. Proper selection will save time and money; also, proper selection means a better opportunity for proper use of the bit, and so cost savings.

Bit performance can be known and estimated in advance either by checking former records of by consulting the bit manufacturer, or by use of the following procedure.

EXAMPLE 6.11: Bit Performance

In deciding which bits to use, either a more costly diamond bit or a less costly conventional bit might be used. Data might be presented as follows (all figures are hypothetical). The reader should note the variables used in each column.

Step 1—Previous bit performance:

Size of bit, in.	Feet drilled	Rotating hours of drill	Trip time of rig, hr	Depth out, ft
8.5	42	10.25	11	7,583
8.5	82	14	9	7,665
8.5	66	6.5	12	7,731
8.5	43	5	12	7,774
Total:	233	35.75		

Let us assume that rig cost is $180/hr, the cost of 8-1/2" of a conventional bit is $350 for each bit, and the average life of a bit in use is 9 hr. Approximately 4 bits were used. So 35.75 hr ÷ 9 hr average life of bit = 3.97 bits at $350 each, or a total cost of $1,400 ($350 × 4) for bits. Average footage of bit in use is 58 ft, average trip time of rig is 11 hr, and net diamond bit cost (estimated 50% of price) is $3,600. Then:

$$\text{Drilling cost} \atop \text{per foot of} \atop \text{bit in use} = \frac{\left({\text{rig cost} \atop \text{per hour}}\right)\left({\text{average life} \atop \text{of bit in} \atop \text{use} + \text{average} \atop \text{trip time} \atop \text{of rig}}\right) + {\text{cost} \atop \text{of bit} \atop \text{in use}}}{\text{average footage of bit in use}}$$

$$= \frac{180(9+ 11) + 350}{58} = \$68$$

So it costs $68 for every foot of drilling when 8.5-in. conventional bits at $350 each are used. This $68/ft can be compared with other bits, or with performance of other oil companies, in determining the value of bit performance.

Step 2—The ratio of rig cost to cost per foot of bit in use is $180:$68 or 2.6. In other words, a rig cost of $180 is 2.6 times more than the cost per foot of bit being used. This kind of performance can be compared with other past performances to determine its value.

Step 3—The ratio of average trip time of the rig times the rig operation cost per hour plus the net diamond bit cost to the drilling cost per foot of bit being used (and as given in Step 1) is

$$\frac{(180 \times 11) + 3,600}{68} = \$82$$

This $82 represents the rig cost. This cost, the trip time of the rig, and the net bit cost, when added to actual drilling hours times the ratio of the rig cost to cost per foot of the particular bit being used, gives total footage drilled as follows. Assuming that the number of drilling hours involved is 20, then 20 × 2.6 = 52 + 82 = 134 total footage drilled.

The penetration rate of the drilling and bit is then found by dividing the total footage drilled by the number of drilling hours, or 134 ÷ 20 = 6.7, which is the penetration rate in feet per hour.

Step 4—A graphical presentation, similar to the one illustrated in Figure 6.12, which is drawn from data in Table 6.11, might be worked up for several drilling hours of the 8.5-in. bit.

Although total footage increased with an increase in drilling hours, the penetration rate in feet per hour dropped. The basic economic principle, the law of diminishing returns, is acting here in the first 20 hr of drilling, after which the diminishing returns became less severe.

Step 5—Next we assume a recommended diamond bit size and style of 8-7/16" medium hard style, with estimated diamond bit performance, from bit records, of 320 ft per bit, and a penetration rate, or total footage ÷ number of drilling hours of 8 ft/hr. Then the drilling hours for the 8-7/16"

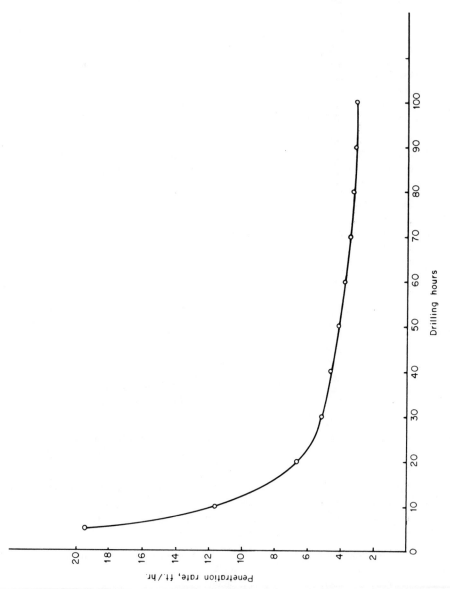

FIG. 6.12.   Penetration rate in feet per hour to drilling hours, from data in Table 6.11.

TABLE 6.11

Drilling Performance

Drilling hours	Ratio of rig cost to cost per ft of bit +	Ratio of trip time, rig cost + net diamond bit cost: drilling cost of bit =			Total footage ÷	Drilling hours =	Penetration rate, ft/hr
5	2.6	13	+	82	95	5	19.5
10	2.6	26	+	82	108	10	10.8
20	2.6	52	+	82	134	20	6.7
30	2.6	78	+	82	160	30	5.3
40	2.6	104	+	82	186	40	4.7
50	2.6	130	+	82	212	50	4.2
60	2.6	156	+	82	238	60	4.0
70	2.6	182	+	82	264	70	3.8
80	2.6	208	+	82	290	80	3.6
90	2.6	234	+	82	316	90	3.5
100	2.6	260	+	82	342	100	3.4

medium hard style diamond bit is 40 hr (320 ÷ 8). The 40-hr figure compares more favorably in usage to the 8-1/2" conventional bit, which shows only 186 ft, or a penetration rate of 4.7 ft/hr after 40 hr of drilling with the use of 5 bits, 9 hr to a bit.

Step 6—The cost of the drilling interval with the bit in the hole is equal to the product of the ratio of rig cost to cost per foot of bit in use times the footage of the 8-7/16" diamond bit, or 68 × 320 = \$21,760. The cost of diamond drilling is thus:

(1) Rotating cost of drilling:

$$\text{Rotating cost of drilling} = \frac{\text{total footage}}{\text{penetration rate in feet per hour}} \times \text{rig cost}$$

$$= \frac{320}{8} \times 180 = \$7,200$$

(2) Average trip cost of rig = (average trip time × rig cost per hour)

$$= 11 \times 180 = \$1,980$$

(3) Diamond bit cost (estimated 50% of price) = \$3,600

(4) Rotating cost of drilling + average trip cost of rig

+ net diamond bit cost = $12,780

The difference, $21,760 - $12,780 = $8,980, is the net saving with diamond drilling.

Breakeven with Bits. Figure 6.13 is an example of a breakeven curve used sometimes to predict bit performance. It is plotted from the data in Table 6.12. Generally, if the point describing anticipated diamond bit performance falls well above the basic breakeven curve, the diamond bit should be economical to use. If the point describing anticipated diamond bit performance falls below the curve, the bit should not be used because of doubtful economics. If the point describing anticipated diamond bit performance is very near the curve either, above or below it, then other factors such as hole conditions, mud costs, rig wear, etc., should be studied carefully studied, because indirect savings they provide may influence the choice of bit.

If the breakeven curve indicates that, for the bit under consideration, performance versus cost of bit will not "break even," and if a faster penetration rate cannot be achieved, the bit should be pulled before any

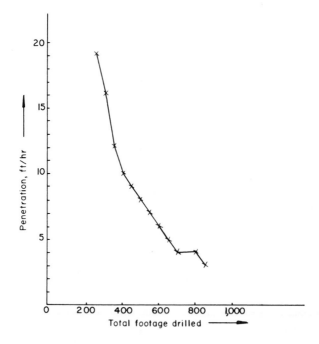

FIG. 6.13. Breakeven curve, from data in Table 6.12.

TABLE 6.12

Bit Performance

Total footage drilled	Penetration rate, ft/hr
250	19
300	16
350	12
400	10
450	9
500	8
550	7
600	6
650	5.5
700	4
750	4
800	4
850	4.3

appreciable diamond wear occurs, or before excessive rig time is consumed. This will minimize losses from "under performing" diamond bit "runs."

Determining Costs. After pulling a diamond bit, cost per foot should be calculated to determine net gain or net loss. Either estimated or actual diamond bid consumption may be used. In most cases an approximation will indicate whether the bit run was an economic success or failure.

The following basic cost equation for rotary drilling is the same for any type of bit performance:

Total drilling cost per foot = (drilling time + average rig trip time)

× (rig cost + support cost + net bit cost)

Thus, if drilling time is 40 hr, average trip time is 10 hr, rig cost is \$200/hr with no support cost, net diamond bit cost is \$3,000, and we assume that total footage is 500 ft, then total drilling cost per foot is

$$\frac{(40 + 10) \times (200 + 0 + 3,000)}{500} = \$26/ft$$

Bit Selection. Choice of bit style, diamond size, and diamond quality, can mean the difference between an economical or a costly diamond bit run.

Some earth formations are more easily drilled with diamond bits than with any other type of bit, but these formations and their "drillability" change from area to area. Some formations, such as sands, sandy shales, salts, and limestones, are easily drilled at fast drilling rates with diamond drilling bits. Also, diamond bits normally perform better in hard formations, because it is easier to keep the bit clean and cuttings are smaller. Individual diamonds cut by a plowing action rather than by chipping and tearing. But diamond bits require hydraulic equivalent to or greater than other bits to stay clean and run cool in softer, stickier formations.

Usually, the smaller the diamond bit, the better the bit performs relative to the penetration rate in feet per hour, mainly because of hydraulics. Larger bits may show greater decreases in penetration rates unless hydraulics are maintained at high-energy levels of 2.0 to 3.0 hydraulic horsepower per square inch.

Factors Affecting Penetration in Drilling

Studies made by experts from drilling and equivalent companies indicate that there is a positive effect of weight and speed of rotation on penetration rate, or feet per hour of drilling. This is true whether toothed or carbide-studded bits are used.

Past experience has shown that proper penetration rate of weight on bit rotary speed and hydraulic horsepower can be plotted on a graph to determine optimum drilling at minimum drilling cost. Thus, the penetration rate of a bit varies with weight on bit, rate of rotation, and hydraulic horsepower.

Of course, there are other factors, including the following, which can reduce drilling efficiency and raise footage costs:

1. Poor fluid hydraulics. This can increase the cost of drilling as much as 20% on a 10,000-ft well.

2. An incorrect casing program for a particular well. For example, a 5,600-ft, 9-5/8" casing instead of, say, a 2,500-ft, 10-3/4" casing is not economical.

3. Selection of improper drilling bits. Sizes, type for particular land formation, and use of conventional bits vs. jet-nozzle bits, etc., are just some of the things to consider in bit selection.

4. Too fast a penetration rate for both bit and land formation involved. The object is to increase the penetration per bit without using too many bits.

5. Not using a proper mud program.

6. Increase in abrasive action of drilling fluids, because of high per-
    centage of solids content.

7. Not enough weight per inch of bit diameter used, since too little
    can mean using more bits, and too little is just as bad as too much
    in weight per inch of bit diameter used.

8. Too many changes in drilling hookups, and taking too much time
    for each change.

9. Depth of well is too deep for efficient drilling.

But most of all, laboratory data and oil field results are available for
study of the effect of weight on bit on penetration rate in feet per hour.
Also, hydraulic horsepower is a factor with weight on bit that must not be
overlooked when enhancing or extending the penetration rate. Penetration
rate varies linearly with pump hydraulic horsepower and proportionally with
weight on bit, in pounds per inch of bit diameter, as past results have proven.
Thus, if a 20" bit costing $15,000 per bit is being used, the penetration rate
is 20 ft/hr, the weight is 10 lb/in., the drillability is 20/10 = 2, and the
cost is $15,000/20 or $750/ft/hr. To cut costs of drilling, the weight on
bit must be increased; that is, the drillability must be increased.

So average penetration rate in feet per hour can be plotted on graphs
with weight on bits and average pump hydraulic horsepower. Under most
circumstances, an increase in hydraulic horsepower will generate an increase in penetration rate even when hydraulic horsepower is not accompanied by an increase in weight on bit.

Directional Drilling

When drilling is conducted in inaccessible locations, change of direction
of drilling is often considered. The experience of the drill operator himself becomes important in determining the potential success of a change of
direction and the costs involved in directionally drilled wells. Constant
attention and control are needed with the directional drilling technique, but
efficiency in directional drilling, as in all drilling operations, still depends
on how long the bit is actually drilling on the bottom of the well.

Changes of direction per well in directional drilling can vary from 20°
to as much as 170° for the direction of the hole away from vertical at that
point to the final direction of the hole on completion.

In Canada, one company, Eastman Oil Well Survey Company, Ltd.,
reported the following data in Table 6.13 for 8 different wells in a 1960
study. The study indicates wells with a variety of footages, deviation in

TABLE 6.13

Drilling Data for 8 Directional-Drilled Wells

	Well 1	Well 2	Well 3	Well 4	Well 5	Well 6	Well 7	Well 8
Controlled footage	4,137	3,975	1,010	985	5,155	1,250	1,197	1,490
Deviation in feet	1,512	1,558	205	218	1,905	162	267	280
Maximum angle	31°	32° 30'	15°	20° 15'	34°	14° 30'	20° 15'	14° 30'
Estimated days	21	21	8	9	21	9	7	9

Source: Oil and Gas Journal, October 21, 1961.

feet, and maximum angles of deflection. However, the figures also reveal that it takes about 3 weeks of 24-hr/day drilling, to drill at an angle of 31 to 32°, and 7 to 9 days for an angle of 15 to 20°.

Estimates of cost of partly drilling a well under directional control are in most cases prepared before the well is spudded. Consequently, it has to be assumed that at the point of deflection the hole is vertical.

When petroleum is discovered underwater, a little offshore, one way of tackling the problem of getting the oil out is by directional drilling.

A directional well, or "crooked well," is one that does not extend directly vertical from the center of the derrick where the derrick is located nearby land. This is illustrated in Figure 6.14.

Price Elasticities and Well Drilling

Oil experts agree that the prices of crudes influence crude production, but that development drilling is more sensitive to these economic incentives than exploration drilling (see discussion of price elasticity in Chapter 3). Also, the development of marginal oil fields is the first to be curtailed when crude prices fall. In the case of exploration, or discovery, of new oil fields, price elasticity of available oil reserves is probably substantially less than price elasticity of drilling and other expenditures necessary to produce those reserves.

The best available estimate of price elasticities (or sensitivities of oil drillers, suppliers, relative to selling prices of crudes), of exploration drilling and of development drilling—and also with respect to any other

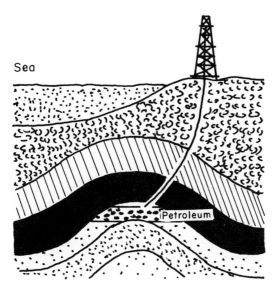

FIG. 6.14. Drilling a crooked well.

factors that affect profits in the same way, such as depletion allowances
for oil resources—places drilling price elasticity in the neighborhood of at
least +2.5 or more, and the reserves price elasticity at considerably less,
or around +0.5. This means that suppliers or development drillers of oil
are usually ready to increase their drilling by 2.5 times as crude prices
increase, while explorers of oil will only increase their test drillings 0.5
times as much as prices of crudes increase. In other words, development
drillers are more sensitive to price rises than exploration oil people. Actually, the price elasticity of supply for development drilling is highly elastic (highly sensitive), while the price elasticity of supply for exploration
drilling is inelastic or not too sensitive to price increase in crudes. An
increase in price of crudes of, say, 5% should cause at least a 12.5% increase in development drilling supply of crude.

$$\frac{\text{Development drilling}}{\text{price elasticity}} = \frac{\text{\% of increase change in wells and footage drilling}}{\text{\% of increase change in price of crudes}}$$

$$= \frac{12.5\%}{5\%} = 2.5$$

An increase in price of crudes by 5% should increase exploration drilling
by 2.5%:

Reserve
(exploration) = % of increase change in test wells and other explorations
price   % of increase change in price of crudes
elasticity
       $= \dfrac{2.5\%}{5\%} = 0.5$

Oil experts further agree that when the price of crude rises by 10%, there is a direct effect of a sharp increase in "wildcat" development drilling. (Wildcat drilling is a U.S. term for indiscriminate well drilling by landowners where oil is discovered, rather than by organized oil companies who obtain concessions for land where oil is believed to be available.) There is also, however, a reduction in the success ratio or number of "successful" wells to total wells drilled; and also, there is a reduction in average discovery sizes per successful wildcat well.

Whether sensitivities (elasticities) of exploration and development drilling are of a high degree or not depends upon many factors outside the context of this analysis. Some of these factors involve policies and objectives of oil companies and policies of governments, both supplying and consuming ones, relative to the oil industry.

While elasticities greater or less than unity, that is, 1.0, provide a convenient terminology, an elasticity of +2.5 may be low for some purposes, and an elasticity of 0.5 intolerably "high" for other purposes. There are also long- and short-run price elasticities as well. These are complicated, and will not be discussed here.

The high cost of petroleum exploration and development is in sharp contrast to the low cost of bringing the crude up to ground surface. Up to 1955, on the average approximately $2 was spent on exploration and drilling in order to realize a successful investment of $1 in producing wells [3]. Today, these figures should be upgraded by at least 50%, according to U.S. cost indexes.

SECONDARY RECOVERY

Primary recovery of oil means that the oil is removed from its reservoir by its own natural pressure, or by "stripper action." When pumps must be used to raise oil up to ground surface, the wells are referred to as "stripper wells." When a stripper well goes dry, usually a large percentage of the original oil in the producing formation immediately surrounding the well still remains. Today, with improved methods for additional recovery, much of the remaining oil from primary recovery is not "lost," but can be brought up to the surface. We call the claiming of the remainder oil left after primary recovery, "secondary recovery."

The basis of secondary recovery is usually one of forcing the oil into the producing well. In general, two secondary recovery methods are used. They are injected gas drive, and water injection or water flooding. Thus secondary recovery is accomplished chiefly by injecting gas or water through selected wells.

Gas injection is more common and popular than water injection because of the low viscosity of gas, and because gas, unlike water, is a rather non-wetting fluid. But gas injection is more expensive than water injection.

In water flooding, a number of water injection wells are drilled around the producing well. For instance, the "five spot pattern" is one model used whereby water injection wells are drilled at the corners of a rectangle with the producing oil well in the center. Chemically treated water under pressure is brought into the water injection well, and the water is then forced through the oil-producing formation. As the water expands in a ring around the oil-producing formation, the oil is flushed out of the sandstone and rocks and pushed toward the producing well.

A good rate of return from the use of water flooding or injection is to recover between 5,000 and 10,000 bbl of oil per producing well. But a high rate of injection, or about 300 to 1,000 bbl of water per day per well, is necessary to obtain a good and "economic" response. Figure 6.15 shows the performance of a flood project in Hutchinson Country, Texas, in 1960, developed on 20 acres with an old well producer, and converted to a water injection well to make one 40-acre, five-space pattern.

Gas under pressure is also used, either instead of or in conjunction with water, in secondary recovery. By injecting gas, the oil and gas from a well reach the surface in a frothy mixture which is routed through the valves of a "Christmas tree" series of valves to tanks where the gas is separated and the oil sent on to storage.

The deciding factor, or whether to use water or gas, is usually one of economics. Thousands of dollars are invested in advancing the water flood and gas drives. The oil company usually uses the least costly method which will also produce or recover more oil at a profit.

The result of water flooding in the example shown in Figure 6.15 was over 50,000 bbl of additional oil in 2 years as a consequence of converting one producing well into an injection well. At the time of this writing, more than 100,000 bbl of water flood oil were expected to be recovered before abandonment. This oil, while adding only 25% to the primary recovery, was being produced at a very nominal cost. This cost is largely for the water, which is being supplied at a cost of less than 10% of the net revenue from the additional oil. By this process, water is pumped into the oil sand, and the water then flushes the oil ahead of it to the wells.

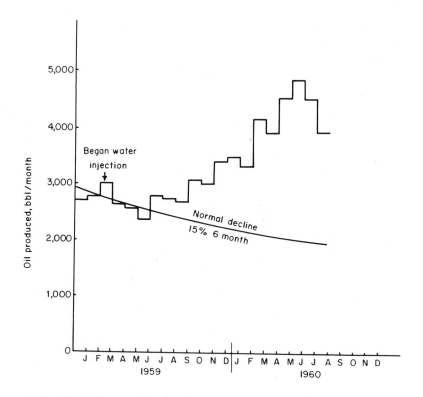

FIG. 6.15. One result of water injection: A 4-well pilot flood.

EXAMPLE 6.12: Secondary Recovery—Water Injection

Consider, for example, the following case of secondary recovery, which occurred in the United States, and which illustrates the use of water injection, or flooding, of a fractured reservoir. By using 8,325,000 bbl of water, a recovery of 880,000 bbl of oil resulted.

Actually, 8,325,000 bbl of water was injected into 46 fractured wells, which were natural low-producing wells, at maximum pressures of 900 psig, and at average pressures of 750 psig. Thus, the gross ratio of cumulative water injected to cumulative oil produced was 9.4 bbl of water to 1.0 bbl of oil (8,325,000 bbl of water to 880,000 bbl of oil).

The 880,000 bbl of oil recovered the second time was, in this case, even more than the primary recovery. Primary recovery was 643,650 bbl of oil, which represented only 15.4% of the original oil in place. So secondary recovery added approximately 20% of the original oil in place. Figure 6.16 illustrates.

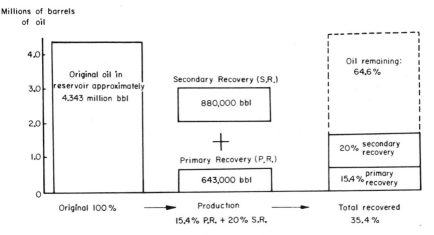

FIG. 6.16. Primary and secondary oil recovery.

The investment of $196,000 resulted in the production of $2,640,000 worth of oil (880,000 bbl × $3.00/bbl of crude). Operating costs of $547,000 over 16 months resulted in a net revenue to all investment interests of $1.9 million (rounded off) from the water flood operation ($1,900,000 = $196,000 + $547,000 − $2,640,000). Thus, in this case, the investment in the use of water injection as a secondary recovery method paid off. Receipts acquired more than compensated for the investment made plus operating costs.

Many wells are fractured before completion, or before possible amount of recovery is made. With rising costs of finding oil and improved techniques for secondary recovery, it is becoming more profitable to reevaluate old oil fields where drilling risks are low, rentals cheap, and where there is an abundance of subsurface information from previous drilling and therefore no costs for exploring.

Most of the primary development of wells drilled usually has occurred within 5 or 6 years after the year of discovery, but if only on the basis of the elimination of exploration and test drill costs, it is worth reviewing former producing wells. Also, improved methods and techniques are not too costly in relation to the amount of oil derived through secondary recovery. Secondary recovery is not always economically advisable, however.

EXAMPLE 6.13: Secondary Recovery

If, for instance, it costs $80,000 to use secondary recovery driving techniques, and this will only recover $40,000 worth of oil which primary recovery would not recover, the oil company could lose at least $45,000 ($80,000 - $40,000 + $5,000 in royalties). For obvious reasons, out of the $40,000 in oil that the oil company would recover, it would have to pay a royalty charge of one-eighth or more. Thus the "royalty owner" gains $5,000 (1/8 × $40,000). The oil company would take in receipts from sale of crude in the amount of $40,000, but would pay out $5,000 for royalties and $80,000 in secondary recovery driving costs, or a total cash outflow of $85,000. The net cash outflow in this example would be $45,000, the difference between $85,000 out and $40,000 in. Obviously, this project would be turned down.

Thus, from Chapter 4, on the financial aspects of oil management, each project involving secondary recovery needs to be analyzed in terms of net cash flow, rate of return, payout, and present value measures.

PIPING FROM THE OIL FIELDS

The United States, where the oil pipeline was invented, has many more miles of pipeline than any other country in the world. The United States was formerly the number one source of petroleum supply, and the United States is still the world leader in oil consumption.

Although pipelines are an American invention, the working principles of the pipeline are ancient. Like the wheel, the pipeline is a basic principle of transport. Its working history probably predates that of the wheel simply because, physically, it is easier and less hazardous to move liquids in steady streams within protective walls of pipes than to move the same liquids by carrying them in buckets, jugs, or backpacks, or to haul them in trucks, wagons, railroads, planes, or any other conveyance.

The value of a pipeline is in its economy of operation and in its consistency of operation. Today, there is great diversity in size of pipe used to carry crude oil and refined oil products, ranging from 6" to as much as 36", and in some cases in the Middle East, larger than 36" piping. Lines are single or multiple, laid on top of the surface or buried in the ground, with booster pumps spaced anywhere from approximately every 25 miles to as much as 200 miles apart.

Costs of Pipelines

Pipeline costs vary, of course, with capacity, the character of the terrain which the lines will traverse, and the type of product which the line is intended to carry.

EXAMPLE 6.14: Pipeline Investment

If the investment cost in pipelines in flat terrain, for instance, is $60,000/ mile (or $11.36/ft, $60,000 ÷ 5,280 ft), for 350,000 bbl of crude per day, a pipeline of from several hundred miles to 1,000 miles in length, or up to $60 million total investment, could more than be recovered in a year's revenues (operating expenses) even with a very low price of crude per barrel of $1.00/bbl. At $1.00/bbl, total gross revenues received in one year would amount to $350,000 × 365 × $1.00 or $127,750,000. This example shows the immense costs of a pipeline, but also the quick return on investment. In general, only the tanker can better the pipeline as a low-cost mass transporter of oil.

Figure 6.17 illustrates the transport of oil by pipelines which run into millions of pipe feet and tonnage per oil field, as well as per refinery. From each individual well head in an oil field, the crude oil is collected in small-diameter gathering pipelines which then converge on a collecting center. At the collecting center, the crude oil passes through gas separators, where gas is "liberated" from the crude oil. Usually, there are a number of collecting centers in different parts of the oil field.

From the collecting center, pipes of extremely large diameter lead the crude oil to a tank farm, a center or group of large circular enclosed storage tanks. From here, the crude is conveyed either to a refinery, or to storage tanks at terminals for overseas delivery by sea tankers or long-distant pipeline. Large-diameter pipe is used where volume is large, where it is practical, and where long distances are involved, for the greater the diameter of the pipe, the less the fall in pressure and thus the fewer pumping stations required.

Estimated average pipeline investment for any amount of piping involves millions of dollars. Size of pipe in diameter, length of the line in distances of miles and feet traveled, and type of pipe used all contribute to total investment in pipelines.

Economic Balance in Piping

When pumping of a specified quantity of oil over a given distance is to be undertaken, a decision has to be made as to (1) whether to use a large-diameter pipe with a small pressure drop, or (2) whether to use a smaller-diameter pipe with a greater pressure drop. The first alternative involves a higher capital cost with lower running costs; the second, a lower capital cost with higher running costs specifically because of the need for more pumps.

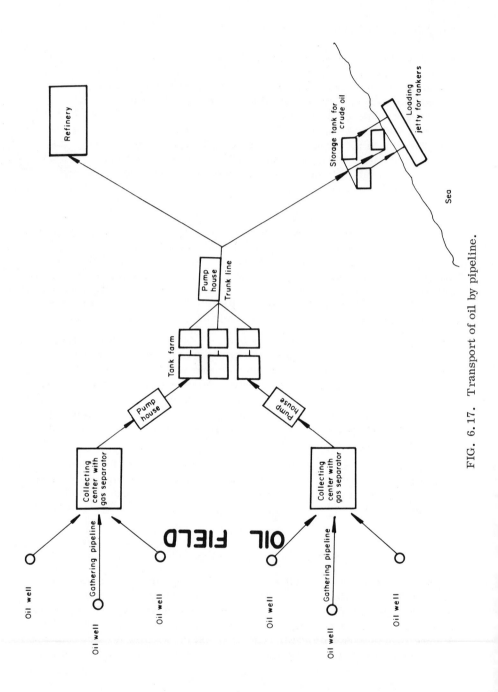

FIG. 6.17.   Transport of oil by pipeline.

So, it is necessary to arrive at an economic balance between the two alternatives. Unfortunately, there are no hard and fast rules or formulas to use; every case is different. Costs of actual pumping equipment undoubtedly must be considered, but also the area in which the pipes will "run" is important. For instance, to obtain the same pumping effort in the desert as opposed to a populated area could involve much higher costs in the form of providing outside services and even the creation of a small self-contained township.

EXAMPLE 6.15: Economic Balance in Piping

This is an example of the principle of economic balance as applied to piping involving two alternatives. One alternative is the use of a large-diameter pipe with a small pressure drop; the other alternative is a smaller-diameter pipe with a greater pressure drop and more pumps. Pumps and pump room installation are considered part of the investment in pipelines.

Assume that the requirement is to transfer 100,000 bbl/day of crude oil for a distance of 200 miles by pipe. In order to arrive at the optimum conditions where total annual costs of an assumed $1.00/bbl will be minimized, fixed costs, or installation costs, and corresponding operating costs for the pipeline for different diameters must be determined and the optimization technique then applied. This is illustrated as follows:

First, fixed charges (installation costs) include the following:

1. Piping: The cost of piping usually varies between $2,500 and $9,000 per inch of diameter per mile of length [4]. However, for a given case [5], the following figures are quoted.

Nominal diameter, in.	$1/mile*
6	$18,000
8	22,000
10	31,400
12	39,000
14	46,000
16	54,000
18	61,000
20	69.000

*Costs for the piping include material and labor.

2. Pumps and their installation: For a distance of 200 miles and for such a quantity of oil (100,000 bbl/day), the number of pump stations varies

between two to three. The installed cost of the pumps is assumed to be
$450/bhp (brake horsepower).

In order to convert the total fixed costs to an annual basis, a payout
time has to be assumed. This is taken to be 5 years, plus 5% annual maintenance. Therefore, the annual "fixed charges" are $0.20 + 0.05 = 0.25\%$
of the fixed costs.

Second, operating expenses include the following:

1. Labor, supervision, and salaries are assumed to be total $50,000/
    year.

2. Electrical power is estimated to be about $20/bhp/year.

Using the above data and taking into consideration the pressure drop
(DP) for each diameter of pump, one can estimate the number of stations
needed and the brake horsepower used in pumping the oil. The ultimate
solution leading to the optimum diameter is found from the graph presented
in Figure 6.18.

The complete solution of the problem is left as a practice for engineering students with a background on fluid flow. An optimum diameter of 14"
is anticipated as a final result for this particular problem.

It is to be noted that the figures given here would vary for other cases.

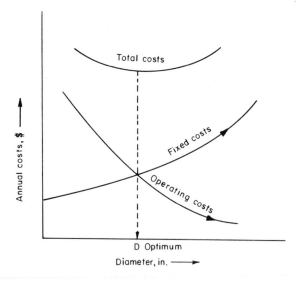

FIG. 6.18. Calculation of optimum diameter of pipe.

Economic Balance in Yields

Recoveries of oil are somewhat more complicated than yields in other industries because of variations in the raw material, the crude oil, and variations in drilling conditions such as temperature, pressure, type of ground layers of soil, etc., which can considerably affect the yield or recovery.

The principle of economic balance in yields can also be applied in oil fluid flow. In the flow of oil in pipes, the fixed charges are the cost of the pipe, all fittings, and installation. All these fixed costs can be related to pipe size to give an approximate mathematical expression for the sum of the fixed charges.

In the same way, direct costs, or variable costs, comprising mostly the costs of power for pressure drop plus costs of minor items such as repairs and maintenance, can be related to pipe size. For a given flow, the power cost decreases as the pipe size increases. Thus direct costs decrease with pipe size. And total costs, which include fixed charges, reach a minimum at some optimum pipe size. This factor can be expressed roughly in a series of simplified equations which express relations in terms of weight rate of flow and fluid density, then weight (or mass) rate of flow, and annual cost per foot for most cases of turbulent flow.

Optimum Economic Pipe Diameter

In choosing the inside diameter of pipe to be used, either in the oil field or in a refinery, selection should generally be based on costs of piping vs. costs of pumping.

Small-diameter pipe, which usually involves quicker drops in pressure than large-diameter pipe and therefore must be supplemented with more pumping equipment when laid for long distances, costs less than large-diameter pipe, but cost of pumping can add considerably to total costs of transferring a given amount of oil. Conversely, large-diameter pipe will have a high fixed capital charge, even though pumping costs are minimized since natural pressure drops are less than with small-diameter pipe. Thus, an economic balance is desirable.

EXAMPLE 6.16: Optimum Pipe Diameter

This example illustrates determination of diameter of pipe (D), and the difference between pipe diameters (d).

Assume the following formulas:

$$\text{Operating cost} = F_1 \frac{1}{D_{pipe}} \quad \text{and} \quad \text{fixed costs} = F_2 (D_{pipe})$$

Then total annual costs for transferring oil will be

$$F_1 \frac{1}{D_{pipe}} + F_2 D_{pipe}$$

Optimum economic diameter of the pipeline is reached when the total annual costs are at the minimum, that is,

$$\frac{d}{dD_{pipe}} \text{(total costs)} = \frac{d}{dD}\left[F_1 \frac{1}{D_{pipe}} + F_2(D)\right]$$

and letting this product equal zero, solving for the value of $D = D_{optimum}$.

To illustrate the principle of $D = D_{optimum}$ in a simplified manner using a and b as constants, let us assume that

$$F_1\left(\frac{1}{d}\right) = \frac{a}{D} + b \quad \text{and} \quad F_2(D) = CD + d$$

Then

$$\text{total annual costs} = \frac{a}{D} + b + CD + d = \text{total costs (T.C.)}$$

and

$$\frac{d(T.C.)}{dD} = -\frac{a}{D^2} + C = 0$$

(where C is a constant representing rate of flow)

$$\frac{a}{D^2} = C \quad \text{or} \quad D_{optimum}$$

$$D^2 = \frac{a}{c}$$

$$D = \frac{a^{1/2}}{C^{1/2}} \quad \text{or} \quad D = \frac{a^{1/2}}{c}$$

where c is a constant and is equal to the square root of C.

The exact equation for predicting $D_{optimum}$ for turbulent flow of incompressible fluids inside steel pipes of constant diameter is given if we assume the following:

$$D_{\text{optimum}} = \frac{2.2\,W^{0.45}}{\rho^{0.32}}$$

where

$D > 1''$

W = thousands of pounds mass flowing per hour

ρ = density, or lb-mass/ft^3

Then, to calculate D_{optimum}, if we are considering the transfer of 500,000 bbl/day of oil of an average API of 33° C (with ρ = 53,70 lb/ft^3) across a distance of 1,000 miles, we have as follows:

First, calculating for $D_{1_{\text{optimum}}}$,

W = 500,000 bbl/day × 300 lb/bbl = 15 × 10^7 lb/day

= 15 × 10^7 × 1/24 = 6.25 × 10^6 lb/hr = 6.25 × 10^3 lb/hr

$\rho^{0.32}$ = $(53.7)^{0.32}$ = 3.58

$D_{1_{\text{optimum}}}$ is then

$$\frac{(5,250)^{0.45}}{3.58} = \frac{50}{3.58} = 14''$$

Therefore 14'' is the optimum economic pipe diameter in this particular case.

Second, calculating the cost of pipeline of 1,000 miles of the 14''-diameter pipe, if we assume that the construction of the 14''-pipeline is $50,000/mile, then total costs for the 1,000-mile 14'' line are $50,000 × 1,000, or $50,000,000.

If a pumping station is needed every 100 miles, or an overall total of 10 stations for the 1,000-mile pipeline, and the cost of each pump station is estimated at $2 million, the total cost of using a 14'' pipeline is $50 million for the pipe plus $20 million for 10 stations, totalling $70 million.

Then, if crude is selling at $3.50/bbl, 20 million bbl will have to be moved through the line before the total investment of $70 million is recovered ($70,000,000 ÷ $3.50). This recovery is simplified, since maintenance and repair expenses plus cost of money invested was not considered.

At a rate of, say, 50,000 bbl/day, or 17,250,000 bbl/year, it will take 1.15 years (20,000,000 bbl ÷ 730,000 bbl) before payback.

Crude Oil Pipeline Capacities

Each pipeline must be considered as an individual problem. In general, the economic capacity of each of the various diameters of pipelines lies between the limits given in Table 6.14.

During the last few years, the maximum diameter of pipe available has been increased progressively to 24", 30", and finally to 36" and today even 48".

These large sizes of pipe are costly to ship, because the space they occupy, relative to their weight, is high, therefore freight costs are up. To reduce freight costs, it has become the practice today to design these large pipelines for equal quantities of two slightly different sizes of pipe, so that they can be "nested" for shipment; for example, one length of 20" pipe is placed inside each length of 22" pipe.

Summary: Pipelines

There are three types of pipelines. These are as follows:

1. Those which run from the oil field to loading ports, and are complementary to ocean transport. Without these, there would be no transport by tankers at all, so they are not competitive with transport by tankers.

2. Those long-distance pipelines which naturally shorten the alternative sea route. They can be competitive with ocean transport tankers if tanker rates are high. But in times of low tanker rates, such pipelines are not competitive with transport by tankers. A good example of this type of pipeline is Tapline, the 1,100-mile pipeline from Ras Tanura through four countries to Sidon, Lebanon. Transport by Tapline saves approximately 3,300 miles each way of ocean transport, and also did save Suez tolls when the Suez canal was open. At this writing the Suez canal has just reopened.

3. Those pipelines which transport oil from ports of discharge to inland refineries located in industrial areas, remote from a seaport. They can be competitive with domestic railroad and motor carriers. Examples of this type of pipeline are the pipelines of Rotterdam on the Rhine, and Wilhemshaven on the Ruhr.

TABLE 6.14

Crude Oil Pipeline Capacities

Diameter, in.	Useful range, million tons/year	Usual pump station spacing, miles
6	0.4-0.7	30-80
8	0.7-1.3	
10	1.3-2.5	
12	2.0-4.1	40-100
16	4.1-8.0	
20	7.0-13.0	
24	12.0-18.0	60-200
30	15.0-25.0	
36	20.0-40.0	

(40 million tons or 300 million bbl)

Source: British Petroleum Company, Ltd., Our Industry, 1947.

When moving oil and oil products, such operating costs as the following, based on a per-ton mile basis, will be important:

1. Construction costs of pipeline and equipment.

2. Amortization of investment.

3. Interest on invested capital.

4. Energy costs for operating pumping stations, etc.

5. Personnel and maintenance costs.

6. Royalties to governments of countries crossed by the pipeline.

NOTES

1. Middle East Economic Survey, Beirut, June 1972.

2. Oil and Gas Journal, December 25, 1972.

3. Baker, A., "Depletion as a Key to Reasonable Returns on Petroleum Production," Humble Oil Company, 1954.

4. Meyers, R., "Planning of Crude Oil Production, Transportation and Refining," 7th World Petroleum Congress, Mexico, 1967.

5. Ibid.

Chapter 7

OIL PROCESSING IN REFINERIES

INTRODUCTION: REFINING PROCESSES

Petroleum is of little use as it comes from the ground. It is only a raw material, much as newly felled trees are raw materials for furniture, construction, etc. Thus crude oil must be put through a series of processes in order to be converted into the many hundreds of finished oil products that are derived from the crude. These processes, collectively, are known as refining.

As mentioned earlier in Chapter 3, molecules in crude oil, known as hydrocarbons, contain varying number of hydrogen and carbon atoms. Some molecules contain more atoms than other molecules. Those molecules containing many carbon atoms make up the thicker and heavier components of petroleum, such as asphalt. Other molecules, containing relatively few atoms, make up the lighter and more volatile components of petroleum, such as gasoline.

The first step in refining is distillation. This step roughly separates the molecules in crude according to their size and weight. The process is analogous to taking a barrel of gravel containing stones of many different sizes and running the gravel through a series of screens to sift out first the small stones, next those slightly larger, and so on up to the very largest stones. As applied to crude oil, the distillation process "sifts out" progressively such components as gas, gasoline, kerosene, home heating oil, lubricating oils, heavy fuel oils, and asphalt.

In modern distillation processes, crude oil is run through a series of pipe lining a large brick furnace. After being heated to about 800° F, the crude enters the bottom of a cylindrical container, called a fractionating tower. There all but the heaviest portions are flashed into vapor.

The various components of crude oil have different boiling points—that is, they change from liquid to vapor, or condense back from vapor to liquid, at different temperatures. By taking advantage of this fact, it is possible to separate the oil into different fractions, or cuts.

Distillation is a physical process. It can separate crude into various cuts, but it cannot produce more of a particular cut than existed in the original crude. Unfortunately, too, consumers' demands for different oil products do not necessarily parallel the natural proportions of the crude. For example, if we had to depend on the amount of gasoline naturally present in crude (about 20%), we would not have enough to run all the automobiles presently on the roads. This leads us to the next step in refining, the chemical process called cracking.

Cracking amounts to breaking big molecules into little ones. It is comparable to crushing the larger stones in a barrel of gravel into smaller stones. There are two common methods of cracking: thermal cracking and catalytic cracking, the latter being the method most widely used today.

In catalytic cracking a catalyst is used. A catalyst is a substance that causes other substances to change chemically without being changed itself. In the oil cracking process, the catalyst is a clay-like material which may be in the form of lumps, pellets, grains, or superfine powder.

Cracking has made it possible to produce more than twice as much gasoline from a barrel of crude as could be obtained by simple distillation.

There are also other chemical processes which may be used to take petroleum molecules apart, put them together, and rearrange their atoms. Polymerization is the linking up of similar molecules to make larger molecules; alkylation is the joining of dissimilar molecules; isomerization is the alteration of molecules so that the atoms are arranged differently, and reforming is a method for making high-octane gasoline.

Petroleum products resulting from distillation or cracking require further refining to make them suitable for use. These further processes are known as treatment processes. Hydrogen, sulfuric acid, and caustic soda are some of the treatment agents used, by which thousands of oil products can be made from crude oil.

Figure 7.1 summarizes the modern refining process, indicating both physical and chemical methods.

ECONOMIC ANALYSIS IN PETROLEUM REFINING

Economic analysis is used in refining in various ways—to determine the most economical refining operations, to determine whether to use new or existing equipment, etc. Economic analysis, including cost analysis, is

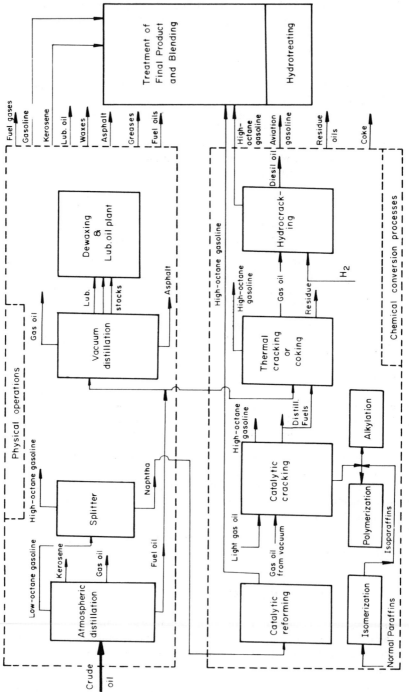

FIG. 7.1. A flow diagram of the physical and chemical processes of crude oil refining.

complicated in a refinery. The reason is that an operation in a refinery with lower operating cost is not necessarily the most desirable procedure, and similarly, an operation giving higher yields, or production rates, is not necessarily a more economical one. Highest yields with lowest costs is what the refiner would like to achieve.

Applying economic analysis is further complicated by the fact that several hundred different products may be produced from one basic raw material, crude oil. There are also other complications. The basic crude may consist of a number of different crudes which have considerably different characteristics and different selling prices (to independent refiners only). Furthermore, it is becoming increasingly more difficult for refiners to determine which products are prime products and which are by-products. The economic analyst faces the problem of establishing reasonable differentials between the costs or values of various products and raw materials consistent with the amounts of one product which can be produced from another product or from crude oil.

Economic analysis helps a refiner determine whether he can meet competition. Application of the results of economic analysis by a petroleum refiner will result in an improved competitive position in the industry and may result in increased profits.

Basic tools used in an economic study of a refinery are operating costs of existing unit operations in a refinery. These can be obtained from normal accounting records. Direct operating costs which are controllable consist of direct supervision, operating labor, maintenance and repair labor, plus materials, chemicals, fuel utilities, auxiliary services, royalties, and employee benefits.

There are also costs of an auxiliary operating nature, such as overhead and burden items, that are generally not controllable. These include depreciation, taxes, administrative and general expenses, and other items not charged directly to operation units.

For a given refinery, total money spent on materials and labor must be distributed to operations in as fair a manner as possible. The economist thus needs to be acquainted with, and use, operating costs in a manner consistent with the method by which the costs were determined. In addition to total operating costs which are controllable, it is necessary for the economic analyst to estimate marginal operating costs (the costs of increasing the charge rate or production from a unit already in operation).

Economic analysis also aids in determining the most economical means of operating a given process unit, and involves factors usually considered in setting up operating standards, such as standards of variations in throughput in temperature and pressure in chemical and catalyst consumption, conversion, and other variables for determining optimum operating conditions.

This part of economic analysis is known as economic balance. In this chapter several hypothetical situations, covering the range of operating conditions to be investigated, are presented. Material balance and operating costs for each of these situations will be evaluated to determine the operation which yields the greatest economic credit. For example, controllable, total, and marginal operating costs for a modern crude oil pipe are shown in Table 7.1. This table indicates those items which contribute to each of the three types of operating costs:

Controllable or direct costs of operation.

Total operating costs, including overhead and burden items, not necessarily controllable.

Marginal operating costs.

In economic analysis, very often only two products are considered; but an oil refinery may be able to produce a multiplicity of products. Linear programming is the method by which an oil refiner is guided in the choice he makes among the various alternatives open to him.

LINEAR PROGRAMMING

The following simplified example illustrates the basic fundamentals involved in linear programming.

EXAMPLE 7.1: Linear Programming Problem on Blending [1]

It is required to obtain 100 gal of 4% acetic acid solution. Two stocks are available for mixing. Solution 1 contains 4.5% acetic acid by volume and is valued at $0.32/gal. Solution 2 contains 3.7% acetic acid by volume and is valued at $0.25/gal. Water can be used in blending at no cost.

The problem is to determine what volume of solution 1, solution 2, and water will require the minimum cost to produce 100 gal of 4% acetic acid solution, bearing in mind the final blend should contain at least 10% by volume of solution 1.

Solution

There are three variables involved: V_1, V_2, and V_w, where V_1 is the volume of solution 1, V_2 is the volume of solution 2, and V_w is the volume of water in the final solution. The constraints are as follows:

TABLE 7.1

Suggested Operating Costs in a Crude Oil Pipe Still
(All Costs Are Estimated)

	Total cost, $/bbl	Percent for marginal	Marginal cost, $/bbl
Labor salaries and wages			
Operating	0.40	0	0
Maintenance	0.75	50	0.37
Total	1.15		0.37
Materials			
Maintenance materials	0.40	50	0.20
Chemicals for operations	0.05	100	0.05
Petroleum and general supplies	0.02	100	0.02
Total	0.47		0.27
Fuel			
In operations	0.80	100	0.80
In utilities and auxiliaries	0.25	100	0.25
Total	1.05		1.05
Utilities			
All depreciation and taxes excluded (excluding fuel)	0.25	100	0.25
Auxiliary services (cooling towers, pumping industrial units, etc.)	0.60	50	0.30
Employee benefits	0.28	—	0.09
Controllable or direct costs	3.80	Marginal cost	2.33
Burden and overhead expenses	1.50		
Depreciation and taxes	1.00		
Total cost	6.30		

1. $V_1 + V_2 + V_w = 100$ gal.

2. $V_1(4.1) + V_2(3.7) + V_w(0.0) = 100(4.0)$.

3. The volume of solution 1 must be greater than 10 gal, that is, $V_1 \geq 10$ and $V_2 \geq 0$, $V_w \geq 0$.

The objective is to minimize the loss of product:

$$\cos V = (V_1)(0.32) + V_2(0.25) + V_2(0.0)$$

$$= \text{minimum}$$

Now, the set of constraints should be combined with the objective function to reach the optimum solution. This is usually achieved by linear programming, for which a computer is usually required.

For this particular simple example, the intuitive approach can be used along with the linear programming as follows:

We have a set of linearized basic equations:

$$V_1 + V_2 + V_w = 100 \tag{7.1}$$

$$4.5V_1 + 3.7V_2 = 400 \tag{7.2}$$

$$0.32V_1 + 0.25V_2 = C \quad \text{(minimum)} \tag{7.3}$$

Combining Equations 7.1 and 7.2 to eliminate V_2 gives

$$V_1 = 4.625V_w + 37.5 \tag{7.4}$$

Combining Equations 7.1 and 7.3 gives

$$\cos V = C = 25 + 0.07V_1 - 0.25V_w \tag{7.5}$$

Then, by intuition, it can be seen that C is a minimum for $V_w = 0$. Applying this result to Equation 7.4, it is found that V_1 optimum $= 37.5$ gal and, consequently, V_2 optimum $= 62.5$ gal for Equation 7.1.

Let us consider next an actual problem in a refinery where the real programming techniques are applied.

EXAMPLE 7.2: Refinery Yield

An oil refinery company processes crude oil into refined products of gasolines, distillates and fuel oils.

The company can buy its crude oil from two sources which differ in yields and quality. These yield characteristics are given in Table 7.2, and the feasible purchase possibilities are graphed in Figure 7.2.

TABLE 7.2

Percentage of Yields of Products from Different Oils

Product	Source 1	Source 2	Purchase limitations
Gasolines	0.20	0.30	1.8
Distillates	0.20	0.10	1.2
Fuel oils	0.30	0.30	2.4
Relative profit	5	6	

Observe in Table 7.2 that from source 1, there is a 20% yield of gasolines, a 20% yield of distillates, and a 30% yield of fuel oils; the remaining 30% is unrecoverable waste or gas. From source 2, there is a 30% yield of gasolines, only a 10% yield of distillates, and a 30% yield of fuel oils, with also a 30% unrecoverable waste.

Thus the yield of 20% of distillates is relatively higher from source 1 than from source 2; but the yield of 30% of gasolines is relatively higher from source 2 than from source 1. The figures for fuel oils and waste are 30% for each source.

The question is: How many barrels of crude oil should the company purchase from each source?

Solution

The answer to the question depends, in part, on the relative profit contributions of the sources. These relative profit contribution figures are calculated by adding the sales revenues associated with the yields for the separate crude oil products and then subtracting the costs of purchasing these crude oils from the sources.

To simplify the linear programming problem, other variable expenses such as sales and distribution costs are ignored. We shall therefore use the term "relative profit" contribution.

The relative profit contribution depends on the refined products produced from the type of crude oil purchased and not on the sources themselves. Now, suppose that the relative profit contribution is $5 for source 1 and $6 for source 2. Even though source 2 is more profitable than source 1, it does not follow that the company should purchase all of its crude oil from source 2, because relative profit contribution does not affect buying allocation decisions, only what kinds of refined products are produced from the crude oil of each source.

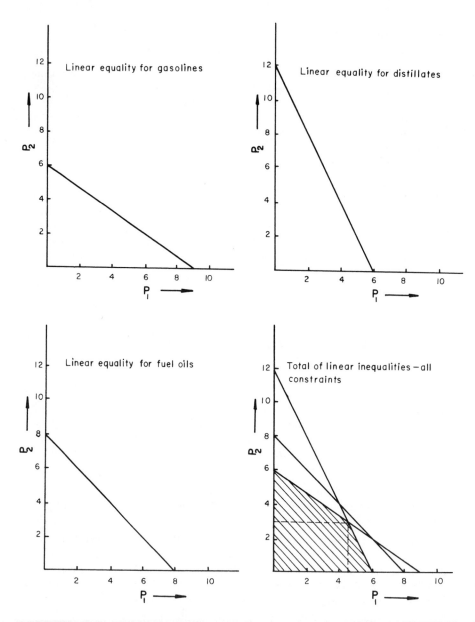

FIG. 7.2. Feasible purchasing policies, given two suppliers of crude oil of differing yields and qualities.

Also two other factors, at least, are relevant to the purchase decision. These two factors are (1) the maximum amount of each product that the company can sell, and (2) the maximum amount of each product that the company can process, given its production facilities, or the capacity of the refinery.

Constraints (restrictions): To keep this exposition simple, suppose that the two factors given above, in concert, imply that total production (the same as purchase limitations here) cannot exceed 1.8 (180%) for gasolines, 1.2 (120%) for distillates, and 2.4 (240%) for fuel oils, where these constants are measured in terms of an appropriate unit of measurement such as millions of barrels. These restrictions can then be expressed mathematically as follows:

Let P_1 denote the number of barrels of crude oil that will be purchased from source 1, and P_2 the number of barrels of crude oil that will be purchased from source 2.

Then the values for P_1 and P_2 are constrained (restricted) by the linear inequalities of:

$$0.2P_1 + 3P_2 \leq 1.8 \text{ for gasolines}$$

$$0.2P_1 + 1P_2 \leq 1.2 \text{ for distillates} \tag{7.6}$$

$$0.3P_1 + 3P_2 \leq 2.4 \text{ for fuel oils}$$

(The numbers on the right are taken from Figure 7.2.) We assume non-negativity; that is, $P_1 \geq 0$ and $P_2 \geq 0$. These non-negativity restrictions are given since a value, such as $P_1 = 04$ would have no physical significance.

All the values for P_1 and P_2 that satisfy Equation 7.6 are shown in the shaded area in Figure 7.2. Notice that each line in the figure is represented by a restriction in Equation 7.6 expressed as an equality. Furthermore, the arrow associated with each line shows the direction indicated by the inequality signs in Equations 7.6.

Optional values for P_1 and P_2 are found by making the relative profit contribution as large as possible, consistent with the constraints. Therefore the optimization problem is to maximize $(5P_1 + 6P_2)$ subject to Equation 7.6.

In this simple problem, the solution can be exhibited graphically, as in Figure 7.2. Each of the parallel straight line segments represents a different combination of P_1 and P_2 that gives the same value for the linear objective function, the maximizing of $5P_1 + 6P_2$. The highest segment still having a point in the feasible constraint region is the optimal value of the objective function, and such a point is an optimal solution. In Figure

7.3 there is only one optimal solution, and it occurs at the intersection of the gasoline and distillate constraints.

Consequently, optimal values can be calculated by solving the associated linear equations as follows:

$$0.2P_1 + 0.3P_2 \quad = 1.8 \quad \text{for gasolines}$$

$$0.2(4.5) + 0.3(3) = 1.8 \tag{7.7}$$

$$0.2P_1 + 1P_2 \quad = 1.2 \quad \text{for distillates}$$

$$0.2(4.5) + 0.1(3) = 1.2 \tag{7.8}$$

Verify that the optimal answers are $P_1 = 4.5$ and $P_2 = 3$ as shown in Figure 7.3, giving an objective function value of 40.5. (Or show that maximizing $5P_1 + 6P_2$ gives $5(4.5) + 6(3) = 40.5$.

The solution is, therefore, that 22.5 bbl should be purchased from source 1 and 18 bbl from source 2.

Example 7.2 illustrates what is termed a linear programming model. Real applications of linear programming usually involve hundreds of constraints and thousands of variables. Learning how to formulate and solve such models as they might pertain to petroleum problems is reserved for an advanced course in oil economics.

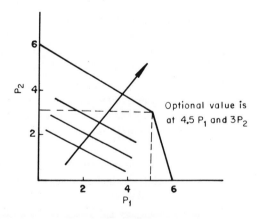

FIG. 7.3. Relative profit

RECIPROCAL COST ALLOCATION, USING LINEAR ALGEBRA

Allocation of costs from a service department of a refinery to one or more
operating department is normally done individually by department, either
by using algebra or by making successive approximations after trial and
error.

When the number of departments, or the extent of the reciprocal interactions, becomes very large, the problem of cost allocation becomes hard
to handle by ordinary methods. Then cost allocation of service departments
may be expressed as a series of linear equations, using matrix algebra.

EXAMPLE 7.3: Allocation of Costs of Service Departments using
                Linear Algebra

Suppose that we have four operating departments in a refinery: distillation,
cracking, reforming, and product treatment. Assume, also, that these
four operating departments are supported by four service departments:
steam and water, maintenance, power, and compressed air. Thus, we
assume a total of eight departments, with interactions among departments
as indicated in Figure 7.4.

Then let total costs for the four operating departments, including allocations from the service departments, be as shown in Table 7.3. Note
that the sum of each column, which represents both operating and service
departments, equals 100%. All service departments are thus completely
allocated to operating departments. Use of such simple linear algebra to
a refinery's problem of reciprocal cost allocation will provide some organized manner for performing this necessary economic analysis.

REFINERY PRODUCT COSTING

Costing of products is an important function of economic analysis as well
as of cost accounting. The establishment of the total cost of individual refined products in a modern refinery is complicated by the great number of
different products produced simultaneously from crude oil. Thus refinery
product costing is not a simple matter.

A procedure must be established for assigning fair raw stock (crude)
costs and operating costs to the various products. This requires technical
knowledge of refinery yields operations and the crude oil stocks involved.
Furthermore, the total cost of all refined products produced must equal the
cost of crude oil plus the total cost of the refinery. The problem is to determine an equitable distribution of crude oil stock and refinery operations

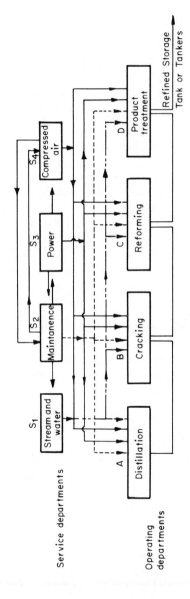

FIG. 7.4. Interactions among operating and service departments.

TABLE 7.3

Cost Allocations (%)

Department	Steam and water	Maintenance	Power	Compressed air	Dept. expenses	Total
Distillation	20	10	10	10	100	150
Cracking	30	20	10	5	100	165
Reforming	15	20	30	30	100	195
Product treatment	15	20	30	25	100	190
Steam and water	—	10	—	—	100	110
Maintenance	10	—	10	25	100	145
Power	10	10	—	5	100	125
Compressed air	—	10	10	—	100	120
Totals	100	100	100	100	800	1200

among the various refined oil products produced. There is no universally acceptable and accurate method of determining the true costs of the many refinery products produced simultaneously from the single raw stock of crude oil. Many methods are possible by making some arbitrary assumptions. Here is one, explained by use of an example.

EXAMPLE 7.4: Refinery Product Costing

Assume that a refinery is operating and producing five products as follows (see Table 7.4): gasoline, kerosene, heating oil, fuel oil, and fuel gas, from crudes costing $2.90/bbl delivered, and that the operating cost of the refinery is $0.60/bbl of crude oil charged (see Table 7.4).

Profit, after tax, is 10.125¢/bbl in this hypothetical example. But the main problem arises as to what is the cost of the individual products and what is the profit being realized on each product before income tax. Various arbitrary methods of arriving at the cost of the individual products produced simultaneously from crude oils are possible. Table 7.5 illustrates some possible methods.

For the average unit cost method (column 2 of Table 7.5), it is assumed that all refined products "cost" the same, except fuel gas which is given an arbitrary natural gas value in this example. By this method, however, costs bear little resemblance to market values. Gasoline and

TABLE 7.4

Costs of Refined Products

Product	Yield, volume %	Middle East market value, in ¢/gal
Gasoline	40	12
Kerosene	10	9
Heating oil	30	8
Fuel oil	15	44
Fuel gas	5	2.14 (15¢/1000 Btu)
Total average	100	$8.87/gal ($3.725/bbl)
Crude oil cost	$2.90/bbl	6.90¢/gal
Operating costs at refinery	$0.60/bbl	1.43¢/gal
Total costs	$3.50/bbl	8.33¢/gal
Refinery profit before income	22.5¢/bbl	0.54¢/gal
Income tax at 55%	12.375¢/bbl	0.30¢/gal
Refinery profit after income tax	10.125¢/bbl	0.24¢/gal

Note: 1 bbl ≅ 42 gal for conversion purposes.

kerosene, for instance, cost less than their market value (selling prices), while heating oil and fuel oil actually cost more than their market values. Fuel oil could have a market value assigned by this method, but then there would have been an increase in the costs of gasoline, kerosene, and heating oil.

In column 3 of Table 7.5, it is assumed that the same profit as a percentage of cost is realized on all products except fuel gas. Thus this method shows the costs of the four major products, excluding fuel gas, as being proportional to market values.

In the by-product method (column 4), all products except gasoline are assigned market values. The cost of gasoline is calculated by difference, so all profit appears on gasoline.

In column 5, where the total replacement cost method has been used, costs are calculated by first processing crude to ultimate or maximum yield of gasoline with zero yield for kerosene and heating oil in a hypothetical operation. Total replacement cost of gasoline is then calculated by debiting the cost of crude oil and the hypothetical operating costs required, and crediting the by-product fuel oil and fuel gas produced at market values.

TABLE 7.5

Methods of Evaluating Costs of Products (Before Income Tax), in ¢/gal

Product	Yield, % (from Table 7.4)	Average unit cost method	Same percent profit for all products method	By-product method	Total replacement cost method	Incremental total replacement cost method	Middle East market value, in ¢/gal (from Table 7.4)
	(1)	(2)	(3)	(4)	(5)	(6)	(7)
Gasoline	40	8.66	11.27	10.66	11.91	10.87	12.0
Kerosene	10	8.66	8.46	9.00	8.50	8.40	9.0
Heating oil	30	8.66	7.51	8.00	6.50	7.92	8.0
Fuel oil	15	8.66	4.13	4.40	4.40	4.40	4.4
Fuel gas	5	2.14	2.14	2.14	2.14	2.14	2.14
Total and average	100%	8.33	8.33	8.33	8.33	8.33	8.87

The costs of kerosene and heating oil are then calculated in terms of the gasoline cost by establishing the yield of gasoline, fuel oil, and fuel gas that can be produced from the product in question. Fuel oil and fuel gas are also assigned market values, and hypothetical operating costs required for conversion are debited.

Finally, in column 6, in the incremental total replacement cost method, costs are calculated in a manner similar to total replacement costs except that in this method enough hypothetical virgin gas oil fraction in the heating oil boiling range is cracked to achieve zero net heating oil production rather than to crack actual refractory stocks to get ultimate yield. There is therefore an increased ultimate or maximum gasoline yield from crude, and so lower unit gasoline cost. This method is applicable only to costing incremental production.

In summary, the comparison of costs in Table 7.5 calculated by the various methods with market values indicates that, where market values are used to distribute costs, the costs bear a direct relation to market value. However, since costs by these methods vary with market values, the costs represent allocated and not calculated costs.

COST ANALYSIS

There are certain costs in the oil industry which are approximately the same regardless of the amount of refined products produced. These costs are known as "overhead," or fixed costs. Some of these costs are interest, pensions, taxes, depreciation and depletion, payroll, goods and services purchased from others, etc.

The composition of these costs will differ in some respect for different oil companies. Also, oil refiners may have different percentages for the components of their fixed costs depending on the degree of integration and their capital structure. However, regardless of their composition, such costs are relatively large for major producers in the petroleum industry and, with typical low operating rates, are a substantial percentage of total costs.

Costs over and above fixed costs represent "additional costs" incidental to the production of each additional barrel of refined oil products (marginal costs), assuming the oil refinery is in operation. Actually, addition to total costs arising from the production of each additional barrel of refined oil products is the same regardless of the operating rate at which the additional output is obtained, as long as the other factors affecting costs remain constant. This phenomenon of constant additional costs covers a range of output from 20% of capacity to about 90% of physical limit of output. As the physical limit of capacity, or 100%, is reached, the equipment becomes overtaxed and for various reasons operates less efficiently and at greater cost. In such cases, additional costs incidental to production (marginal costs) of an additional unit of output cease to be constant and probably rise sharply. So the basic economic law of diminishing returns enters to make further production uneconomical.

AVERAGE COSTS

Average costs are the sum of additional costs plus an amount equal to fixed costs divided by the number of units produced. Average costs must necessarily be higher than the additional cost for nearly the whole range of operations if the refinery is to operate efficiently.

The components of the average cost of producing a barrel of refined oil lie, to a certain degree, largely outside the control of oil refiners; wage rates tend to be inflexible, lags in adjustment prices paid for goods and services are often fixed by outside agencies, interest is determined by factors in the money market, taxes are established by law, and depreciation and depletion changes cannot be disregarded.

PRICE-VOLUME-COST RELATIONSHIP

The inelasticity of total demand for refined oil products and the aforementioned characteristics of cost in the oil refinery industry place definite limitations on the financial ability of the oil refining industry to increase production by decreasing prices. Assuming that each 1% decrease in price would increase consumption of refined oil products by 1% (but prices do not decrease, and the oil industry is not worried about increasing demand by cutting selling prices since substitutes for refined oil products are no problem and world demand continues to rise faster than supply or reserves), a 10% decrease in average level of refined oil product prices, even though offset by a 10% increase in demand for oil products, would only serve to cut the profits of the oil refining industry.

Despite this overall price-volume-cost relationship in the industry, potential elasticity of demand for oil products of an individual refinery and the internal problems within the individual refinery, because of the characteristic cost pattern, can affect the market for refined oil products. Except in periods of high operations and in times of slack demand, which is of course purely theoretical, there would be a tendency to cut prices below average cost so long as the selling price for the additional unit (bbl) sold is above the "additional cost" (marginal cost) necessary to produce this additional barrel of refined oil.

Large individual orders for crude or refined oil products could also accentuate such a tendency; also, because of the inelastic nature of total demand for refined oil products, the problem for the individual oil refiner is to obtain a share of the going business, so he will tend to increase his output.

There are problems in applying cost analysis. Cost relationships, which economists wish, are not always available from conventional accounting records. Accounting records are maintained for other purposes, and the needs of refinery management go beyond those of accountants and economists.

The oil refiner is expected, however, to combine his input factors of production, such as labor, capital goods, and materials used, in such a manner that their marginal productivities are equal. That is, the marginal productivities, or contributions (output) to oil refinery production are equal in value to the prices paid for these input factors. This is a basic principle of production economics.

The following illustrates the effect of factor price changes on cost curves of a refinery. A change in underlying conditions of production, such as (1) in prices of input factors or (2) in nature of the product, brings a shift in the overall cost curves as shown in Figure 7.5. (This is analogous to changes in the underlying conditions of demand which cause shifts in demand curves.)

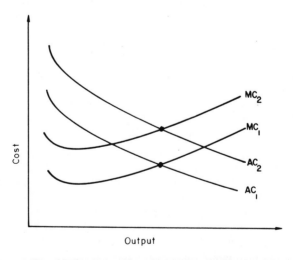

FIG. 7.5. Effect of factor price changes on cost curves for a refinery, where MC = marginal cost and AC = average cost.

Finally, in order to make important decisions relative to input factors as they affect output of a refinery, the refiner needs to know the several relationships previously described, namely, (1) the costs of building differ-ent-sized refineries, (2) the operating costs which will be experienced in these different-sized plants, (3) the way in which cost responds to changes in output produced with a given-size refinery, (4) the relationship of unit costs to alternative techniques of production, and (5) the effect of refined products.

INCREMENTAL COSTING

Incremental costs are referred to as "out-of-pocket" or differential costs. Any change in a process, or in the operations of any refinery equip-ment, which changes the production rate of the refinery or any one of its departments and so affects costs accordingly, brings a net change in pro-duction and in costs.

For example, consider a gas-fired boiler that has been used to supply steam at a rate of 100,000 lb/hr at an average cost of $0.60 per 1,000 lb of steam. The gas-fired boiler is replaced by a waste-heat boiler that sup-plies steam at a rate of 95,000 lb/hr at a cost of $0.50 per 1,000 lb of steam. Our first thought may be that a savings of $100 ($1,000 × $0.60 - $0.50) has been made. We do indeed have an incremental decrease in costs, but the steam-generating rate is still 100,000 lb/hr. If we need 100,000 lb/hr, it would cost $2.50 more or a total of $50,000, even though we are saving $10 per 1,000 lb of steam.

BREAKEVEN ANALYSIS

Breakeven analysis is an important economic and business concept for refineries. It is important for refinery management to know what level of production is needed to break even with regard to sales dollars received vs. costs of production (see Fig. 7.6).

The breakeven point is the refinery capacity at which gross income from refined oil products just equals the cost of sales and production when all refined oil products produced are sold. The breakeven point is a useful guide for oil management in analyzing its operations. Above the breakeven point a profit results, below breakeven point there will be a loss (Fig. 7.7).

Mathematically, breakeven can be expressed as total fixed costs, or investment, divided by selling price per unit-variable cost per unit, which gives a breakeven figure in number of barrels. Breakeven is best illus-trated by the following example.

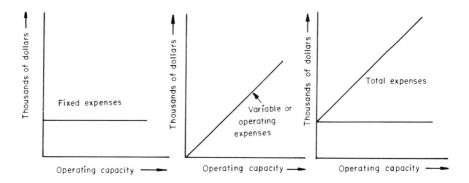

FIG. 7.6. Components of total cost curves.

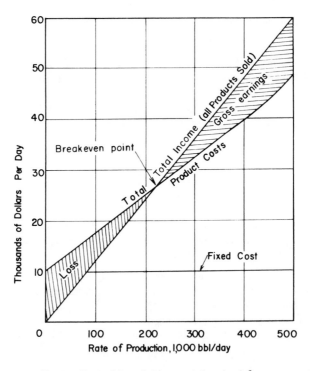

FIG. 7.7. Illustration of breakeven point principle.

EXAMPLE 7.5: Breakeven Point

For a given refinery, the average selling price of all refined oil products combined is $1.50/bbl, variable costs of the refinery are $0.70/bbl, and nonvariable, fixed investment is $12,045,000 or about $1 million a month. The designed capacity of the refinery is 100,000 bbl/day. How much production is necessary for the refinery to break even?

Solution

It will be necessary for the refinery to sell 15 million bbl/year on approximately 41,250 bbl/day (41% of capacity) to break even. Figure 7.8 illustrates this breakeven point. The reasoning by which these figures are obtained is as follows.

Each additional barrel of product sold increases the profit by $0.80 (or reduces the loss by $0.80), because each barrel adds $1.50 to revenues and $0.70 to expenses. To cover the nonvariable, fixed investment of $12,045,000, or $33,000/day, sales of 41,250 bbl, with each barrel providing about $0.80 toward these expenses, are required.

Selling price per barrel	$1.50
Less: variable cost per barrel	$0.70
"Contribution" per barrel to fixed costs	$0.80

Then $33,000 (fixed costs) ÷ $0.80 = 41,250 bbl, breakeven point.

Proof that the breakeven point is 41,250 bbl/day is as follows:

Total costs per day		$61,875
Less: fixed costs per day	$33,000	
Less: variable costs per day (21,250 × $0.70/bbl)	28,875	
Revenues on 41,250 bbl at $1.50/bbl		$61,875

Breakeven charts or graphs, such as those in Figures 7.6-7.8, can be constructed for total refinery operations, for subdivisional operations such as separation, cracking, and treatment processes, or even for the relationships between expenses and revenues pertaining to a particular salable refined product, such as gasoline or jet fuel, or to a refined product line such as a line of gasoline products. All that is required is knowledge of expense and revenue behavior in relation to volume of production. Unfortunately it is not always possible to identify these elements with any degree

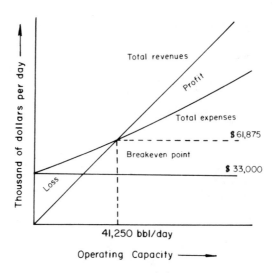

FIG. 7.8.   Solution to Example 7.5.

of accuracy.   Nevertheless, when expenses and revenues can be reasonably estimated, breakeven analysis is helpful to oil management in visual-
izing the effect of their relationship on earnings at different volumes.

Refinements in Breakeven Analysis

If the market is such that a refinery is required to operate below the breakeven point (for instance, at 41,250 bbl/day or 41% of capacity as in Example 7.5), the refinery will operate at a loss.   This does not necessarily
mean, however, that the refinery should be shut down, because fixed costs
must still be paid, even if variable costs are cut off completely.   The shutdown point occurs only when revenues received do not exceed total fixed
costs.   In Example 7.5, for instance, total fixed costs were $33,000/day.
Thus, at a selling price of $1.50/bbl, at least 22,000 bbl would have to be
produced to obtain at least enough revenues ($33,000) to cover fixed costs,
even though variable costs would not be covered by 22,000 bbl.   Variable
costs are zero only at zero production.

On the side of expenses of operation, there are many uncertainties.
One uncertainty is whether the marginal costs of this year's and last year's
production figures also apply to next year's sales.   Certain adjustments to
productive capacity are made from time to time in response to changing
market conditions.   Changes from one level of production to another always
affect marginal costs, which will in turn change the breakeven point.   Finally,

continuing needs for increases in output may be impossible to satisfy with existing plant, and the purchase of new or additional refinery equipment may be necessary. This, of course, will affect marginal costs and the breakeven point.

Breakeven analysis is a technique somewhat akin to full-cost pricing, in that it sets a level of capacity utilization at which the refinery wishes to operate in order to break even, thus allowing profit to be a true residual which arises after a certain number of barrels are produced and sold. By full-cost pricing, we mean an attempt to set prices for refined oil products such that total revenue equals total cost. Then when output exceeds this figure (e.g., 41,250 bbl/day in Example 7.5), a profit is made. If output falls short of the required capacity, a loss is incurred.

Breakeven analysis yields a determinate solution for oil refiners so so long as the breakeven capacity utilization is known. It thereby yields stable prices, apart from changes in expenses and fluctuating profits. But it is only a technique for arriving at a vague target of a satisfactory level of profit year by year.

REPLACEMENT OF REFINERY ASSETS

Excessive operating costs, and especially increased maintenance costs, indicate that a replacement study is needed. When costs for operating become greater than average costs expected from a new asset, replacement of the old asset is necessary.

Operating and maintenance costs for most refinery equipment tend to increase with age. At the same time, annual capital recovery costs decrease, because greater age spreads the capital costs over a longer period of time.

By totaling maintenance costs and capital recovery costs on a yearly basis for a number of years, a particular year is easily identified at which average annual cost is minimized.

A simple illustration will help explain the matter of timing the replacement of oil equipment.

EXAMPLE 7.6: Timing of Equipment Replacement

A piece of equipment is to be replaced at some time during a 5-year period. The equipment costs $5,000, and operating costs and potential resale prices are as follows:

	Year 1	Year 2	Year 3	Year 4	Year 5
Operating and maintenance					
costs	$550	$875	$1,550	$2,375	$3,100
Resale price	$2,800	$2,000	$1,500	$1,000	$500

The resale price, or realized cash value if sold, is obtained by subtracting the market value of the machine from the original cost each year. Zero interest, or the fact that no interest could be earned on another investment, is assumed.

Average annual costs including maintenance and capital costs, but excluding depreciation, for the purpose of determining when to replace the machine are given in Table 7.6. From Table 7.6 it would appear that the third year would be the last year to use this machine, because average annual costs including operating costs and capital recovery is lowest at $2,158 for this year. After the third year $3,500 will be "lost" on the $5,000 machine investment (at the end of the third year the machine will be worth $5,000 - $3,500 or $1,500) and cumulative maintenance costs for three years will be $2,975, for a total of $6,475 for three years (average annual costs of $2,158).

Setting up such tables for each major asset or classes of assets in the refinery can lead to more efficient replacement of assets at the "right" times.

ECONOMIC BALANCE IN REFINERIES

Economic balance in the refinery is defined in the same way as economic balance in the oil field (see Chapter 6). Basically, economic balance means that costs are balanced with revenue, inputs with outputs, and crudes with refined products. The object is to find the combination of least cost with the "greatest" contribution.

Economic balance may refer to (1) the period before installation of equipment, in which case it consists of a study of costs and values received on design of equipment; or (2) the period after installation of equipment, in which case it is a study of costs and values received on processing operations. The latter means on one side an economic balancing of costs against optimum yield or optimum recovery, and on the other side of elimination of as much waste as possible.

Economic balance applies to both physical operations (unit operations) and chemical conversion processes.

TABLE 7.6

Summary of Replacement of Refinery Assets

Age of machine, in years	Cumulative maintenance and operation costs	Capital costs each year	Total costs	Average annual total costs
1	$550	$2,200	$2,750	$2,750
2	1,425	3,000	4,425	2,213
3	2,975	3,500	6,475	2,158
4	5,350	4,000	9,350	2,338
5	8,450	4,500	12,900	2,580

Economic Balance in Design

Design of equipment for process operations is a complex problem because of the many variables involved and the fact that broad generalizations about these variables cannot be made. Economic balance is not discussed in detail here, as much of it is beyond the scope of this book. A number of cases of economic balance in design, however, will be discussed:

1. Economic balance in evaporation is a problem of determining the most economical number of effects to use in a multiple-effect evaporation operation. There is economy in increasing the amount of steam used because direct costs are reduced, but at the same time there is an increase in fixed costs when an increasing number of effects are used. So selection of which number of effects will balance direct costs is desirable.

2. Economic balance in vessel design may involve specific design problems such as heating and cooling, catalyst distribution, design of pressure vessels for minimum cost, etc.

3. Economic balance in fluid flow involves the study of costs in which such direct costs as power costs for pressure drop and repairs, as well as fixed costs of pipe, fittings, and installation, are related to size of pipe. For example, power costs decrease as pipe size increases, and total costs are at a minimum point at some optimum pipe size.

4. Economic balance in heat transfer requires an understanding of how fixed costs vary with a selected common variable used as a basis for analysis. Variable costs must also be related to this same variable. Thus both fixed costs and variable costs are required for economic balance.

In any study of either design or operations, only the variable costs—often referred to as direct costs—which are affected by variations in operation are included. Those not affected by variations in operations, such as selling expenses and costs of warehousing, inventory charges, royalties, etc., are excluded. And for a study of existing equipment, fixed costs are usually not involved; only operating costs connected with that equipment are considered. Furthermore, all costs that remain essentially constant are usually excluded from any analysis of economic balance in design or operation, since they do not affect the economic balance. They are, of course, included in final costs where a total cost is required.

The following examples illustrate the principle of economic balance in design in a refinery.

EXAMPLE 7.7: Economic Balance in Design [2]

In designing a bubble plate distillation column (Fig. 7.9), the design engineer must calculate (1) the number of plates, (2) the optimum reflux ratio, and (3) the diameter of the column. It is well established that if the reflux ratio is increased from its minimum value, R_m, the number of plates must be decreased to attain the same desired separation. This means lower fixed costs for the column. The other extreme limit for the reflux could be reached by further increase in R with corresponding decrease in the number of trays until the total reflux, R_t, is reached (case of minimum number of trays, N_m). Attention is now directed to the effect on the diameter of the column of increasing the reflux ratio, that is, increasing vapor load.

As R increases, the vapor load inside the column increases; consequently, the diameter of the column must be increased to attain the same

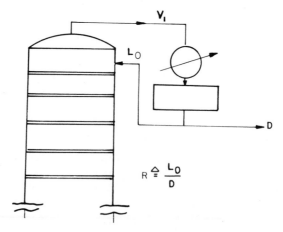

FIG. 7.9. A bubble plate distillation columns.

vapor velocity. A point is reached where the increase in column diameter is more rapid than the decrease in the number of trays. Hence the only way to determine the optimum conditions of reflux ratio that will result in the right number of trays for the corresponding column diameter is to use economic balance. For different variable reflux ratios, the corresponding annual fixed costs and operating costs must be combined and plotted vs. the reflux ratio.

Annual fixed costs are defined as the annual depreciation costs for the column, the reboiler, and the condenser, where the cost of a column for a given diameter equals the cost per plate of this particular diameter times the number of plates. Therefore, the operating cost equals the cost of the steam plus the cost of cooling water.

It is required to find the optimum reflux ratio (moles of liquid returned to the column per mole of distillate product, L_0/D), for a bubble plate distillation column to be designed to handle 700 lb moles/hr of a feed mixture of benzene and toluene. The composition of the feed mixture is 45 mole % benzene.

The feed is liquid at its boiling point. The separation required by this column is 92 mole % of benzene (top product) and 95 mole % of toluene (bottom product). The column operates at atmospheric pressure. The cost of steam is taken at $0.50 per 1,000 lb and that of cooling water as $0.03 per 1,000 gal or $0.036 per 10,000 lb. The sum of costs for piping, insulation, and instrumentation can be estimated to be 60% of the cost of the installed equipment. Annual fixed charges amount to 15% of the total cost for installed equipment, piping, instrumentation, and insulation.

Assuming costs for the installed equipment, including delivery and erection, the final result of calculation is summarized in Table 7.7.

By plotting the annual costs (fixed, operating, and total) vs. the reflux ratio as given in Figure 7.10, it is seen that the minimum annual total cost occurs at a reflux ratio of 1.25, at which value the number of trays in the column should be about 27 (by interpolation and not by exact calculation).

EXAMPLE 7.8: Fixed Costs vs. Operating (Pumping) Costs for
                        Heat Exchangers [3]

When the velocity of a stream in a heat exchanger is increased (but not exceeding 20 ft/sec in order to avoid erosion), the following beneficial results are obtained:

1. The heat transfer rate is improved (higher overall transfer coefficient U).

2. A smaller heat exchanger is required (less surface area).

TABLE 7.7

Summary for Example 7.7

Reflux ratio	No. of plates required	Column diameter, ft.	Annual costs					Total annual costs
			Fixed			Operating		
			Column	Condenser	Reboiler	Cooling water	Steam	
1.14	—	6.7	∞	$1,870	$3,960	$3,780	$44,300	∞
1.2	29	6.8	$8,930	1,910	4,040	5,940	45,500	$66,320
1.3	21	7.0	6,620	1,950	4,130	6,200	47,500	66,400
1.4	18	7.1	5,920	2,000	4,240	6,470	49,600	68,230
1.5	16	7.3	5,490	2,050	4,340	6,740	51,700	70,320
1.7	14	7.7	5,290	2,150	4,540	7,290	55,700	74,970
2.0	13	8.0	5,210	2,280	4,800	8,100	61,800	82,190

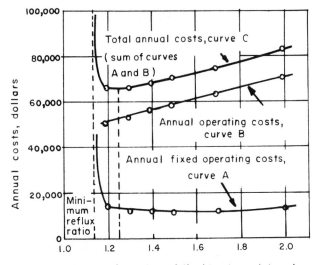

FIG. 7.10. Optimum reflux ratio in distillation columns. [From
Peters, M. S., and Timmerhouse, K. D., Plant Design and Economics
for Chemical Engineers (New York: McGraw-Hill Book Company, 1968.)]

3. Structural costs are reduced (less plot plan area).

4. Increased length of time between cleaning shutdowns (less fouling
   and plugging).

However, this increase in velocity is accompanied by a decrease in
pressure or an increase in ΔP. Consequently, pumping costs increase to
compensate for this pressure drop. For the following example of pressure
drop inside an oil cooler, data for three cases are presented in Table 7.8.

One has to make an economic balance between these prices and ΔP. If
we design our exchanger (cooler) according to case 3, for instance, we get
a saving in the price of the unit totaling $1,800, which, however, might be
offset by the cost of increased pumping due to the higher ΔP.

In practice, this fixed economic operating balance is seldom made,
because the optimum ΔP is high. Operators usually sacrifice this apparent
economy for dependability. Therefore the allowable pressure drop, ΔP
operating, is generally specified, and is not a design variable.

TABLE 7.8

Data for Example 7.8

Case	1	2	3
Shell side ΔP, psi	5	10	15
U, overall heat transfer rate, Btu	80	89	95
Tubular surface area (ft^2)	1,200	–	1,010
Diameter of heat exchange (in.)	26	–	24
Percent less surface required based on case 1	–	–	15
Cost of exchanger	$11,400	–	$9,600

EXAMPLE 7.9: Optimum Temperature of Heat Exchange of Crude Oil
(with Products Prior to Heating Inside a Pipe Still
(Furnace)

It is the practice in oil industry to utilize the heat coming with the products
from a distillation column to preheat the cold oil going into the unit. This
is accomplished by installing heat exchangers. This proposal is not as
simple as it sounds. It could be considered a "mixed blessing" for non-
optimum conditions. The following facts are given:

1. Establishing heat exchangers in conjunction with a furnace will cut
    the amount of heat given in the furnace, which will lead to a small
    furnace heat duty.

2. Following the sequence of events in (1), the flue gases will leave
    the furnace at a higher temperature than is the case without an
    exchanger, which means more loss in heat with the gas.

3. Using heat exchangers for crude oil products will bring savings in
    terms of coolers and cooling water required before storing the
    products.

We must calculate the right amount of heat exchange in terms of surface area and temperature in order to reach optimum conditions of maximum
monetary savings.

One way to treat this case is by suggesting a comparison between
(1) the total cost of absorbing heat in the exchangers and the furnace, and

(2) the cost of removing the heat by cooling using water coolers in the absence of the heat exchangers.

Nelson [4] presents costs for absorbing heat in a 20 million Btu furnace, showing the savings for different cases of heat exchange.

The result could be presented in percentage form as follows:

Heat exchange of crude oil	Annual percent savings based on cost of heating the oil with no exchange*
Up to 300° F	$\dfrac{20,600}{109,000} = 18.9$
Up to 500° F	$\dfrac{39,400}{109,000} = 36$
Up to 550° F	$\dfrac{37,600}{109,000} = 34.4$

*Cost of heating with no exchange = 20 million Btu × 0.665 ($/million Btu) × 24 hr/day × 340 day/year = $109,000.

Eventually the optimum temperature of heat exchange of entering crude oil would be 500° F, where we get maximum saving as indicated in Figure 7.11.

By adding insulation to pipe systems and vessels carrying fluids, it is possible to reduce heat losses. An economic balance must be reached because, by increasing the thickness of insulation, the cost of maintaining the insulation increases, while the cost of heat loss decreases. Eventually, an optimum thickness (economic insulation) is to be calculated which involves maximum savings. The following example illustrates this principle.

EXAMPLE 7.10: Economic Insulation Thickness

It is required to calculate the optimum thickness to choose for a heat exchanger from 1-, 2-, 3-, or 4-in. insulation. The cost of heat can be taken as $0.30 per million Btu. The return in terms of savings to be expected from the capital invested into this insulation is about 15%. Fixed charges are 10%. Given data are summarized in Table 7.9.

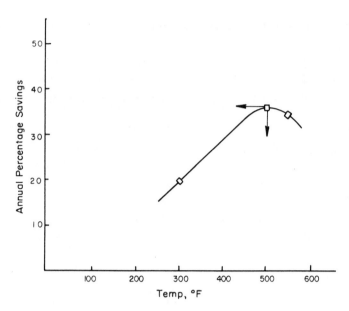

FIG. 7.11. Maximum savings in head exchanging of crude oil.

Solution

For the 1-in. insulation (basis: one year of 300 working days), the money
saved is

(annual total Btu saved) × (cost per Btu) $= 216 \times 10^7 \times 0.3 \times 10^{-6}$

$$= \$648$$

Fixed charges amount to 10% of the fixed capital:

$$\frac{1,200 \times 10}{100} = \$120$$

Therefore net savings is

$$\$648 - \$120 = \$528$$

By similar calculations, we can show that the net savings for the 2-in.
insulation is $596, for the 3-in. insulation is $620, and for the 4-in. insulation is $634.

TABLE 7.9

Data for Example 7.10

| | Thickness of insulation | | | |
	1-in.	2-in.	3-in.	4-in.
Btu/hr savings	$300,000	$350,000	$370,000	$380,000
Capital cost involved for the installed insulation	$1,200	$1,600	$1,800	$1,870

Now, if we take 1 in. as a minimum thickness and compare the return on investment to that for 2-in. insulation, we will find that the increase in capital cost is $400, the increase in savings is $68, and the return on investment for the 2-in. insulation is 17%. The 2-in. insulation is a better choice than the 1-in. insulation.

Now we compare the 2-in. insulation with the 3-in. insulation. By the same reasoning we used above, we find that the return on investment for the 3-in. insulation is 13%. Thus the 2-in. insulation is preferable to the 3-in. insulation.

Finally, we compare the 2-in. to the 4-in. insulation: We find that the return on investment is 14% for the 4-in. insulation. Again, the 2-in. insulation is the preferred choice.

EXAMPLE 7.11: Optimization Technique in Reactor Design

In a refinery, it is required to calculate the size of a reactor, the feed rate, and the conversion for the production of 100 moles/hr of product B. The feed to the reactor consists of a saturated solution of component A of concentration $C_A = 0.1$ g mole/liter. This reaction goes according to the equation $A \rightarrow B$. In addition, it is also required to calculate the cost of producing one unit of the product B.

Given:

1. The rate of reaction of component B is

$$r_B = (0.2 \text{ hr}^{-1}) C_A$$

2. The cost of raw material, component A, at this concentration is

$$P_A = \$0.50/\text{g mole}$$

3. The price of fixed capital of the reactor (including installation and auxiliary equipment) plus the cost of overhead, labor, and depreciation is estimated to be

$$P_R = \$10.01/\text{hr/liter}.$$

Solution

If Q is the volumetric flow rate in liters/min, V is the total volume of the reactor, and S is the space velocity of the reactant inside the reactor, then one can say that

$$Q = (V)(S)$$

For a given volumetric flow rate Q, one can choose larger V with smaller S or vice versa.

Consequently, in terms of economic consideration, one can consider the problem as an optimization problem between high conversion rate using a larger reactor (which involves higher equipment cost, P_R) vs. low conversion rate using a smaller reactor.

Solution is attained through differentiating an expression for the total costs, P_t, with respect to the conversion X_A, and then setting the result equal to zero. This will lead to the optimum conditions, that is, minimum total costs.

$$P_t = VP_R + Q_A(\text{feed rate})P_A$$

Q_A is related to Q_B by

$$Q_B = (Q_A)(X_A) = 100 \text{ moles/hr}$$

Also,

$$V = \frac{Q_A X_A}{kC_A(1 - X_A)}$$

Hence,

$$P_t = \frac{(100)(0.01)}{(0.2)(0.1)(1 - X_A)} + \left(\frac{100}{X_A}\right)(0.5)$$

$$= \left(\frac{50}{1 - X_A}\right) + \frac{50}{X_A}$$

$$\frac{dP_t}{dX_A} = \frac{50}{(1 - X_A)^2} - \frac{50}{X_A^2} = 0$$

Solving for X_A, it is found to be 0.05

Hence, the feed rate of A is

$$A = \frac{Q_B}{X_A}$$

$$= \frac{100}{0.5}$$

$$= 200 \text{ moles/hr}$$

The optimum reactor size, V, is calculated and found to be $10,000$ liters.

$$\text{Cost per unit of the product} = \frac{\text{total costs}}{\text{number of units produced}}$$

$$= \frac{\$200}{100}$$

$$= \$2 \text{ for one g mole of B}$$

Economic Balance in Processing and Production

There are two corollaries of great significance to the oil refiner that follow from the principle of diminishing productivity: namely, the principle of variable proportion, and the principle of least cost combination.

The principle of variable proportion enters into all decisions relative to combining economic factors (inputs) for full production. In chemistry, we know that elements combine in definite proportions. For instance, the combination of 2 atoms of hydrogen with 1 atom of oxygen will produce 1 molecule of water: $H_2 + O \rightarrow H_2O$ water. No other combination of hydrogen atoms and oxygen atoms will produce water. What is true in this instance is also true in all other chemical combinations, and in oil production as well. In other words, a law of definite proportions governs the combination of the various chemical elements and the various factors of production, such as amount of labor and capital in a plant investment.

EXAMPLE 7.12: Economic Balance in Agricultural Yield

Different quantities of nitrogen (commercial fertilizer as capital) were applied to five plots of land of equal size and equal fertility:

1	2	3	4	5
43 lb	86 lb	129 lb	172 lb	None

The resulting yields for the various plots are shown in Table 7.10. Yield diminished for each successive dose of 43 lb of nitrogen. As shown in Table 7.10, the return per 43-lb dose of nitrogen on plots 3 and 4 was almost constant; while the output on plot 5, which received no nitrogen, was much less than for any of the others. The gain on plot 4 over plot 3 was only 5/8 bu/acre, so plot 4 was discontinued after 8 years.

EXAMPLE 7.13: Economic Balance in Refinery Yield

In a refinery, one of the streams produced is a 22° API residue which comes mixed with about 5% by volume of light hydrocarbons (heptanes and heavier). It is required to recover these light hydrocarbons from the heavy residue by steam stripping (see Fig. 7.12).

The amount of light hydrocarbons (HC) removed is a function of the quantity of steam used (rate of steam). Let us consider four cases where the conditions are exactly the same in terms of charge composition, charge rate, etc., except for one variable—the rate of stripping steam (Fig. 7.13).

Let us consider the value of gain for each of the four cases to decide which will give the optimum profit on the basis of the value of the amount of light hydrocarbons recovered and the cost of steam added.

TABLE 7.10

Plots in Order of Yield

Plot	Average annual yield, bu/acre	Gain for 43 lb of nitrogen
5	19	—
1	27-7/8	8.875
2	35-1/2	7.625
3	36-7/8	1.375
4	37-1/2	0.625

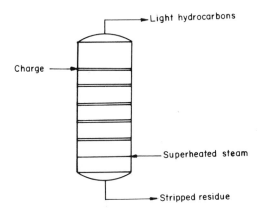

FIG. 7.12. Steam stripper.

The cost of steam added is assumed to be \$0.50/1,000 lb. The cost of value of light hydrocarbons is assumed to be \$3.78/bbl. The data are summarized in Table 7.11. For simplicity this economic analysis is based on the assumption that the feedstock has nearly the same value as the residue coming from the still and that the amount of residue is nearly constant from case to case.

It is obvious that the refinery working under conditions of 6,000 lb/hr of steam is the optimum. Working under a higher rate of steam is not justified by the marginal increase in profit, given the unforeseen (hidden) operating costs due to the necessity for a larger separating vessel (rundown tank) for the condensed water from the hydrocarbons.

EXAMPLE 7.14: Principle of Variable Proportion

Consider two refineries, one with a capacity of 5,000 bbl/day and the other with a capacity of 10,000 bbl/day. Factors of production involved in this problem are labor and capital (equipment). Labor is \$25/unit, capital is \$50/unit. Therefore the ratio of capital to labor is 2 to 1.

The larger refinery (10,000 bbl/day) will use 10 units of capital and 5 units of labor; the 5,000 bbl/day refinery will use 5 units of capital and 10 units of labor. Then for a throughput of, say 5,000 bbl/day, the results are as shown in Table 7.12.

Refinery 2 is working to full capacity, and Refinery 1 at only 50% capacity. This example demonstrates the principle of variable proportion, in which different combinations of labor and capital were used for each refinery. The principle of variable proportion enters into every decision involving combinations of factors of production.

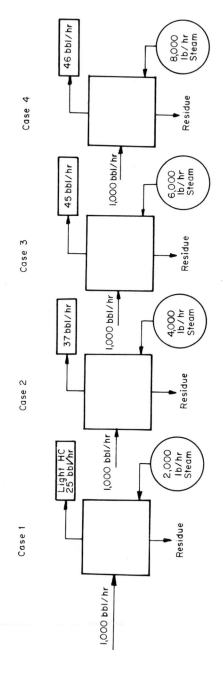

FIG. 7.13. Four cases of steam stripping.

TABLE 7.11

Summary of Data for Example 7.13

Case	Gain of light HC, bbl/hr	Value, $	Quantity of steam, lb/hr	Cost of steam, $	Profit, $
1	25	95	2,000	1.00	94
2	37	140	4,000	2.00	138
3	45	170	6,000	3.00	167
4	46	174	8,000	4.00	170

TABLE 7.12

Data for Example 7.14

	Refinery 1 10,000 bbl/day capacity	Refinery 2 5,000 bbl/day capacity
Capital	$500 (10 units × $50)	$250 (5 units × $50)
Labor	125 (5 units × $25)	250 (10 units × $25)
Total capital-labor costs for 5,000 bbl/day	$625	$500
Cost per barrel	$0.125 (625 ÷ 5,000 bbl)	$0.10 (500 ÷ 5,000 bbl)

Obviously, when the 10,000 bbl/day refinery produces to full capacity, the result will be different—there will be a lower cost per barrel and a higher productivity per day.

EXAMPLE 7.15: Principle of Variable Proportion

Five different-sized refineries with different outputs are compared in Table 7.13 on a cost-per-barrel basis using different combinations of labor and capital.

With capital costs assumed to be $25 per 500 bbl/day and wage rates of labor assumed to be $25/day per man, the 10,000 bbl/day refinery—with a $10 million capital investment operating at 90% capacity, or 9.000 bbl/day—has the lowest cost per barrel at $0.07/bbl. The least-cost combination using the principle of variable proportion indicates 15 men (labor) at a

TABLE 7.13

Output and Cost of Different–Sized Refineries Operating at Different Levels of Capacity

No. of workers and corresponding pro- duction capacity, bbl	Plant cost $500,000, 500 bbl/day capacity	Plant cost $1 million, 1,000 bbl/day capacity	Plant cost $2 million, 2,000 bbl/day capacity	Plant cost $5 million, 5,000 bbl/day capacity	Plant cost $10 million, 10,000 bbl/day capacity
10 men	100	500	1,000	2,000	5,000
12 men	200	700	1,200	2,500	6,000
13 men	400	800	1,500	3,000	7,000
15 men	450	900	1,700	4,000	9,000
20 men (full capacity)	500	1,000	2,000	5,000	10,000
			Costs and Output		
Wages for 10 men	250	250	250	250	250
Fixed capital costs	25	50	100	250	300
Total costs	275	300	350	500	550
Output (bbl)	100	500	1,000	2,000	5,000
Cost/bbl	$2.75	$0.60	$0.35	$0.25	$0.11
Wages for 12 men	300	300	300	300	300
Fixed capital costs	25	50	100	250	300
Total costs	325	350	400	550	600
Output (bbl)	200	700	1,200	2,500	6,000
Cost/bbl	$1.625	$0.35	$0.33	$0.22	$0.10

Wages for 13 men	325	325	325	325	325
Fixed capital costs	25	50	100	250	300
Total costs	350	375	425	575	625
Output (bbl)	400	800	1,500	3,000	7,000
Cost/bbl	$0.875	$0.47	$0.28	$0.19	$0.09
Wages for 15 men	375	375	375	375	375
Fixed capital costs	25	50	100	250	300
Total cost	400	425	475	625	675
Output (bbl)	450	900	1,700	4,000	9,000
Cost/bbl	$0.89	$0.47	$0.28	$0.16	$0.07
Wages for 20 men	500	500	500	500	500
Fixed capital costs	25	50	100	250	300
Total costs	525	550	600	750	800
Output (bbl)	500	1,000	2,000	5,000	10,000
Cost/bbl	$1.05	$0.55	$0.30	$0.15	$0.08

total cost of \$375 and fixed capital costs of \$300. Then total costs of \$675
÷ by 9,000 bbl = \$0.07/bbl.

Actually, this example shows that in all the refineries except one having the 5,000 bbl/day capacity, the costs per barrel increase as full capacity
is attained. This illustrates the basic economic concept of the law of increasing costs with increase in productivity.

Economic Balance in Yield and Recovery

Principles of economic balance must be applied to different processes in
the oil refinery for the purpose of determining how variations in yield, as
affected by design or operation, will produce maximum profit. The effect
of changing the crude feed and refined oil product compositions on the overall profit for a refinery process can best be illustrated, in most cases, as
follows.

Assume a barrel of crude oil of a particular grade to which all other
costs can be realted. Thus if a certain cut of refined oil is worth X dollars
per barrel, the value of one barrel of another grade, expressed as a fraction f, is Xf. The value of X, however, also usually varies with the concentration.

A typical study of economic balance in yield and recovery reveals that
obtaining a higher-grade product from a fixed amount of given feed means
an increase in variable costs because of costs of increased processing. The
final refined oil product, of course, has a higher value; but for some product grades the costs may equal the selling price, with the result that it becomes uneconomical to exceed that particular "specification."

At some optimum grade of a product, however, a maximum gross
profit, or difference between the sales dollars curve and the total costs
curve, may be obtained per barrel of pure material (crude) in the feedstock.

In general, capacity is reduced as grade is increased, with the result
that the maximum profit per barrel of pure material (crude) may not cor-
respond to the maximum annual profit.

Although graphical analysis is the best procedure to use for such problems, there are also some useful mathematical relations. For example, if
D is total refined product, F is total feed (crude), and Y is a conversion
factor relating feed (crude) and product (refined), then, under physical
operations,

$$Y = \frac{D \text{ bbl of total refined product}}{F \text{ bbl of total crude feed}} \quad \text{or} \quad \frac{\text{output}}{\text{input}}$$

or recovery in percent form. Also, if fixed costs are constant for a given process, then fixed costs will be constant for a given value of F or total feed (crude). However, as is usually the case, equipment costs will be higher for a higher-grade product, with the result that the annual fixed costs per unit of refined product increases.

For a given crude feed rate, raw material costs are constant, but refinery processing costs usually increase for a higher-grade product to give a variable cost curve that also increases.

The value of the finished product, like that of fixed costs and variable costs per unit of refined oil, will vary with the grade of product.

Figure 7.14 is a typical economic chart with curves illustrating economic balance curves in a refinery. Recovery, or ratio of output to input, in the oil refinery is greater than recovery in the oil fields. It is to be noted that to make a profit the refiner must stick to the product grades marked between A and B on Figure 7.14.

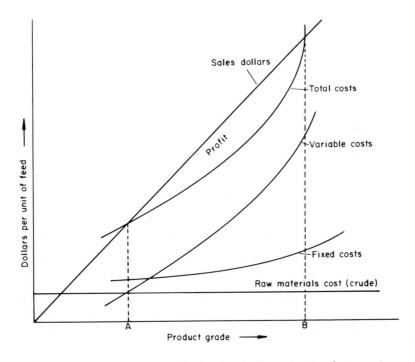

FIG. 7.14.   Economical level of refined oil production from a given crude feed.

EXAMPLE 7.16: Economic Balance in Yield and Recovery

Assume the following facts:

1. Crude oil is processed to refined oil cuts of 25° API (fuel oil),
    40° API (kerosene), and 60° API (gasoline).

2. Refined oil production is 6,000 bbl/day for 25° API, 5,000 bbl/day
    for 40° API, and 3,500 bbl/day for the 60° API grade.

3. Recoveries (where recovery means output as a percentage of re-
    fined oil from crude) are 99% for the 25° API concentrate; 90% for
    the 40° API concentrate, and 75% for the 60° API concentrate.

4. Some refinery equipment is used on all concentrates with fixed
    charges of $1,000/day. (Allocation of this $1,000 fixed cost is
    made to each on a percentage basis of the amount of crude oil
    feed each concentrate is taking.)

5. Variable costs, excepting crude raw materials, are constant at
    $1.50/bbl.

6. Crude feed grade is 38° API and is worth $1.875/bbl.

7. Refined oil prices are $3.125/bbl for 25° API, $4.56/bbl for
    40° API, and $6.25/bbl for the 60° API concentrate.

We want to find the profit per barrel of crude oil and the profit per
barrel of refined products, if production varies inversely as the ratio of
product grade for the higher grades.

The economic analysis is given in Table 7.14. The table may be analyzed as follows:

1. Figures of crude oil required based on recoveries is regarded as
    100%.

2. Total costs include a percentage of fixed costs: for 25° API with
    6,060.6 bbl or 39% of $1,000, for 40° API with 5,555.5 bbl or 31%
    of $1,000, and for 60° API with 4,666.6 bbl or 30% of $1,000. Then
    each concentrate is charged $1.50/bbl for variable costs.

3. If crude is purchased from another source, the figures shown ap-
    ply; if a refinery uses its own crude, the profit per barrel is as
    follows: for 25° API, $1.561; for 40° API, $3.004; and for 60° API
    concentrate, $4,686. The latter is less than if the crude was pur-
    chased.

TABLE 7.14

Analysis for Example 7.16

Product grade	25° API	40° API	60° API
Recovery	99%	90%	75%
Refined oil product	6,000 bbl/day	5,000 bbl/day	3,500 bbl/day
Total costs involved per barrel of crude oil per day	$1.564	$1.556	$1.564
Value per barrel of refined oil	$3.125	$4.560	$6.250
Profit per barrel of crude oil	$0.311	$0.319	$0.311
Profit per barrel of refined oil	$1.250	$2.680	$5.375

ECONOMIC APPRAISAL OF A REFINERY

The most important economic aspects in an appraisal of a refinery are the following:

1. Factors considered in locating a reinery.

2. Capital investment (fixed and working).

3. Refining costs and total product costs.

4. Profit and payout time (economic analysis).

Locating a Refinery

If a refinery is not located in the most economically favorable site, any competitive advantages gained as a result of a careful process research and study can be wiped out. For optimum site location of a refinery, the following (A) primary and (B) specific factors need to be considered.

A. Primary factors apply to the choice of a region:

1. Raw materials (crude oil): Availability and future reserves, distance from proposed location.

2. Markets: Demand vs. distance—growth or decline—storage requirements for marketing centers.

3. Power and fuel supply: Availability of electricity, type and reserve of fuels—costs.

4. Water supply: Quantity and quality—dependability, costs.

5. Climate: Installation costs—humidity and temperature—winds and tornados.

B. Specific factors are decisive for choosing the exact site location within the region:

1. Transportation: railroads—highways—water—pipeline—air.

2. Labor: Availability of skilled labor and rate of wages.

3. Waste disposal.

4. Site characteristics: Structure—room for expansion—costs of land.

Capital Investment

The total capital investment involved in a refinery [5] includes (A) fixed capital and (B) working capital.

A. Fixed capital includes:

1. Installed process equipment (principal items), which is equal to the delivered cost $\times$ 1.43.

2. Process piping—30 to 60% of item 1.

3. Instrumentation—10 to 20% of item 1.

4. Utilities plants and storage facilities—40 to 100% of item 1.

5. Buildings and site preparation—15% of item 1.

6. Electrical auxiliaries—10 to 15% of item 1.

7. Insulation—8% of item 1.

The subtotal fixed capital cost (physical plant cost) is therefore the sum of items 1 to 7. In addition, the costs of engineering and construction generally equal 20% of the subtotal fixed capital cost, and the usual allowance for contingencies is 15% of the subtotal fixed capital cost. Hence the total fixed capital cost is the subtotal fixed capital cost multiplied by 1.35.

Another way of presenting the above data is as follows. If the cost of installed equipment is denoted by EC (equipment cost), then the physical plant cost, PC, is

$$PC = EC \times 2.6$$

where the process piping is taken as 40% of item 1 in the list on page 250, instrumentation is taken as 15% of item 1, utilities and storage is taken as 70% of item 1, electrical auxiliaries are taken as 12% of item 1, and the total fixed capital cost, FC, is [6]

$$FC = PC \times 1.35$$

Comments on Items Comprising Fixed Capital Cost, FC:

1. Installation of equipment usually covers three types of costs: foundations, platforms, and erection.

2. Utilities include installation for generating or supplying steam, electricity, refrigeration, filtered water, softened water, cooling water (cooling towers), compressed air, and inert gases.

3. Electrical auxiliaries includes items such as electrical substations and feeders and wiring.

4. Insulation, needed for piping and equipment.

5. Engineering and construction takes care of costs of design, engineering, plant supervision during construction, and inspection.

6. Contingencies compensate or take care of unexpected expenses due to process changes, cost changes, etc. [7].

B. Working Capital is the funds needed for starting up the refinery, usually determined as the costs of one month's supply of crude oil, plus one month's supply of in-process inventory, plus one month's supply of product inventory.

It is estimated as 15% of the total fixed capital cost (FC).

Refining Costs [8]

Refining costs include variable and direct refining costs, fixed charges, overhead expenses, and general expenses.

A. Direct refining costs include costs of utilities, costs of fuel, operating labor and supervision, maintenance and repairs costs (the latter is usually estimated to be 8% of the capital investment).

B. Fixed charges include depreciation costs, local taxes and insurances (taxes and insurances costs are estimated to be about 3% of the capital investment).

The depreciation costs account for the charges associated with the amortizable investment. One method is to calculate it on fixed annuities on the basis of amortization of 15 years and a certain interest rate, say 8%, on capital outlay. (See Chapter 5 for other methods of calculating depreciation.)

C. Overhead expenses account for employee benefits, medical service, etc., and is estimated to be 50 to 75% of the operating labor and supervision.

D. General expenses account for administration, sales, and research expenses, estimated to be about 10% of sales.

Profit and Payout Time (Economic Analysis)

$$\text{Gross profit or revenue} = \text{annual value of salable products (income)}$$
$$- \text{annual total product costs}$$

$$\text{Net profit} = \text{gross profit} - \text{income taxes}$$

$$\text{Payout time} = \frac{\text{fixed capital investment}}{\text{gross (or net) profit}}$$

EXAMPLE 7.17: Refinery Design

The following example is drawn from a plant design course for Applied Chemical Engineering students at the College of Petroleum & Minerals in which students were asked to design a refinery of a capacity of 30,000 bbl/ day to be built in the area of Al-Azezia, near Al-Khobar, on the Arabian Gulf (see Fig. 7.15). This location was proposed on the assumption that Al-Azezia might develop as an industrial center in this area near Dhahran, since a water desalination plant had already been constructed there and practically all utilities were to be available (fresh water, fuel gas, steam, electricity, etc.). Supply of crude oil and shipping of finished products would be done ultimately through pipelines because of the shallowness of the sea at Azezia. The proposed refinery is to include an atmospheric distillation unit, a vacuum distillation unit, and a catalytic reforming unit. The feed to this reforming unit, which is naphtha, has to be hydrodesulfurized. Crude oil specifications are those of the local Saudi type. The economics of the proposed project are as follows.

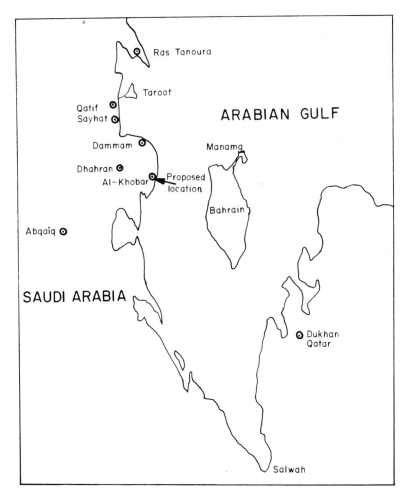

FIG. 7.15. Location of proposed refinery site.

Total Capital Investment:

Process Units	
Unit	1956 construction costs, million $
1. Atm. distillation ⎱ 2. Vac. distillation ⎰	1.500
3. H_2 desulfurizer	0.400
4. Catalytic reforming unit, capacity about 4,000 bbl/day	1.500
Total	3.400

Using the standard cost indices, the total up-dated costs of process units for 1971 are

$$(3.400)\left(\frac{403}{195.3}\right) = \$7.00 \text{ million} \quad \text{(physical plant costs, PC)}$$

These costs are quoted for complete process units, and not for equipment. The costs are to include installation and construction and provide for items listed under capital investment, except for the utilities plants and the storage facilities. However, provision must be made for buildings and site preparation, and engineering and supervision for building the refinery.

The costs of the process equipment alone could be computed as follows:

Physical plant costs, PC, are equal to the equipment costs EC × 1.60, as given previously. For this example, the PC quoted above does not include the costs of the utilities plants and storage facilities, which is estimated to be 75% of the installed process equipment. Therefore:

$$\$7.00 \text{ million} = EC \times (2.6 - 0.75)$$

$$= EC \times 1.85$$

hence,

$$EC = \frac{\$7.00 \text{ million}}{1.85}$$

or cost of the process equipment is \$3.78 million.

Fixed capital can be computed as illustrated below:

FIXED CAPITAL

Item	Percent of process equipment (as given before)	Costs, million dollars
Process units complete and installed	—	7.000
Utilities and storage facilities	70	2.640 (70% of \$3.78)
Buildings and site preparation	15	0.566 (15% of \$3.78)
Physical plant cost		8.206

Then:

Fixed capital cost, FC $= PC \times 1.35$

$\qquad\qquad\qquad\quad = 8.206 \times 10^6 \times 1.35$

$\qquad\qquad\qquad\quad = \$11 \times 10^6$

Working capital cost $= 15\%$ of FC

$\qquad\qquad\qquad\quad = 0.15 \times 11 \times 10^6$

$\qquad\qquad\qquad\quad = \$1.65 \times 10^6$

Total capital investment $= \$12.65$ million

Annual Refining Costs (in millions of dollars):

---

1. Direct refining costs
   Utilities
Steam	$0.380
Electricity	0.380
Water	0.184
Fuel	0.011
Labor and supervision	0.260
Maintenance and repairs	1.100 (10% of fixed capital)
Catalysts	0.062
Total	$2.377

2. Fixed charges
Depreciation	0.733 (fixed capital to be depreciated over 15 years)
Taxes and insurances	0.220 (2% of fixed capital)
Total	$0.953
Subtotal (refining costs)	$3.330

3. Overhead expenses $0.130 (50% of labor and supervision)
   Subtotal refining costs $3.460

4. General expenses $3.437 (10% of the sales)
   Total refining costs $6.897

5. Cost of crude oil
   30,000 bbl/day $\times$ 340
   days/year $\times$ \$2.00/bbl $20.4

Hence, total product costs are

$\qquad$ \$6.897 million
$+$ 20.400 million
$\qquad$ \$27.297 million

Value of Salable Products (Income):

Product	Quantity, bbl/stream day	Selling price, $/bbl	Value (million $)
Gasoline	4,740	6.80	11.00
Light gas oil	4,000	3.44	4.67
Heavy gas oil	3,930	3.30	4.40
Gas oil from vacuum column	6,000	3.35	6.80
Asphalt	12,000	1.90	7.50
Total sales			34.370

Profit and Payout Time

$$\text{Gross profit (revenue)} = \text{income} - \text{total product costs}$$
$$= \$34.370 \text{ million} - \$27.297 \text{ million}$$
$$= \$7.073 \text{ million}$$

Now, from this revenue, a depreciation allowance of 0.733 should first be set aside to give the taxable income as follows:

$$7.073 - 0.733 = \$6.340 \text{ million}$$

If income tax is theoretically deduced (40%), then the net profit is

$$6.340 \times 0.60 = \$3.804 \text{ million}$$

The annual percentage return on fixed capital investment is

$$\frac{\text{Cash flow}}{\text{Capital investment}} \times 100 = \frac{3.804 + 0.733}{11.0} \times 100 = 41.2\%$$

One of the main reasons for having such a high percentage return is the low price of crude oil.

$$\text{Payout time} = \frac{\$12.65 \text{ million}}{\$4.537 \text{ million}} \text{ or } 2.8 \text{ years}$$

$$\text{Cash flow} = \text{profit of } \$3.804 + 0.733 \text{ depreciation}$$

Finally, it could be concluded that building a refinery to process 30,000 bbl/day of crude oil at Al-Khobar would require a fixed capital, PC, of $11.00 million, and a total capital investment of $12.65 million. The total annual refining costs was calculated and found to be $7.073 million. A gross profit of $7.473 million and a payout time of 2.8 years are to be anticipated for this project.

Breakdown Costs of Processing Units in a Refinery on Item-by-Item Cost Estimation

In the previous example, the cost of an entire unit, distillation, reforming, etc., was quoted as a "lump sum." Another way of making a cost estimation for a processing unit is to make a list of the main equipment comprising a unit, find the costs of this equipment, and then apply the "percentage technique" to find the overall cost of the unit, that is, "the installed price," as illustrated in Table 7.15.

Comments on Table 7.15

1. The piping factor is high for the processing units which handle large quantities of liquid or gas for recycling, refluxing, and pumping. This is the case when the processer is primarily a fractionation unit.

The piping factor includes all processers, compressors, fuel gas, air, water, steam, drainage, and relief pipe as well as valves and fittings necessary for a complete plant.

2. The structural material factor includes concrete and supports, ladders, platforms, etc. It includes necessary reinforcing steel and form lumber, and all concrete for equipment foundations, grounding supports, piers, building floors, curbs, pipe trenches, sidewalks, and underground envelope encasing and electrical conduit. The steel includes supports, ladders, safety cages, stairs, platforms, walkways, grating, hand rails, bracing brackets, clips, bearing plates, anchor bolts, etc.

3. The electrical material factor includes conduit, wire, conduit and wire fittings, lighting fixtures, receptacles, lighting transformers. It does not include starters, push buttons, or lighting panels, which are all part of the principal items.

4. The insulation factor includes the insulation of vessels, insulation for pipes, and internal insulation for reactors. But steel lining, a part of the fabricated vessel, is included with the principal items.

5. The painting factor involves the painting of all structural steel building interiors where needed and painting of all uninsulated vessels, equipment, and piping.

6. The "miscellaneous" factor is a "catch-all," including crushed stone for paved areas, expansion joint materials for floors for curbing for foundations and for sidewalks, tie wires, etc.

TABLE 7.15

Breakdown of the Percentages of the Costs of Processing Units in a Refinery

Items	Crude vacuum unit	Gas plant	Catalytic cracking unit	Catalytic Polymerization unit	Catalytic reforming unit	Alkylation unit
Piping	27.1	29.0	28.2	22.0	16.0	29.0
Structural concrete	5.9	6.7	6.1	6.3	3.5	3.6
Steel supports, ladders, platforms	2.1	3.4	3.1	1.8	1.1	2.2
Electrical	2.3	5.0	4.5	5.3	3.6	2.5
Insulation	15.8	8.3	14.6	7.4	6.3	8.0
Painting	1.5	1.5	1.6	1.7	1.2	1.8
Miscellaneous	0.5	0.5	0.5	0.6	0.5	0.6
Total materials above	55.2	54.5	58.6	45.1	32.2	47.7
Total materials above + direct materials	155.2	154.4	158.6	145.1	132.2	147.7
Total direct labor	39.8	36.1	39.5	28.5	26.0	40.0
Total indirect costs	23.8	23.9	24.0	23.4	18.0	25.1
Total direct and indirect costs of installed units	218.8	214.4	222.1	197.0	176.2	212.8

7. Direct field labor includes all manual and craft labor, as well as supervisory labor on structural steel and equipment erection, field pipe fabrication, inspecting, and testing labor.

8. Indirect costs include labor of field engineers, accountants, field clerks, and other field personnel, and materials such as small tools, construction supplies, field office supplies, telephone and telegraph costs, travel and subsistence costs of supervisory field personnel, and such costs as workmen's compensation and insurance.

9. Engineering, overhead, and fee contingency include preliminary and final working drawings, flow diagrams, etc., and specifications, bills of materials, engineering layouts, and plot plans as done by the engineering and consulting firms.

This method can best be demonstrated by an example.

EXAMPLE 7.18: Item-by-Item Cost Estimation

Suppose that the cost of a catalytic reforming unit (capacity 3,000 bbl/day) is required.

From a flow diagram of a catalytic reforming unit (Fig. 7.16), a list of needed principal items of equipment can be made and classified as to type of equipment. The following is an overall possible list of principal items of equipment for a catalytic reforming unit.

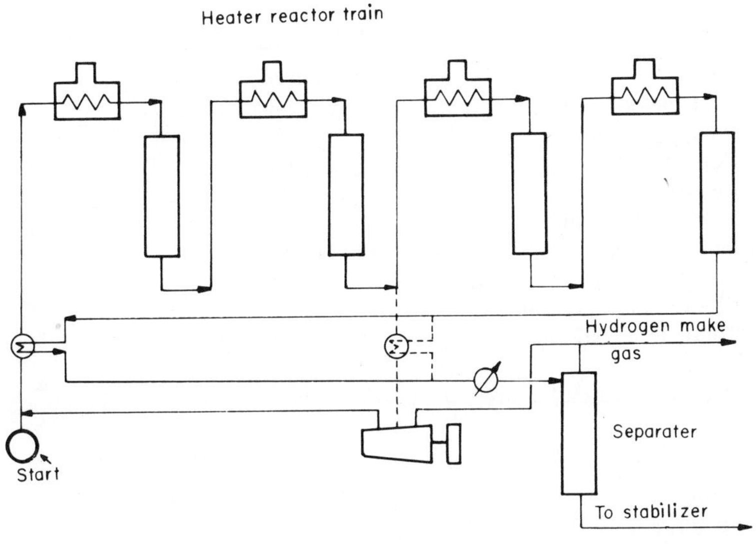

FIG. 7.16. Flow diagram of a catalytic reforming unit.

1. Vessels—columns, towers, tanks, drums, internal linings, trays, caps, etc.

2. Heat exchanger—shell and tube, finned tube, double pipe, pipe coils, atmospheric sections, air-cooled sections, including driver.

3. Pumps—including drivers, such as motors, turbines, and engines. Centrifugal, turbine, rotary, reciprocating and power.

4. Compressors, blowers, and fans, including drivers.

5. Heaters (furnaces).

6. Cooling towers—atmospheric, forced draft, induced draft, including fans and drivers.

7. Mechanical, miscellaneous—elevators, cranes, conveyors, cyclones, ejectors, eductors, etc.

8. Instruments—temperature, pressure, flow indicators, recorders, and controller instruments, instrument panel boards.

9. Structures—main supporting structures for equipment.

10. Electrical equipment—starters, push buttons, etc.

The result of establishing principal items of equipment for the catalytic reforming unit is next illustrated.

The principal items needed and their costs as given by manufacturers of such equipment are as follows:

Item	Cost (1956)
Vessels	$47,000
Heat exchangers	101,000
Pumps	8,000
Compressors	45,000
Heater	59,000
Misc.—Mechanical	1,000
Instruments	17,000
Structures	3,000
Electrical equipment	4,000
Total principal items	$288,000

"Other" materials, applying percentages given in Table 7.15:

Item	Cost	Percentage
Piping	$46,000	16
Structural:		
Concrete	10,000	3.5
Supports, ladders,		
platforms	3,000	1.1

Item	Cost	Percentage
Electrical	$10,000	3.6
Insulation	18,000	6.3
Painting	3,000	1.2
Miscellaneous	1,000	0.5
Total "other" materials		32.2 = $91,000
Total direct material (principal items + other materials)	$379,000	61
Total direct field labor, 26%	75,000	12
Total indirect costs, 18%	52,000	4.5
Engineering, overhead fee, and contingency costs, 23%	116,000	
Total job cost (installed)	$622,000	

The updated costs of this unit will be

$$(622,000) \times \frac{380}{190} = \$1,244,000 \text{ (using a Nelson cost index [9])}.$$

More on Investment in Capital Equipment in Refineries

New Equipment. An estimate of investment for new equipment is necessary in an economic analysis to establish the estimated operating cost of the new equipment. Also, since the economic problem involved in installing new equipment is generally to estimate the rate of return on investment in years, or as a percent per year, it is necessary to know the new equipment investment required.

Existing Equipment. Investment for existing equipment affects total operating costs of existing equipment because of maintenance and repair, with depreciation and taxes being included in total operating costs.

Thus, in preparing an economic analysis on investments both present and future, equipment operating costs, equipment investment, raw material costs (prices of crudes), product values (refined oil product prices), etc., must all be considered.

Fuel for Refineries

How much fuel does a refinery use? It is estimated that the amount of heat needed in the processing of oil varies between 555,000 and 700,000 Btu/bbl of crude.

If crude oil has a heating value of 6 million Btu/bbl, the above figures indicate the equivalent of 9.2 to 11.7% of the crude oil. The heat requirements include the burning of coke from catalyst as well as such common fuels as refinery gas, natural gas, residual fuel oil (pitch), and acid sludge and coal.

Oil fuel	13.7%
Coal	1.2%
Natural gas	37.4%
Refinery gas	43.0%
Coke (catalyst)	4.2%
Acid sludge	0.5%
	100.0%

Fuel is used for (1) sensible heat and latent heat in direct-fired heaters, (2) generating steam and electricity, (3) gas and diesel engines, and (4) heating catalyst or solid material in the catalytic cracking processes.

Not more than 65% of the heat, or about 4,500,000 Btu, actually enters the barrel of oil, the remaining 35%, or about 1,575,000 Btu, of heat goes up to the stack and out, or remains as chemical energy in carbon monoxide (in regenerator flue gas).

PLANNING REFINERY OUTPUT AND SHIPPING

Estimating the future demand for refined products is important and vital in the petroleum industry. Since oil refining has been developed to such a degree that crude petroleum is processed into many different products, it is essential to have long runs of the principal types of fuels to keep costs low. This is especially true in large refineries with heavy investments.

To stop one process, change the equipment to produce a different mix, and start up again can become very expensive if there are many process changes within a short period of time. Yet competition in the oil industry is so keen and refining processes so complex that production estimates must be carefully determined—output must be enough to meet customer demand but not so much that extra storage tanks will be needed.

Oil management cannot afford to guess or to rely entirely on previous years' figures. But statistics are needed as a basis for making estimates of consumer requirements.

An oversimplified example illustrating the importance of predicting annual demand for fuel oil, before production, so that efficient production plans can be made, is given as follows.

EXAMPLE 7.19: Predicting Demand for Fuel Oil

The problem is to project estimates of annual demand for fuel oil for the years 1975-1982 from the known demand for 1967-1974. Here we are using demand for the past 8 years, but any number of past years might be used.

Table 7.16 shows the demand for distillate fuel oil, the kind used in home heating, for the years 1967-1974 inclusive. In order to plan refinery production, we must predict the demand for the years ahead. The figures in Table 7.16 show that while the demand has continued generally upward, the rate of increase has varied greatly.

Normally, the first step in graphical analysis would be to mark off the x (horizontal) axis in years on a coordinate grid. This step can be simplified in calculations by calling the midpoint zero and marking the time in years as plus (after) and minus (before) that midpoint date.

These calibrations are shown in Figure 7.17, with the midpoint being between years 1970 and 1971. Thus, all years to the left of zero, or between 1967 and 1970, have minus numbers, and all years to the right of zero have plus numbers.

The year 1971 is set at +0.5 rather than at zero because the number of observations, or years, is 8 and is an even number. When the number of observations is even, there is no single middle observation; thus a point midway between the two central figures, 1970 and 1971, is selected as the midpoint and is zero. If the number of observations was odd, the middle year would be zero and all other numbers would be plus or minus an integer.

TABLE 7.16

Total Domestic Demand for Distillate Fuel Oil,
in million bbl

Year	Barrels
1967	226
1968	243
1969	298
1970	341
1971	329
1972	395
1973	447
1974	477

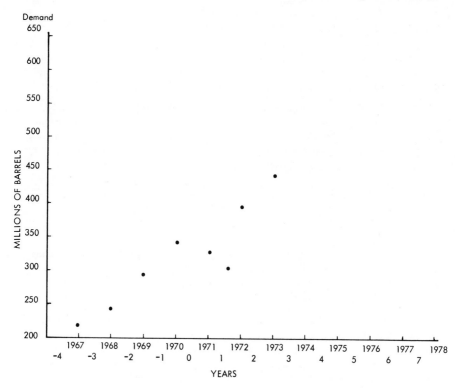

FIG. 7.17. Coordinate grid for Example 7.19.

Using the data from Table 7.16, the next step is to find the regression or trend line. The method of least squares to compute a and b for the line $y = ax + b$. By referring to the expressions for a and b, it is clear that the following will need to be found:

$$\sum_{i=1}^{8} x_i, \quad \sum_{i=1}^{8} y_i, \quad \sum_{i=1}^{8} x_i^2, \quad \text{and} \quad \sum_{i=1}^{8} x_i y_i$$

Table 7.17 presents data for the trend line, as plotted in Figure 7.18, by the method of least squares. What does the trend line predict for 1975? All we need to do to answer this question is substitute the proper number for x. If 1974 corresponds to x = 3.5, then the x that corresponds to 1975 is 4.5. Reading up vertically at 4.5 to the trend line and reading across to the y vertical, or demand, gives a predicted demand in 1975 of 500,000 bbl of fuel oil.

In 1976, the trend predicts about 540,000 bbl.

TABLE 7.17

Calculation of Trend Line for the Data of Table 7.16
by the Method of Least Squares

Year	i	x_i	y_i	x_i^2	$x_i y_i$
1967	1	-3.5	226	12.25	-791
1968	2	-2.5	243	6.25	-607.5
1969	3	-1.5	298	2.25	-447
1970	4	-0.5	341	0.25	-170.5
1971	5	0.5	329	0.25	164.5
1972	6	1.5	395	2.25	592.5
1973	7	2.5	447	6.25	1117.5
1974	8	3.5	477	12.25	1669.5
		0	2756	42.00	1528.0

$$a = \frac{8(1,528) - 0}{8(42) - 0} = \frac{1,528}{42} = 36.4$$

$$b = \frac{2,756}{8} - 36.4\frac{(0)}{(8)} = 344.5$$

From these computations, since $y = ax + b$, the trend line is

$$y = 36.4x + 344.5$$

Planning refinery output and shipping involves, basically, the balancing of refinery operations with overall product demand and with tanker scheduling. Tanker schedules are usually received in the refinery by the 20th of each month for the following month. A tanker schedule usually shows the names of the tanker and tanker company, the capacity of the tanker in deadweight tons, the date of arrival, the destination, and the product or grades of crude oil to be carried. Once the tanker schedule is received, plans for the crude and for the refining of the crude to meet these shipping dates can commence.

Most refineries have a product slate which is a list of products that the particular refinery will produce. Product slates of small refineries are naturally smaller than those of large-size refineries. The latter may have a slate showing anywhere from 15 to 20 different products, ranging from naphthas to asphalts, including a number of grades for each product.

Refined products have certain crude equivalents; for example, 1 bbl of crude will yield approximately 0.2 bbl of naphtha, 0.12 bbl of kerosene,

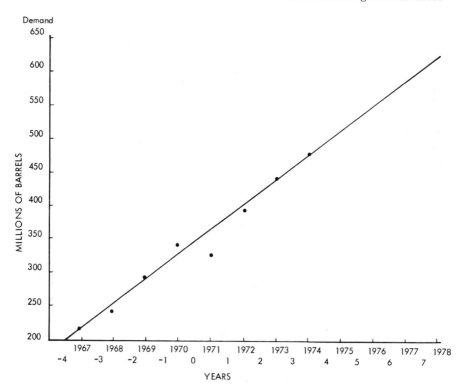

FIG. 7.18.   Trend line for Example 7.19.

0.15 bbl of diesel, and 0.5 bbl of fuel oil, 0.03% representing gas and
losses. Processing is to a great extent a "matching" procedure. The type
of crude used influences the finished products produced. A refinery might
get, for instance, four cuts from crude runs, such as naphtha, kerosene,
diesel, and fuel oil. These then become basic stocks which are blended,
and sometimes treated, to get the desired finished products.

Refinery management determines crude (input) to run. This might be,
for example, 600,000 bbl/day, but it might include 300,000 bbl of light,
250,000 bbl of heavy, and 50,000 bbl of medium crude. Figures and data of
crude input and of product (output) requirements are then fed into a computer
to determine maximum yield and adherence to grade specifications of each
product to be produced from the crude input. Results are forwarded to the
sales department. If there is any deviation from the customer order, it is
up to sales to contact the oil product buyer. The refinery attempts to meet
specifications of buyers as much as possible, but production is governed by
the mixture or type of crude involved. The refinery does work with some

tolerance or range on grade specifications. In the Middle East, the tolerances are greater than in the United States, where the buyer visits and usually gets the grades he wants.

In other words, demand says that this is what we want. Supply says that this is the crude we have and then proceeds to determine just how well we can match crude input with the requirements of demand. Thus, the refinery computes "product mix," in which maximum yield and quality is determined by computer, before the refinery run. The refinery takes this information as given by the computer and assigns crudes to various refinery crude processing units in terms of basis unit capacity and product yield capability.

These, essentially, are the steps in planning crude processing on a monthly basis. Next, the planning of day-to-day operations becomes imperative.

A schedule of tankers, noting their time of port arrival, is prepared and carefully observed day by day and hour by hour.

Operating orders, pumping orders, blending orders, and tank assignment sheets are all sent out to the various departments involved. Operating orders and blending orders are given to refinery units, pumping orders go to pump stations, and tank assignment sheets are sent to the terminal, as well as to the laboratory, since the laboratory must analyze the blending to determine if the processed products meet the required grade specifications. So the refinery and terminal know what is needed in terms of which tanks will be available and which tanks will be tested, etc.

Thus, the processing procedure is generally one in which the already-prepared base stocks, whether naphtha, kerosene, diesel, or fuel oil, are blended in refinery tanks. It may include some downstream treatment such as hydroforming, reforming, and/or sweetening, or the mix may be blended downstream in the refinery tanks on the way to the ship terminal.

After the refined products have passed laboratory inspection, the terminal is notified. This allows the terminal to plan for loading of tankers upon their arrival in port.

Everything must work on a close time schedule. Tankers must not be allowed to wait too long for their oil load. This is economically bad, since awaiting time costs money. Ships must be kept moving with a minimum amount of cargo queue time and berth occupancy time. Basically, the ideal condition at the terminal is to have the order ready when the tanker moves into its terminal berth for loading, so that little or no awaiting time is involved. Success depends on both available tank capacity and regular arrival of tankers at scheduled times. When orders are not ready, or when tankers do not arrive as scheduled, cutbacks in production and processing can result at the oil plant.

CONCLUSIONS

Thus the modern oil refinery is a chemical blending and compounding plant which attempts to fulfill requirements given the petroleum industry by consumers for various products.

Many of the refined products found on the market are made by blending stocks together to meet particular industrial requirements. As an example, the various grades of motor oil and machine oil are prepared by compounding different amounts and grades of neutral stocks and bright stock. Neutral oils might be of the type of nonviscous sewing machine oil, and bright stock might be heavy airplane oils. At any rate, although it is theoretically possible to produce any type of refined product from any crude oil, it is not usually economically feasible to do so. But sometimes the demand for a particular product or the dropping of a low-value product becomes so prevalent, or necessary, that it requires a major conversion of a chemical nature.

In keeping abreast of the demand for greater efficiencies, lower costs, and improved products caused by increased competition, modern armaments, and rapid advancements in the automation industries, various petroleum refining processes are used. Among these are thermal or catalytic cracking to increase the yield of gasoline from crudes, a catalytic cracking process to reduce the yield of residual fuel oil, polymerization to recover olefinic waste gases, solvent extraction processes by which unsuitable hydrocarbons are eliminated from lubricating oils, diesel fuels, etc., and catalytic refining to produce superior motor oils and fuels. So it is up to the refinery to get the best possible and most economical combination of refined oil products from each crude without forgetting that each crude must be processed to supply its own particular market environment.

As Figure 7.19 illustrates, the first step in a great number of complicated steps in the refining process is the breaking down of the crude into fractional parts. A number of steel tubes of special quality are connected within a firebrick-lined furnace so that crude oil enters at one end and flows back and forth through the furnace. This is often known as a pipe still. Crude oil enters at one end of the pipe still at a low temperature and emerges at a temperature of 700° F or more. The hot crude is discharged about one-fourth of the way up from the bottom of a fractionating tower. The tower is usually 75 ft or more in height and about 4 to 6 ft in diameter.

A number of steel plates, located about 3 ft apart in the tower, baffle and break up the rising stream of vaporized crude oil. The tower acts like a large nest of test sieves to separate various parts of the crude oil. The "cuts" are as follows.

Gasoline is tapped off of the top at about 100° F, naphtha is taken off lower on the tower, kerosene is withdrawn still lower on the tower at about

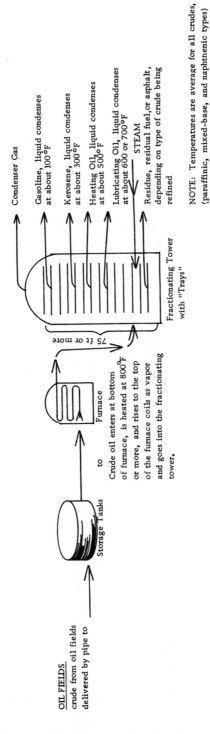

Condenser Gas

Gasoline, liquid condenses at about 100°F

Kerosene, liquid condenses at about 300°F

Heating Oil, liquid condenses at about 500°F

Lubricating Oil, liquid condenses at about 600 or 700°F

STEAM

Residue, residual fuel, or asphalt, depending on type of crude being refined

NOTE: Temperatures are average for all crudes, (paraffinic, mixed-base, and naphthenic types) as reported by API.

Fractionating Tower with "Trays"

75 ft or more

to   Furnace

Crude oil enters at bottom of furnace, is heated at 800°F or more, and rises to the top of the furnace coils as vapor and goes into the fractionating tower.

Storage Tanks

OIL FIELDS
crude from oil fields delivered by pipe to

FIG. 7.19.   Intermediate refined products.

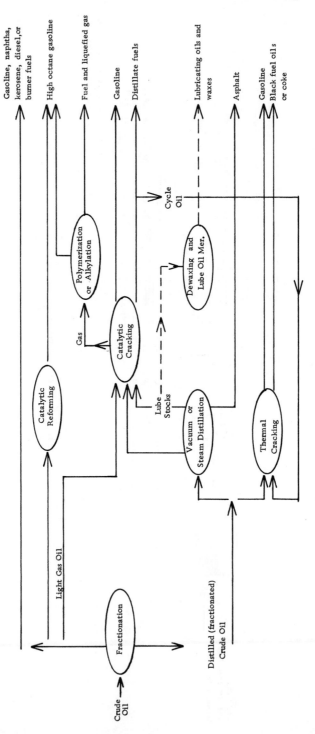

FIG. 7.20. Flow of basic refinery operations.

300° F, and so on down the tower. Heavy residues are taken from the bottom, and in this cut is found the paraffin or asphalt from which the crude type derived its name.

The latest crude distillation units are designed as three-stage units, including a prefractionator, an atmospheric tower, and a vacuum tower with side units leading from these to handle waste products. As crude oil with a larger percentage of impurities occurs, the crude must be further refined to make marketable products. This, of course, increases refinery costs, and makes selling prices of products higher.

The cracking process, along with other steps of processing, follows distillation processing as shown in Figure 7.20. Cracking usually refers to a combined operation of topping and thermal cracking, and in most refineries the gas oil is catalytically cracked. No lubricating oils are produced in cracking processing. Catalytic cracking results in smaller yields of residual fuel oil. But by thermal cracking, the total yield of gasoline may be increased from about 20 to 30% by simple topping (or fractionating) to as much as 70% by topping and cracking.

In lubricating oil processing (or topping with the manufacture of lubricants from the residue of the crude oil), paraffin-base crude oils are always processed for lubricants because paraffin-base crude oils are particularly suited for the manufacture of lubricating oils.

Finally, most large refiners practice topping, viscosity breaking (thermal cracking), catalytic cracking, catalytic reforming, and lubricant manufacture simultaneously in an enormously complicated process.

NOTES

1. Modified from Peters, M. S., and Timmerhouse, K. D., Plant Design and Economics for Chemical Engineers (New York: McGraw-Hill Book Company, 1968).

2. Modified from Peters and Timmerhouse, Plant Design and Economics.

3. Baker, William J., Petroleum Refiner, July 1959.

4. Nelson, R., Petroleum Refinery Engineering, 4th ed. (New York: McGraw-Hill Book Company, 1949).

5. Vilbrandt, F. C., and Dryden, C. E., Chemical Engineering Plant Design, 4th ed. (New York: McGraw-Hill Book Company, 1959), p. 194.

6. Ibid., p. 194.

7. Summation of items 1 to 7 in the list on p. 250.

8. The difference between refining costs and total product costs in the cost of crude oil.

9. Nelson, R., Petroleum Refinery Engineering.

Chapter 8

MARKETING OIL AND OIL PRODUCTS: PRICING

INTRODUCTION

Price is one of several elements which determine profits in the oil industry. Oil industry earnings depend not only on the prices of products, but also on the volume of sales and the extent operating expenses.

We have discussed production volume, costs, and operating expenses at some length in previous chapters. We turn now to the pricing of crude oil and refined oil products. The prices of crudes and oil products are exceedingly important in determining the behavior of both buyers and sellers, each in their own interest. Falling crude prices are generally in the interest of buyers, who in this case are generally independent refiners. Rising crude prices favor the sellers of crude oils. So the oil price mechanism has to satisfy the divergent interests of both buyers and sellers. It is the means by which buyers and sellers, whose economic gains lie in opposing directions, are brought together.

The supply of crudes and of refined oil products and the demand for crudes and for refined oil products affect price setting in the oil industry. But the demand and supply of both crudes and refined oil products reflect the interdependence of price and quantity as well. For example, the willingness of buyers to purchase more crude from a particular source depends somewhat on price; the buyer of crude expects a price discount or lower price as quantities purchased increase. Similarly, when demand exceeds supply, sellers of crude expect a premium over the "going price" as more buyers bid for limited supplies.

HISTORICAL BACKGROUND

Before World War II, international trade in crude oil was insignificant. Also, most petroleum exports were refined at or near the source of crude. The United States was an important supplier in almost all markets. Prices of refined oil products everywhere tended to equal f.o.b. prices from the United States plus transportation costs. However, since price elasticity of demand differed in different markets, there were some deviations from these prices, depending on the conditions of competition in particular markets. Also, prices charged to affiliates of integrated companies were not always the same as prices charged to independent buyers.

But because the United States was the marginal supplier to most world markets, products were priced f.o.b. from the U.S. coast of the Gulf of Mexico plus the cost of transportation. The prices so calculated were known as "world parity prices for oil," and the Gulf area of the United States was the "basing point."

The system grew out of the underlying comparative cost condition of the industry and not out of monopolistic restrictions in the market. Even under highly competitive conditions, low-cost oil products from the Arabian Gulf were not sold in nearby markets, such as India, at prices less than those of similar products from the United States as long as the United States was the big supplier of market demand. Such a common recognized formula for calculating prices was adopted by all sellers, and the system was thus institutionalized and turned into a restrictive "basing-point" system. This pricing system not only ensured uniform prices by all sellers, but also ensured that low-cost producers (like the Middle East) could not use their lower costs to expand their share of the market by reducing prices. The basing-point system was restrictive when adhered to by all sellers, since it prevented the expansion of low-cost production and price competition. Competition would, in fact, have produced uniform delivered prices in any given market. So the big oil companies recognized parity prices as evidence of the existence of competition.

When and to what extent the basing-point method of pricing became more or less formally adopted by the oil companies is not clear. It is clear, however, that delivered parity prices played an important role in the pricing policies of each company. There were often informal meetings of representatives of the large companies to ensure collusive adherence to the system. However, the basing-point system actually protected the Eastern Hemisphere: Under effective price competition the production and use of low-cost oil from the Eastern Hemisphere would have increased to the point where no further expansion of output would have been profitable at prevailing prices.

The basing-point system, however, was not wholly effective. It failed to provide scope for expansion of markets for new sources of supply. Price wars resulted, with new sources of supply breaking away from the system.

Finally, as an increasing proportion of Eastern Hemisphere markets came to be supplied from the Middle East, delivered prices of crudes and oil products fell in these markets. Also, f.o.b. prices in the Middle East fell relatively to those from the U.S. Gulf Coast, since the marginal costs of additional Middle East oil were much lower than those prevailing for the relevant rates of supply in the United States.

Thus the decline in delivered prices for oil products from the Middle East broke the hold on prices of the major companies, eventually leading to the establishment of a second basing point, the Arabian Gulf, for fuel oil and later for crude oil. However, governments in the Middle East accepted the demand of the oil companies for equalization of f.o.b. prices Arabian Gulf with f.o.b. prices from the U.S. Gulf Coast.

In reality, two basing points, the Arabian Gulf and the U.S. Gulf Coast, were established. The adoption of two basing points meant that delivered prices could be calculated on the basis of f.o.b. prices plus a standard freight from either the Arabian Gulf or the U.S. Gulf Coast to any given destination. Therefore, there was an area in which prices f.o.b. Arabian Gulf were less than prices f.o.b. U.S. Gulf Coast. A boundary was needed at which delivered prices from the two areas could be set. The two-basing-point system, as such, continued into the 1950s for crude and into the 1960s for refined oil.

The amount of crude oil entering international trade rose rapidly after World War II, and the relative important of trade in refined oil declined, since there was now a discernible movement toward establishing refineries in consuming areas. In consequence, the price of crude began to take on more significance from the point of view of both exporting and importing countries, although most refineries were still owned by the integrated oil companies producing the crude oil, and the price of crude was still an internal transfer price.

In 1948, the relative decline of Middle East oil prices intensified. Also, by this time the Caribbean had emerged as the Western Hemisphere basing point, since the United States had become a net importer and Venezuela supplied far more oil to Europe than did the United States.

The relationship between Caribbean, U.S. Gulf Coast, and Arabian Gulf prices which had been established in 1948-1949 negotiations between the large oil companies and the Economic Cooperation Administration (ECA) remained roughly the same until after the Korean War.

Thus, prices in the oil industry were determined by supply and demand, but not in the same way that prices are determined by competition, since, producers of crudes restricted supply in order to maintain the price of crudes.

The influence of competition on prices may be seen by the extent to which prices tend to approach the long-run costs of production, and not so much discrimination among different markets. In the controversy over oil prices today, the issue is not whether the major oil companies have control over prices, but the fact that each of the companies has always attempted to increase its share of the market as far as possible by means other than price competition. Supply has been restricted by these companies, but not by the unprofitability of selling more at existing price levels. Supply has been restricted by each company's estimates of the amount it could sell without initiating increased price competition in either crude or refined product markets. The condition of restricting supply is beginning to change, however, as producing countries have become owners or part owners of oil companies and demand has rapidly escalated.

OPEC AND THE TEHERAN AGREEMENT [1]

The year 1970-1971 was one of the most successful years in the history of the Organization of Petroleum and Exporting Countries (OPEC), which was established in 1960 primarily to coordinate oil policies so as to obviate the deflationary trend in crude oil prices and to restore the prices of crude oil of 1958 which had registered a drastic reduction in 1959 and 1960. After its establishment, OPEC was able generally to maintain oil prices at the level of 1960 and to arrest the declining trend. However, 10 years after its establishment, it has been remarkably successful in achieving its objectives by a united stand in the face of the oil companies.

The success of OPEC members in 1970-1971 is reflected in the Teheran Agreement, which was signed on February 14, 1971, by the six governments of Saudi Arabia, Iran, Iraq, Kuwait, Abu Dhabi, and Qatar, on the one hand, and 22 international oil companies on the other. The agreement provides for an increase of 33 ¢/bbl in posted prices of 40° API crude in the Arabian Gulf, plus 2 ¢/bbl in settlement of freight disparities. For crude oil of gravity less than 40° API, a reduction of only 1.5 ¢/bbl per degree differential became applicable instead of the previous 2 ¢/bbl. The agreement also provided for the elimination of the existing percentage allowance, the gravity allowance, and the marketing allowance as of the effective date of the agreement. To compensate for worldwide price inflation, the two negotiating parties agreed on an annual increase of 2.5% per annum in the posted prices of crude oil in addition to an annual increase of 5 ¢/bbl in these prices, beginning the first of June 1971, to compensate for future rises in the prices of refined products. The agreement was expected to increase the per-barrel receipts of the governments concerned by about 30 ¢ in 1971, rising to about 50 ¢ in 1975. It was also expected that the six governments of the Arabian Gulf would get an additional revenue of about $1.2 billion in 1971, rising to about $3 billion in 1975 assuming an average annual growth of 10% in the oil exports of the Gulf area.

POSTED PRICES

A posted price system has been the basis today for prices of crude and refined oil products, although actual prices (realized prices) have been able
to vary and usually have varied from posted prices in oil trading. Formal
posting of prices did not start until 1956. Before 1950, prices were estimated using published data.

Oil companies, including the large integrated companies in the Middle
East, "posted" or published prices for both crude and refined oil products.
Price quotations for crudes were f.o.b. well head, so costs of gathering
and transportation to destination were not included. Price quotations for
refined oil products were f.o.b. refinery. Under this posted price system,
each oil company posted its crude oil prices and, with some exceptions,
all refinery producers received the same prices for the same grade of
crude, regardless of the volume of oil sold.

Unlike most commodities, crude oil was sold on the basis of buyer-
set prices. This was due to the fact that crude has a derived demand in
which its entire demand depends on the demand of the refineries, who are
the buyers of crude. When demand for refined oil products drops, demand
for crude also drops, and vice versa for increases in demand for refined
oil products.

Posted prices have actually been "posted," or published by buyers,
who are the refineries, and the large integrated oil companies, which now
also include the participating oil-producing countries who are part-owners
in the products of the crudes and also in the operation of the refineries. In
substance, the posted prices acted as a means of notifying crude oil producers of the prices at which buyers will take the oil. Also, these posted
prices acted as a basis for determining income tax and as a basis on which
royalty interest was settled. Furthermore, the posted price system also
provided the means of informing customers of oil and of oil products, as
well as crude oil producers, of the terms on which oil and oil products
could be purchased. So most sales of crudes and sales of oil products by
refineries to refineries were made at prices varying from posted prices.

Large orders of both oil and oil products have always been sold at
discounts of posted price. Thus, the realized price, the market selling
price, has or has not necessarily been the posted price. For example,
projected posted prices may have been $2.30 and $3.00/bbl for two different types of crude oil, but realized prices may have been $1.61 to $1.84
and $2.10 to $2.40, respectively. The realized prices were thus based on
discounts of 20 to 30%.

Cash discounts alone could range from 10¢ to 40¢. Also, such considerations as quantities purchased, long-term commitments on the part

of the buyer, the location of the buyer, etc. , could lower the realized price from the posted price. There were also some nonprice considerations affecting market price, such as unusually favorable tanker rates, special refinery processing deals, easy credit terms, etc. These could effect a lower realized price from posted price.

Discounts from posted prices for long-term commitments and large orders were common and could be justified, since large orders usually meant lower handling costs and reduction of risks for the oil-producing company, whether it was the selling of crude or refined oil products. Discounts from posted prices based on large quantity purchase of crude have been defensible, since realized prices actually reflected short-run conditions of supply and demand, whereas published or posted prices reflected long-run conditions of supply and demand. But companies selling oil products or crude preferred to give temporary discounts rather than to lower posted prices, because posted prices were favored by oil-producing governments as a basis for royalties and income tax calculations, especially in dealing with short-run conditions. Also, of course, it was easier for oil companies to withdraw discounts when they were no longer justified, or in fact raise selling prices.

Before realized (market) prices could reach posted price levels, however, new posted prices were often set, as posted prices continued to remain above realized prices. Finally, both oil companies and their oil-buying consumers could plan more effectively when relatively stable posted prices represented an estimate of fundamental long-run conditions of supply and demand.

Posted prices were noncompetitive. They were not established by bargaining between buyers and sellers in an open market. They were administrative prices, as stated previously, set by oil-producing countries in concurrence with the large integrated oil companies who "lead" the oil industry. In many cases, the integrated oil companies with their huge size and numerous refineries, must be buyers as well as sellers of crudes in order to meet their consumer market requirements of oil products. In addition, there are many independent refiners who buy either from the large integrated oil companies or direct from the oil-producing countries. And since there is no consumer market for crude as far as direct use of crude by ultimate consumers is concerned, the market for crudes is the refineries. Thus, we say that the posted prices for crude have been "buyer-set." Demand for crudes is a derived one; the demand for crudes depends entirely on the demand for refined oil products.

As stated previously, the value of posted prices to oil-producing countries has been a basis for determining oil revenues from royalties and income taxes due to these countries from the production and sale of crudes and refined oil products. In addition, posted prices have been used for statistical purposes. They have been helpful to oil management in determining trends of increase and of decline in realized (market) prices, since market prices have been somewhat affected by posted prices.

Since posted prices for crude have been buyer-set, an increase in posted prices by the buyer has had much the same effect as a decrease in posted prices in a "seller-set" pricing situation. A posted price increase in a buyer-set arrangement has had a strong tendency toward siphoning customers away from competitors, assuming, of course, that the price change was not met by competitor buyers. A price decrease, on the other hand, has had the effect of repelling seller-customers, tending to cause them to sell the oil to other buyers.

There has been a departure from this method of pricing crude. It has not been uncommon for the base (posted) price of crude to cut out one posted by some other oil company or companies for the type or grade of crude involved, or even some other grade. In such instances, the base (posted) price has been beyond the control of the buying company. Contracts between buyers and sellers, in such cases, also have stipulated that the market price to be paid and received would be so much above or below the posted price of the other oil company.

Because market prices of crude usually prevailed on the date of delivery, crude was frequently bought and sold on what was called an accrual, or cumulative, basis. That is, the product sold accrued in equal daily quantities, and the price in effect on each successive day was applied to that volume. The reason was that delivery could thus be made in large quantities during one part of the contract period and in small quantities at another part of the same contract period, whereas the crude oil itself was run and purchased currently from the field, thus becoming part of the risk of price change.

Changes in Posted Prices

Any changes in posted prices of crudes, whether increases or decreases, could affect refining profit margins of oil companies, since royalties and income tax payments were based on posted prices. Changes in posted prices were important even though posted quotations for crudes might not have adequately reflected the prices at which many of the actual sales were made. At any one time, intramarket variations in crude prices might have prevailed in any market because at the same time a considerable amount of crude might have been moving into a market on a basis of discounts from the posted prices.

Changes in posted prices were not usually made unless there was sufficient information available to indicate what the new price should be.

Generally, a change in posted prices resulted only from the combined opinions of many individuals who were thoroughly acquainted with the refinery's crude trading operations and needs and with the market trends of its refined oil products. Any change in posted price had to be made with the view of assuring an adequate but not excessive crude supply. If an upward adjustment in posted price was not sufficient, or if a downward adjustment of posted price was too drastic, the objective of obtaining an adequate crude supply might not be met. Also, any change in posted price, up or down, had also to be made only in an amount that did not unduly affect the profit margin of the refinery. In other words, if prices of crudes became too high and refined oil product prices could not be raised without affecting the demand for refined oil products, then profits would suffer.

Increases in posted prices usually followed successive "premium sales," sales made at selling prices above posted prices because of an increase in demand over supplies. An increase in posted prices could also be due to soaring costs of exploration and development over a period of time, as well as steady and higher oil-field operating costs. Decreases in posted prices could result because of continuous oversupplies of crude oil to refinery needs.

Posted prices, always a major influence on f.o.b. crude and refined oil prices, assumed greater importance with respect to these realized prices after the OPEC meeting in Teheran in 1971. For example, tax-escalation clauses were thereafter written into sales contracts as a standard practice. Consequently, particularly for crudes, profit margins tended to remain frozen at the level existing when the contract was signed. Any movement upward in taxes had a direct effect on selling prices of crudes, and a corresponding increase of prices of refined products to consumers.

Effect of Exchange Rates on Posted Prices

Inflation and exchange rate changes have been interrelated in prices of exported goods, and crude oil and refined oil products come under the category of exported products in all producing countries.

As countries change the value of their currencies to correct fundamental disequilibriums in balance of payments and trade accounts, and as long as countries retain relatively fixed exchange rates and artificial barriers to free trade, changes in exchange rates will affect prices of exported goods, including crude oil and refined oil products. OPEC has spearheaded the need for changes in posted prices, since posted prices have affected the oil revenues that member countries of OPEC receive.

On the basis of a January 20, 1972, Geneva Agreement, posted prices at the Arabian Gulf and the Eastern Mediterranean terminals of six countries were raised by 8.4%. The agreement was supplemental to a February 14, 1971, Teheran Agreement.

Both agreements are complex. For example, according to the Geneva Agreement, posted prices will move up or down in response to changes in exchange rates on a quarterly basis, in accordance with an agreed currency index published daily by Reuter International Agency. This currency index is often referred to as the Reuter Index.

The agreed currency index, in percentage terms, is the arithmetic average appreciation of currencies of nine major industrialized countries— Britain, The Netherlands, Switzerland, Belgium, Italy, West Germany, France, Sweden, and Japan—vis a vis the U.S. dollar since April 1971. But the movement of the index does not bring any change in postings unless and until the index reaches 2 percentage points above or below "the starting average" of plus 11.02%. In other words, if the index rises above or below the "starting average" of plus 13.02%, posted prices will move upward accordingly. If the index declines below 9.02%, posted prices will move downward.

Also, the index, according to the Geneva Agreement, is recalculated on March 1, June 1, September 1, and December 1 of each year. If a variation of more than 2 percentage points up or down from the starting average (11.02%) is shown, then the necessary increase or decrease in posted prices will be made in line with the formula introduced on the first day of the next calendar quarter.

Where payments to governments are made in sterling (which is the case in all Gulf countries except Saudi Arabia), the companies' payment obligations, as regards conversions from dollars into sterling, is calculated on the basis of midpoint parity (formerly £1 = $2.40, now £1 ≃ $2.6057) to avoid value distortions.

EXAMPLE 8.1: Calculating Posted Price

Here is an example of how the currency index operated after the Geneva Agreement. Assume a 2% increase, as of March 1, in the average exchange rates of the nine specified currencies against the U.S. dollar. Then the quarterly recalculation of posted prices for Arabian light oil on April 1 would be as follows:

$2.479 (posted price as of April 1)

+ 0.0849 (raise in posted price due to dollar devaluation)

$$\times \frac{\underset{\substack{\text{(posted price for} \\ \text{January 1971)}}}{2.285} \times \underset{\substack{\text{(now effective} \\ \text{average)}}}{[0.1302} - \underset{\substack{\text{(starting} \\ \text{average)]}}}{0.1102}}{0.1102 \text{ (original starting average)}}$$

$$= 2.479 + 0.0849 \times \frac{2.285 \times (0.1302 - 0.1102)}{0.1102}$$

= 2.479 + 0.0849 × (0.415)

= 2.479 + 0.035

= \$2.514   new posted price

The net effect of a 2% appreciation of the nine specified currencies, according to recalculation of the currency index, is actually an increase of 1.41% (2.479 – 2.514 × 10%) in Gulf posted prices with a new posted price of \$2.514.

The existing formula by which oil prices were adjusted, as a result of the recent (February 1973) devaluation of the dollar, will oblige Western companies to pay an estimated 6.2% more for their Middle East oil, according to the authoritative Middle East Economic Survey (MEES).

The complicated formula was worked out in the Geneva meeting between the oil companies and the Arabian Gulf producers in January 1972, soon after the previous drop in the value of the dollar in late 1971.

EXAMPLE 8.2: Increasing Posted Price as a Result of Dollar Devaluation

The February 1973 dollar devaluation, taking Arabian light crude as an example, according to MEES and Platt's Oilgram, meant that the posted price would rise by 15.1¢ from \$2.591/bbl (raised from \$2.524/bbl to \$2.742 as the new posted price for Arabian light crude at Ras Tanura.

This, as calculated, would entail an increase in the government royalty of 9.4 ¢/bbl and an increase of 6.2% above the present level of \$1.51/bbl; also, this would raise tax-paid costs by 9.1% to \$1.712/bbl in favor of producing countries.

The calculation of the dollar devaluation is as follows:

Posted price as it would have been on April 1

+ raise in % of posted price due to dollar devaluation

$$\times \frac{\text{posted price on January 1} \times \text{now effective average} - \text{starting average}}{0.1102 \text{ (original starting average)}}$$

$$= 2.591 + 0.10 \times \frac{2.285 \times (20.32 - 13.02)}{0.1102}$$

$$= 2.591 + 0.10 \times \frac{2.285 \times 7.3}{0.1102}$$

= 2.591 + 15.1

= \$2.742   new posted price for Arabian light crude, at Ras Tanura

The Middle East Economic Survey publication noted that apart from
the question of the effect of the formula on the price of concessionary oil,
the devaluation problem becomes more acute with regard to sales of national
crude oil by the producer governments themselves. In this market the dollar devaluations have placed a cloud of uncertainty over price clauses of
most of the contracts concluded for deliveries after 1973.

Moves by OPEC on Pricing

On December 12, 1974, with a meeting of the Organization of Petroleum
Exporting Countries in Vienna, OPEC began a move toward a single price
of oil; thus, once again moving toward another shakeup in petroleum pricing.
The 12 member nations, which account for 85% of the world's oil exports,
have been striving to reach a consensus on a new pricing method for some
time. It is clear as of this writing that major changes in worldwide pricing
of petroleum have long been underway, and will probably be settled in a
matter of months.

As the producing nations move toward 100% ownership of all the oil
operations within their borders, a development that, through takeovers, is
destined to be concluded in 1975 or early 1976, posted prices, "buy-back"
prices, "third-party" prices, and all the other tabulations that add to the
complexities of the current system of posting prices will eventually be eliminated. They will ultimately be replaced by a single price. That price will
be what the producing nations' governments choose to charge for their oil.

None of the producing nations is likely to be willing to settle for less
than it now receives in oil revenues. On the other hand, adoption of a
single-price system would make any further price changes by the producing
governments highly visible. They would not be able to blame the oil companies for increasing costs to consumers, and thus would probably be more
cautious in making further price changes.

The OPEC nations will need a new "glue" to hold their cartel together
once the posted-price system has vanished altogether. So there is some
urgency to get on with the establishment of a uniform system to replace the
posted-price mechanism that has been the cement binding the OPEC cartel
together for so long a time.

As stated before, posted prices were established initially by the international companies developing oil concessions in the producing nations and
were intended as quotes at which the companies would sell the oil in the

marketplace. Taxes and royalties paid the producing nations have been based on the posted prices. A worldwide surplus of oil caused market prices to drop some years ago, resulting in a wide spread between the prices at which oil was sold and the price at which it was posted. Efforts by the companies to bring the posted prices in line with market tags alarmed the producing countries. Such a reduction would have meant a drop in their oil revenues, through lower taxes and royalties. It was, in fact, this threat to their posted prices that led several of the producing nations to form OPEC in 1960.

Although they had become artificial quotes, posted prices suddenly assumed new importance late in 1973 when the producing nations seized the initiative and began changing posted prices unilaterally. Postings, as a result, had quadrupled by January 1, 1974. Also, as the producing countries began sharing in the ownership of oil operations within their borders, at first through participation and then through takeovers and nationalization, it led to an "Alice-in-Wonderland" pricing structure and skyrocketing oil costs for consumers.

For one thing, Arabian Light crude from Saudi Arabia is being tested and used as the "benchmark" posted price for OPEC. The bulk of Saudi crude is produced by Arabian American Oil Company, or Aramco, which is in the process, at this writing, of becoming completely Saudi-owned. But at the moment, the Saudis hold 60% ownership; the remaining 40% is held by Exxon Corporation, Texaco Incorporated, Standard Oil Company of California, and Mobil Oil Corporation. The posted price for Saudi crude is $11.25/bbl, which is quadruple the level prior to October 1973. The oil companies are paying ever-rising tax and royalty rates, currently around $9.77/bbl, up from $7/bbl at the beginning of 1973. As for the oil accruing to the Saudi government, the oil companies buy most of that back at whatever price the Saudis decide to charge. Currently, this "buy-back" oil is costing the companies about $10.67/bbl, which is under the posted price of $11.25/bbl.

Thus, the actual cost of Saudi oil, the weighted average of the charges to the Aramco owner companies, is $10.36/bbl, or up $1.08/bbl since 1973. Meanwhile, the Saudis sell a part of their current 60% share of Aramco's output to third-party purchasers, or smaller oil companies, at 93% of the posted price, or $10.46/bbl.

This two-tier, and some call it a three-tier, pricing method would be eliminated under the single-price arrangement. Once the Saudi government has completed its intended takeover of Aramco, it will set a government, or what it calls a "market" posted price, for all their oil. The current Aramco owners would buy most of that oil, paying the same as everyone else. Additionally, they would receive a per-barrel fee, or commission, for continuing to operate Aramco and produce the oil.

The above development would fit in nicely with the uniform posted price planning of others in OPEC. But a hitch in the Saudi government take-over of Aramco could effect the structuring of that new pricing system. Iran has advanced a slightly different angle on the structure for single pricing. Iran suggests, at this writing, a benchmark price of around $9.80/bbl, but to link it with the prices of 20 or 30 of the world's industrial commodities. This would mean that the oil price would rise if the commodity prices rose. Thus, the new benchmark price of oil would be tied to the rate of inflation in the industrialized countries.

Because of OPEC's concern about pricing their petroleum in dollars, for as the dollar's value declines so does the oil revenues of OPEC members in terms of what they can buy in other currencies, a commission of OPEC studying oil prices has also suggested that oil be priced in terms of the International Monetary Fund's Special Drawing Rights. The effect of this would depend on the base date at which OPEC decided to make such a switch from dollars to SBR effective. An SDR represents a weighted basket of 16 major currencies. It is currently worth about $1.25, but its value fluctuates daily with the foreign exchange markets.

COMPETITION AND PRICES

Since crude prices have been buyer-set, the universal practice of buyers (refiners) has been to publish bulletins which announce the price paid for each grade of crude petroleum. As stated previously, demand for crude is a derived one, depending wholly on demand for refined oil products. Buyers of refined oil products carry considerable weight in most markets for refined oil, where suppliers are usually ample and competition fierce. For example, consider the case of buyers of gasoline, who definitely influence prices set for gasoline. When the price of the refined oil product, gasoline, goes up, usually crude prices go up as well, but only in the long run. And the same effect holds in reverse when refined oil product prices decline due to oversupply of refined oil products: Prices of crudes drop, but usually in the short run. Refineries look for ways to cut costs to maintain profit margins. Reducing crude prices is usually foremost in the minds of refiners when the squeeze is on profits.

Hundreds of different qualities of crude exist, so crude oil is not a homogeneous commodity. Crudes vary in quality largely in amount of sulfur content and in specific gravity. Crudes with high sulfur content, called "sour crudes," are less valuable than those crudes with a low amount of sulfur content, known as "sweet crudes." Sour crudes are less valuable, or less desirable to buyers of crude, because more refining processes are necessary than in refining sweet crudes. Thus sweet crudes have usually higher prices.

With regard to specific gravity, those crudes with higher specific gravity are generally more valuable to buyers of crude than crudes with lower specific gravity. Thus crudes with higher gravity usually command a higher price. In some places, as in the United States, a "specific gravity degree differential" scheme is used, whereby prices of crude are based on a certain fixed degree differential. For instance, posted list prices by gravity designation for each field in Texas and Oklahoma might be based on 2¢ per degree. Thus, if an oil company with crude to sell offers $2.25/bbl for crude of from 20 to 20.9°, it will offer $2.27 for crude of 21 to 21.9°, and so on, up to 40° and above, at $2.65. When a flat price is posted for crude, it indicates that the crude in that area is largely homogeneous in gravity. An example of the use of this scheme is the oil fields of East Texas in the United States.

Actually, competition among the buyers of crudes, particularly the independent refiners who have no oil fields of their own and must depend on the big oil companies for their crude needs, causes crude price changes up and down. Were it not for this competition for crude supply, the crude price level might be fairly static except for adjustments over a long period of time resulting from cost changes in exploring and finding oil.

For example, posted prices of crude oil (f.o.b. well head) for 36° gravity midcontinent crude (Oklahoma) for the years of 1934 through 1950 inclusive were as indicated in Table 8.1. The table illustrates the uniformity of crude prices, with changes only over long periods of time. When there was a jump in price, it was only after several years of uniform price (except for 1947).

CRUDE PRICE STRUCTURE

As previously stated, the price of crude depends on refinery income. When refineries lose money, crude prices suffer. As the refiner's estimate of the spread between crude price and total realization at the refinery is lowered, the only recourse for refiners is to cut costs if prices of refined oil products cannot advance. Thus the price of crude is one of the first items on which refiners attempt to economize, along with cutbacks in production runs. When refined oil prices advance, crude oil prices do not advance in the same way that they went down. The increase in crude price is slower.

Some crude price differences may be due to quality differences in the overall crude mix, with lighter crudes usually commanding a higher price than heavier crudes. Other crude price differences may be due to freight costs, with consuming countries located farthest from crude supply sources naturally paying the higher rate. However, some anomalies do exist which are difficult to explain, except in terms of differences in crude prices and freight billing practices of integrated companies. For example, the high

TABLE 8.1

Posted Price, 36° Midcontinent Crude

Year	Posted price (f.o.b. well head)
1934	$1.00
1935	1.00
1936	1.10
1937	1.21
1938	1.18
1939	1.02
1940	1.02
1941	1.12
1942	1.17
1943	1.17
1944	1.17
1945	1.17
1946	1.37
1947	1.90
1948	2.57
1949	2.57
1950	2.57

Source: Cassady, Ralph, Jr., Price Making and Price Behavior in the Petroleum Industry (New Haven, Conn.: Yale University Press, 1954).

import cost of crude for Germany and Switzerland, as indicated in Table 8.2, might be due to their demand pattern of refined oil products which require lighter crudes at their refineries. But the high import cost of crudes in the United Kingdom is a mystery, since domestic requirements in the United Kingdom are chiefly for heavy refined products in fuel oil. The United Kingdom thus imports mostly heavy crudes.

Many times, significant differences in crude import prices have occurred from country to country, as Table 8.2 indicates. These differences are due mainly to the time involved when contracts for purchase of crudes were made, the duration of purchase contracts, and the freight rates at which charters were concluded to carry the oil.

REFINERY PRICES FOR OIL PRODUCTS

Generally, the refiner is interested in maintaining a margin available for crude transportation and refining costs which includes an amount for profit

TABLE 8.2

Import Prices of Crude for Second Half of 1971

Country	Dollars/metric ton	Approximate dollars/bbl
Switzerland	$24.86	$3.38
Portugal	24.10	3.28
Germany	23.47	3.19
United Kingdom	22.91	3.12
Denmark	22.35	3.04
Austria	22.30	3.04
Netherlands	22.15	3.01
Sweden	21.98	2.99
France	21.63	2.94
Belgium	21.53	2.93
Italy	20.87	2.84
Spain	19.26	2.62
Japan	17.13	2.33

Source: Parra, Ramos, and Parra, S. A., "International Crude Oil and Product Prices, 15 April 1972," Middle East Petroleum and Economic Publications, Beirut, Lebanon, 1972.

and return on refinery investment. Thus, a table such as Table 8.3, which includes (1) annual average wholesale value (price) of refined oil products, (2) posted prices of crude oil (f.o.b. well head), and (3) margin available for crude transportation and refining costs, including profits, should be of interest to refineries as a source of information for controlling costs and protecting needed refinery margin.

Table 8.3 shows how average wholesale prices of refined oil products remained fairly stable from 1934 to 1938, then bounced up in 1941 and remained much the same until 1948. But in each case, except 1949 but certainly so on a long-run basis, the posted price of 36° crude followed the ups and downs of refined oil product prices. Thus, refinery margins to cover transportation and refining costs were maintained in most cases, except 1949, by as much as 30 to 40%.

Oil refiners indicate margins needed to cover their costs and profits, over and above the prices they pay for crudes. When refined oil product prices drop, refiners must be quick to cut costs, and costs of crude are usually cut first. Transportation and refinery costs are primarily fixed costs, and not subject to much variation.

TABLE 8.3

Comparison of Market Prices and Posted Prices

Year	Market price for 1 bbl of midcontinent crude	Posted price for 36° gravity midcontinent crude (f.o.b. well head)	Margin available for crude transportation and refining costs
1934	$1.37	$1.00	$0.37
1935	1.55	1.00	0.55
1936	1.69	1.10	0.59
1937	1.73	1.21	0.52
1938	1.61	1.18	0.43
1939	1.52	1.02	0.50
1940	1.49	1.02	0.47
1941	1.73	1.12	0.61
1942	1.77	1.17	0.60
1943	1.76	1.17	0.59
1944	1.75	1.17	0.58
1945	1.77	1.17	0.60
1946	1.95	1.37	0.58
1947	2.74	1.90	0.84
1948	3.59	2.57	1.02
1949	3.04	2.57	0.47
1950	3.32	2.57	0.75

Source: Cassady, Ralph, Jr., Price Making and Price Behavior in the Petroleum Industry (New Haven, Conn.: Yale University Press, 1954).

The implication is one in which the short-run prices of most refined products can be raised without much effect on demand (inelastic demand), but in the long run price rises can be self-defeating.

Governments realize that short-run inelasticity of demand makes some refined oil products ideal subjects for the raising of government revenue through indirect taxation. For example, gasoline is heavily taxed, thus raising prices of gasoline to consumers, yet in the short run this does not cut demand.

The oil industry also assumes that a general reduction of pretax prices will not increase purchases significantly by consumers in the short run and, in the long run a reduction of pretax prices might even have an adverse effect and cause a decrease in demand for purchases.

Elasticity of demand for industrial sellers is not the same as elasticity of demand for the industry as a whole. Yet when producers and marketers are few, reductions in market prices will not bring more sales.

In such cases, no seller will attempt to increase his total market share by reducing prices. This would happen only in an oligopoly situation, where one company might act as a price leader and cut prices, and others would follow by cutting prices so as to maintain a competitive equilibrium.

Examples of both posted prices and selling prices to show the differentials between the two are given in Tables 8.4 through 8.7.

Table 8.4 indicates posted prices of crudes for the Middle East for the period January 1, 1971, to January 20, 1972; Table 8.5 gives the posted prices of crudes for the Mediterranean and Nigerian areas; Table 8.6 gives the average c.i.f. (cost, insurance, and freight included) prices in 1972 actually paid by Japanese consumers for various sources of crudes around the world. These are realized prices, not posted prices. Table 8.7 gives posted prices at three export centers (basing points), namely the Caribbean, Bahrein/Ras Tanura, and Pulau Bukom in Southeast Asia. Posted prices are for four refined oil products—motor gasolines of 83 and 95 octanes, kerosene, and gas oil. Table 8.8 gives the average market prices of seven refined oil products in the Arabian Gulf for the period January-September 1972.

Looking at Table 8.4, showing posted prices for crudes in the Middle East, and reading across, we find that f.o.b. posted prices increased throughout the year in every Middle East country. Abu Dhabi, with the lightest oils of 39.00-39.09 and 37.00-37.09°API gravity, appears to have the highest posted prices of crudes, because lighter oils carry the highest prices. Posted prices of Iranian light f.o.b. Kharag and Saudi Arabian light °API f.o.b. Ras Tanura 34.00-34.09 appear to be similar, with the Iranian light 1¢/bbl cheaper. The Arabian light 34.00-34.09 f.o.b. Sidon is higher than the Arabian light 34.00-34.09 f.o.b. Ras Tanura, reflecting Tapline running costs for the 1600 km of pipeline and pump operations.

Nigeria competes with Saudi Arabia on °API 34.00-34.09 (light) and 27.00-27.09 (medium) crudes, but Nigerian light and medium f.o.b. posted prices for these comparable crudes are considerably higher than Saudi Arabian light and medium. The difference is reflected in higher exploration, development, and production costs in Nigeria in comparison with Saudi Arabia.

The list of crude imports to Japan from January to December 1972 in average c.i.f. dollar price per barrel, as given in Table 8.6, indicates that the highest prices paid by Japan for crudes, on an average of 4,283,000 bbl/day imported in the period January-December 1972, were to Venezuela, Australia, Angola, Libya, and Indonesia sources. All but Indonesia were

long-distant export points, thus freight was probably a large part of the
c.i.f. price per barrel. Indonesia's crude costs, however, are higher than
other areas, which helps explain the high import price from Indonesia.

Japan purchased 965,000 bbl/day of Arabian light, medium, and heavy
crudes from the Arabian Gulf area in 1972. The average c.i.f. price of
$2.39 which Japan paid for these crudes was rather low if one compares
this price of $2.39 to f.o.b. Ras Tanura crude prices in January 1972 (see
Table 8.4) of $2.479, $2.373, and $2.239 for Arabian light, medium, and
heavy, respectively. However, the 965,000 bbl were obtained from offshore
wells of the Arabian Oil Company, a Japanese-Saudi Arabian Company in the
Neutral Zone, and shipped in Japanese tankers, thus accounting for the price
differential.

Table 8.7 gives posted prices on four refined oil products: motor
gasoline of 95 and 83 octane, kerosene, and gas oil. Three basing points
for posted oil products are the Caribbean, Bahrein/Ras Tanura, and Pulau
Bukom in Southeast Asia.

Market prices at Arabian Gulf ports in 1972 were substantially lower
than posted prices in the area, as indicated by Table 8.8, which gives a
listing of average quotations for the first half of 1972 of both posted and
market prices for seven refined oil products from January to September
1972. Prices of refined oil products, without variations in crude prices,
are usually relatively stable; also, refined oil product prices are influenced
only slightly by seasonal fluctuations in demand. The Middle East, that is,
the Arabian Gulf area, has lower prices, year by year, or date by date, than
the other two export centers of the Caribbean and Pulau Bukom. The high-
est-price export center is understandably, the newest center, Pulau Bukom
in Southeast Asia, with posted prices of oil refined products in most cases
at least $0.50/bbl more than the Bahrein/Ras Tanura export center.

Average market quotations, or prices of refined oil products in the
Arabian Gulf for the period from January to September 1972, were stable
in the early part of 1972 for the six refined oil products listed. Market
prices increased, however, in the latter part of the first half of 1972 which
is the summer season and the heaviest season for refined oil products such
as motor gasoline, naphtha, and kerosene.

CONCLUSIONS AND SUMMARY

Use of API gravity as a posted pricing base for crude has come under at-
tack. Researchers claim that a more realistic method of assigning practical values (posted prices, or reference prices) to crudes is needed. They
propose an evaluation based on certain hydrocarbon fractions contained in
the oil regardless of the oil's origin or oil's gravity. The use of API

TABLE 8.4

Middle East Crude Oil Prices: Gulf Postings, $/bbl

Crude oil	f.o.b. point	°API gravity	Jan. 1	Year 1971 Feb. 15	June 1	Jan. 20 1972
Abu Dhabi						
Murban	Jebel Dhana	39.00–39.09	1.88	2.235	2.341	2.540
Abu Dhabi Marine	Das Island	37.00–37.09	1.86	2.225	2.331	2.529
Dubai						
Dubai	Fateh	32.00–32.09	1.68	2.130	2.233	2.423
Iran						
Iranian light	Kharg Island	34.00–34.09	1.79	2.170	2.274	2.467
Iranian heavy	Kharg Island	31.00–31.09	1.72	2.125	2.228	2.417
Bahrgansat	Bahrgan SBM terminal	32.00–32.09	—	2.135	2.238	2.428
Bahrgansat	Bahrgan MBM terminal	32.00–32.09	—	2.120	2.223	2.412
Nowruz	Bahrgan SBM terminal	20.00–20.09	—	1.840	1.936	2.100
Darius	Karg Island	34.00–34.09	1.63	2.160	2.264	2.264
Cyrus	Cyrus terminal	19.00–19.09	1.34	1.830	1.926	1.926
Sassan	Lavan Island	34.00–34.09	1.70	2.172	2.276	2.469
Rostam	Levan Island	38.00–38.09	—	2.232	2.338	2.536
Iraq						
Basrah	Kohr al-Amaya	35.00–35.09	1.72	2.155	2.259	2.451
Kuwait						
Kuwait	Mina Al-Ahmadi	31.00–31.09	1.68	2.085	2.187	2.373

Neutral Zone						
Khafji	Ras al-Khafji	28.00-28.09	1.55	1.970	2.069	2.245
Hout	Ras al-Khafji	35.00-35.09	1.81	2.185	2.290	2.484
Eocene	Mina Saud	16.5-17.4	1.28	(1.280)*	(1.280)*	(1.280)*
Burgan	Mina Saud	23.5-24.4	1.48	(1.480)*	(1.480)*	(1.480)*
Ratawi	Mina Saud	23.5-244.4	1.41	(1.410)*	(1.410)*	(1.410)*
Oman						
Oman	Mina Al-Fahal	33.00-33.09	1.82	2.205	2.310	2.506
Qatar						
Qatar (Dukhan)	Umm Said	40.00-40.09	1.93	2.280	2.387	2.590
Qatar (Marine)	Halul Island	36.00-36.09	1.83	2.200	2.305	2.501
Saudi Arabia						
Arabian light	Ras Tanura	34.00-34.09	1.80	2.180	2.285	2.479
Arabian medium	Ras Tanura	31.00-31.09	1.68	2.085	2.187	2.373
Arabian heavy	Ras Tanura	27.00-27.09	1.56	1.960	2.064	2.209

*Parentheses indicate postings which have not yet shown effect of Teheran and/or Geneva price agreements.

Source: Parra, Ramos, and Parra, S.A., "International Crude Oil and Product Prices, 15 April 1972," Middle East Petroleum and Economic Publications, Beirut, Lebanon, 1972.

TABLE 8.5

Mediterranean and Nigerian Crude Oil Posted Prices, $/bbl

Crude oil	f.o.b. point	°API gravity	1971				1972		
			Jan. 1	March 20	July 1	Oct. 1	Jan. 1	Jan. 20	April 1
Arabian light	Sidon	34.00–34.09	2.370	3.181	3.158	3.136	3.106	3.370	3.341
Iraq	Tripoli/Banias	36.00–36.09	2.410	3.211	3.188	3.166	3.136	3.402	3.373
Libyan	Libyan ports	40.00–40.09	2.550	3.447	3.423	3.399	3.386	(3.386)*	(3.357)*
Nigerian light	Bonny et al.	34.00–34.09	2.420	3.212	3.195	3.178	3.176	(3.176)*	(3.156)*
Nigerian medium	Bonny	27.00–27.09	2.280	3.104	3.087	3.070	3.068	(3.068)*	(3.048)*

*Parentheses indicate postings which have not yet shown effect of Geneva price agreements.

Source: Parra, Ramos, and Parra, S.A., "International Crude Oil and Product Prices, 15 April 1972," Middle East Petroleum and Economic Publications, Beirut, Lebanon, 1972.

TABLE 8.6

Japan's Crude Imports, January–December 1972, $1/bbl

Source	Barrels/day	Average c.i.f. price per bbl
Sarawak	4,000	$2.84
Brunei	96,000	2.96
Indonesia	554,000	3.40
Iran	1,654,000	2.42
Iraq	5,000	2.51
Saudi Arabia	965,000	2.39
Kuwait	494,000	2.46
Qatar	4,000	2.56
Oman	143,000	2.53
Arab Emirates	244,000	2.57
USSR	7,000	3.00
Egypt	4,000	2.12
Libya	4,000	3.47
Nigeria	62,000	3.22
Angola	30,000	3.54
Venezuela	8,000	3.70
Australia	3,000	3.59
Other	2,000	—
	4,283,000	$2.58 average

Source: Platt's Oilgram Price Service.

gravity as a pricing base for crude now means for example, that two crudes of similar gravity used by a refiner can often realize as much as 20¢/bbl more profit to him from one crude than the other.

There are numerous different f.o.b. prices for crudes depending on quality, quantity, impurities, etc. In principle, such prices are related either to differences in costs of processing, such as the removing of impurities, or to relative costs of refiners for using the different crudes in producing the desired pattern of refinery output (product mix) at expected product prices.

In practice, because of the absence of an efficient market mechanism, certain types of price differentials, especially differentials related to specific gravity, are sometimes deemed appropriate to patterns of demand. times, too, crude oil prices have been influenced by relative prices of refined products, such as fuel oil, kerosene, or gasoline, that are made from these crudes. Also, in some cases, crude oil prices have been influenced

TABLE 8.7

Posted Refined Oil Product Prices: 1970-1972, ¢/gal,
Cargo Lots, f.o.b. Export Centers (Three Basing Points)

Product and date of change	Caribbean	Bahrein/ Ras Tanura	Pulau Bukom
Motor gasoline, 95 octane			
1970 average	9.7	9.7	10.6
	($4.074/bbl)		($4.452/bbl)
1971 average	10.7	10.4	11.8
1971 Sept. 1	11.2	10.6	11.9
1972 Jan. 19	11.7	10.6	11.9
Jan. 20	11.7	10.9	11.9
Jan. 28	11.7	10.9	12.1
Jan. 31	11.7	10.8	12.1
Feb. 1-March 31	11.7	10.9	12.2
	($4.914/bbl)	($4.578/bbl)	($5.124/bbl)
Motor gasoline, 83 octane			
1970 average	7.1	7.2	8.0
1971 average	8.1	7.9	9.2
1971 Sept. 1	8.6	8.1	9.3
1972 Jan. 19	9.1	8.1	9.3
Jan. 20	9.1	8.4	9.3
Jan. 28-March 31	9.1	8.4	9.6
	($3.822/bbl)	($3.528/bbl)	($.03/bbl)
Kerosene			
1970 average	9.2	9.0	9.9
1971 average	10.1	9.7	11.1
1971 Sept. 1	10.0	9.9	11.2
1972 Jan. 17	10.2	9.9	11.2
Jan. 19	10.5	9.9	11.2
Jan. 20	10.5	10.2	11.2
Jan. 28	10.5	10.2	11.5
Jan. 31	10.5	10.1	11.5
Feb. 11-March 31	10.5	10.2	11.5
	($4.41/bbl)	($4.284/bbl)	($4.830/bbl)
Gas oil, 53-57 DI			
1970 average	7.7	6.7	7.6
1971 average	9.9	7.4	9.1
1971 Sept. 1	9.8	7.6	9.2
1972 Jan. 17	10.0	7.6	9.2
Jan. 19	10.5	7.6	9.2
Jan. 20	10.5	7.9	9.2
Jan. 28	10.5	7.9	9.5
Jan. 31-March 31	10.3	7.9	9.5
	($4.326/bbl)	($3.318/bbl)	($3.990/bbl)

Source: Parra, Ramos, and Parra, S. A., "International Crude Oil and
Product Prices, 15 April 1972," Middle East Petroleum and Economic Publications, Beirut, Lebanon, 1972.

TABLE 8.8

Prices of Refined Oil Products in Arabian Gulf, January–September 1972

Refined oil product	Posted prices, ¢/gal	Market Selling Prices		
		Feb. 15, 1972 $/bbl	May 15, 1972 $/bbl	Aug. 15, 1972, $/bbl
Motor gasoline, 95 octane	10.8	8.90	8.90	10.20
Motor gasoline, 90 octane	9.6	7.90	7.90	9.10
Straight-run	5.5	4.30	4.30	5.30
naptha		4.50	4.50	5.30
Kerosene	10.2	9.50	9.50	10.00
No. 2 fuel, low sulfur	7.4	7.10	7.00	7.00
Light fuel oil				
2.5% sulfur	2.07	1.65, 1.75	1.70–1.80	1.80–1.90
3.2% sulfur		1.45, 1.55	1.55–1.60	1.65–1.70
Heavy fuel oil				
3.5% sulfur	2.02	1.30, 1.50	1.50–1.60	1.60–1.70
4.5% sulfur		1.25, 1.45	1.40–1.50	1.50–1.60

Source: Platt's Oilgram Price Service.

by indirect taxation or the tax component of the price, since consumers
pay this price. Sometimes the tax component of price even exceeds the
cost component of the product itself.

New technological developments have significantly increased the ability of refiners to vary the product mix obtainable from a given crude, but
it is not yet possible to produce refined products in any proportions desired.
The basic reason for crude prices being buyer-set is that the demand for
crude is derived from the demand for refined products. Except for the
limited use of burning crude as fuel, crude is of little use to anyone except
those who can refine it into other products. Thus, within the limits of the
technical capacity of existing refineries (or special stockpiling arrangements), the amount and qualities of crude demanded in the short run are
directly related to the products that refiners expect to be able to sell.

In general, short-run elasticity of demand at prevailing aftertax
prices is too low in elasticity, although there are significant variations
between different refined products in different areas. For example, demand for gasoline is usually highly inelastic within most price ranges.
Sometimes, however, where gasoline is used in direct competition with
diesel oil, it can become elastic in demand.

Fuel oil is in competition with other fuels in many uses. The amount
demanded for fuel oil over other substitute fuels is likely to be influenced
by the relative costs of fuel oil and of the costs of the other substitute fuels.
Also, sales of both gasoline and diesel oil are likely to be influenced by
lower prices of substitutes if time is allowed for adaptation of size of engines, replacement of existing fuel-burning equipment, etc., to convert
from the use of gasoline and diesel oil.

NOTES

1. Saudi Economic Survey, "Developments in the Oil Sector in 1970,"
    Ashoor Public Relations Services, Jeddah, Saudi Arabia, December 8,
    1971.

Chapter 9

MARKETING OIL AND OIL PRODUCTS: TRANSPORTING

INTRODUCTION

There are four modes of transportation used in moving crude oil and refined oil products throughout the world. They are the railroad, motor (tank
trucks), sea tankers, and pipeline carriers. Of these, only the airplane is
not as yet used to any great extent for commercial transport of oil in large
quantities. In times of war, however, the military services often resort to
movement of oil cargoes by air transportation to meet extreme emergencies.

Since distribution at the wholesale and retail levels are not discussed
in this book, we shall deal here only with those carriers involved in overseas shipping, that is, sea tankers and pipelines. Further, tankship transportation will demand most of our attention because of its importance to the
movement of oil and oil products throughout the world.

When both production and consumption of crude oil take place in the
same country, as they do in the United States, the location of the refinery
depends on local conditions and transport facilities; it may be near the oil
field, with the oil products then transported to wherever they are needed,
or the crude oil may be shipped to a consuming center where it is refined
for local use. Delivery of oil products to the final consumer is normally in
relatively small batches, so if refining is carried out near the source of the
crude oil, bulk storage and delivery centers are needed in the marketing
area. If, however, the refinery is located in a reasonably compact consuming area, extra handling in the form of storage can sometimes be avoid-
ed and delivery made direct from the refinery.

However, apart from North America and the USSR, there are no large
areas where production in any way matches consumption; hence international,

or more precisely interarea, movements of oil are usually necessary. The transport chain from the oil field to the ultimate consumer thus becomes complex and costly. Europe, Japan, and now the United States, which depend upon imports for the great bulk of their oil needs, require three transport links: (1) oilfield to loading port; (2) sea transport to a seaboard refinery; and (3) delivery of oil products by road or railroad to the consumer.

Although the discussion of oil marketing in this book is limited to the first two links of the transport chain, the reader should bear in mind that each separate handling operation entails added cost. Thus, the basic need in transportation is to use the most economical transport chain that will meet the requirements of users of transport services.

TERMS USED IN DISCUSSING OIL TRANSPORT

Dirty, clean: "Dirty" is the term used by the trade for crude oil cargoes; "clean" is the term used for cargoes of refined oil products. Dirty oil, also called black oil, includes crude oil, heavy diesel oil, fuel and furnace oil. Clean oil includes motor and aviation gasoline, solvents, white gasoline, tractor vaporizing oil, kerosene, gas oil, naphtha, and other refined oil products.

A time charter is a contract of duration longer than a single voyage. The rental rate, or hire, is usually paid on the basis of deadweight tons per month. The rate does not, however, include vessel expenses such as fuel for propulsion, port charges, or canal tolls. These expenses are paid by the charterer.

A spot charter is a contract for a single voyage between two ports. Payment is based on tons of oil delivered. Under this type of contract the owner of the vessel pays for all vessel expenses.

Tankship is a term used to designate supply of tankers and tanker services.

Tanker rates are the prices of tanker services. "Free on board" (f.o.b.), price of oil included, means that the buyer obtains the carrier and is responsible for shipment after the oil is loaded at the export point. "Cost, insurance, and freight" rates (c.i.f.) mean that the seller of oil makes all arrangements, including insurance and freight. Thus the buyer gets a "delivered price," with all costs included in the selling price to him up to the import point. The buyer takes over at the port of entry of the oil shipment, assuming all costs of the oil cargo from then on.

Figure 9.1 illustrates main oil movements by sea, as of 1972. The thicker lines indicate heavy oil exports in tonnage, and show the Arabian Gulf to be the principal exporter. One route is around the Cape of Good

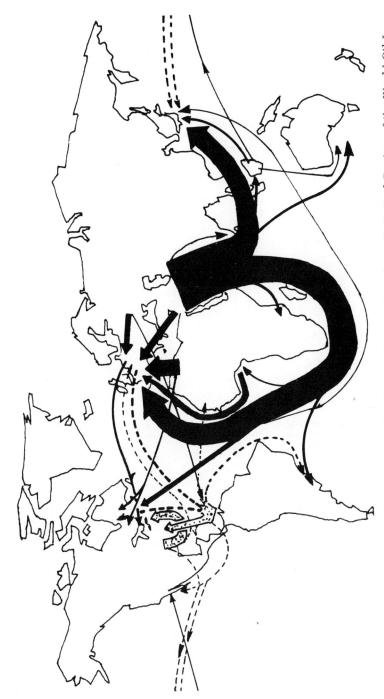

FIG. 9.1. Main oil movements by sea, 1972. Source: British Petroleum Statistical Review of the World Oil Industry, 1972, The British Petroleum Company, Ltd., London.

Hope to Western Europe, with a "less thick" line directed toward the United
States. Still another movement from the Arabian Gulf is through the Indian
Ocean around Southeast Asia to Japan. The Caribbean supply area shows
heavy exports into the Gulf area of the United States, and less heavy exports
up the Atlantic coast of the United States to New York, Canada, and Western
Europe.

Figure 9.2 shows estimated distribution of the world tanker fleet as of
the end of 1970. With the size of tankers in five different deadweight ton
(dwt) classes given, the approximate number of vessels of each deadweight-
tonsize can be estimated, as it was in 1970, by multiplying the number of
tankers by 100 for each deadweight-ton size class.

Tankers up to 25,000 dwt comprised the largest percentage of the total
tanker fleet, or about 2,200 vessels in 1970. The 25,000-50,000 dwt class
was next, with approximately 900 ships. Following this group were the
50,000-100,000 dwt class, with approximately 650 tankers.

The current trend is to larger deadweight-ton size tankers, as Table
9.1 shows, because the oil industry is relying more and more on sea shipping for transporting crude oil and oil products to distant consuming markets. Tanker transportation has the disadvantage of high operating costs,
but capital costs are much lower per barrel and per ton for larger-size per
barrel and per ton for larger-size tankers because of the amount of tonnage
that can be transported at any one time. For this reason, also, there has
been an increase in the rate of construction of even larger sizes, of the
supertanker types. At the time of writing, a 500,000-dwt tanker is being

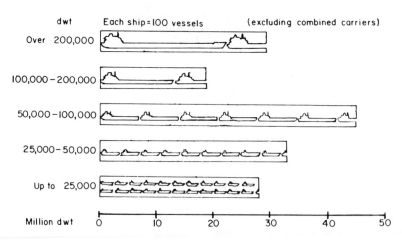

FIG. 9.2. World tanker fleet at end of 1970. Source: British Petroleum Statistical Review of the World Oil Industry, 1970, The Brisith Petroleum Company, Ltd., London.

built, and plans have been made for others, including some as large as 700,000 dwt in capacity.

Tables 9.1 through 9.4 show the extent of the world tanker fleet as of the end of 1972. Table 9.1 shows the world tanker fleet in million tons deadweight by flag (country). Table 9.2 shows the total tanker fleet in classes of deadweight-ton sizes, as well as the total deadweight tonnage by size of tanker that was being built or was on order as of 1972.

As Table 9.3 shows, only the United States has failed to show any significant increase in tanker fleet capacity from 1962 to 1972. In 1962, the United States had 9.0% of the total world tanker tonnage; in 1972, it had 9.7% of the world total, or only a 0.7% increase in oil tanker capacity in 10 years.

TABLE 9.1

World Tanker Fleet by Flag at End of 1972
(excluding 28.5 million dwt combined carriers,
2,000 dwt and over)

By flag	Total, in million dwt	Percent change, 1972 over 1971	Percent of total, 1972
Liberia*	50.2	+7.1	26
Norway	19.9	+0.8	10
England	25.3	+0.1	13
Japan	22.6	+3.8	12
United States	9.7	0	3
Panama*	7.6	+1.9	4
France	8.3	+0.7	4
Greece	10.5	+2.0	5
Other Western European countries	24.2	+0.4	13
Other Western Hemisphere countries	4.9	+1.6	3
USSR, Eastern Bloc, China	6.4	+0.2	3
Other Eastern Hemisphere countries	4.3	0	2
Total	193.9	+18.6	100
Fleet at end of 1971	175.3		
Net increase in 1972	18.6		

*Convenience fleets.
Source: British Petroleum Review of the World Oil Industry, 1972, The British Petroleum Company, Ltd., London.

TABLE 9.2

World Tanker Fleet by Size, at End of 1972
(excluding 28.5 million dwt combined carriers,
2,000 dwt and over)

By size, in dwt

Under 25,000	27.9	1.1
25,000–45,000	28.1	4.2
45,000–65,000	22.5	0.4
65,000–125,000	40.2	11.4
125,000–205,000	13.1	10.5
205,000–285,000	58.6	71.7
285,000 and over	3.5	20.9
Total tonnage	193.9	120.1

Source: British Petroleum Review of the World Oil Industry, 1972, The
British Petroleum Company, Ltd., London.

"Convenience" fleets and various European countries appear to own and
operate the bulk of the world tanker fleet. In terms of deadweight tonnage,
the largest increases have been in the classes between 65,000 and 285,000
dwt.; and 1968 was the year of the largest increase in these size tankers.
The closing of the Suez Canal in 1967 hastened this increase.

Table 9.4 presents the estimated proportions of the world's active
ocean-going fleet by voyage origin and destination. Voyages from the Middle East to Western Europe employed the largest percentage of tankers,
with 44.5%; Middle East to Japan was next largest, with 14.0%.

The big emphasis has been on tankers of 205,000 dwt and over, with
58.6 million tons of capacity available at the end of 1972 and another 71.7
million tons of capacity in the building and "on-order" stages.

With a total of 120.2 million dwt in new tankers building and on order
at the end of 1972, roughly 62% of the 193.9 million dwt in service, or the
supply of tanker service available for oil shipments, represented the amount
that was to be potentially added to world capacity as of the end of 1972. The
increase in tanker supply in 1972 over 1971 was roughly about 18.6%, or
193.9 million from 175.3 million dwt, respectively; the increase in 1971 over
1970 was approximately 25%; and in 1970 over 1969 was approximately 14%.
Thus, it is quite safe to say that with world conditions remaining as they
currently are, world tanker capacity should continue to increase each year
over the previous year by at least 15%.

TABLE 9.3

World Tanker Fleet
by Flag, 1962–1972 (million dwt)

Year	United States	England	Norway	Other W. Eur.	"Convenience"*	Japan	Rest of World	Under 25,000	25,000–45,000	45,000–65,000	65,000–125,000	125,000–205,000	205,000–285,000	285,000 and over	Total
1962	9.0	11.1	10.1	17.6	14.1	3.5	4.9	34.0	24.7	8.4	3.1	0.1	—	—	70.3
1963	9.0	11.8	10.8	18.1	15.7	4.1	5.5	32.7	25.4	12.0	4.8	0.1	—	—	75.0
1964	8.7	11.7	12.2	18.6	18.9	5.0	6.5	31.3	25.4	16.1	8.7	0.1	—	—	81.6
1965	8.8	11.8	13.3	19.0	23.3	6.5	7.4	31.2	25.5	19.3	14.0	0.1	—	—	90.1
1966	8.7	12.5	14.9	21.5	25.2	8.4	8.2	30.0	25.3	21.2	21.8	0.9	0.2	—	99.4
1967	8.7	13.2	16.6	22.8	27.8	9.8	9.0	29.7	25.0	21.8	28.9	2.3	0.1	—	107.9
1968	8.8	15.4	16.4	26.3	31.5	11.4	9.7	29.8	25.6	22.1	33.4	5.4	2.6	0.6	119.5
1969	9.1	18.8	15.7	30.5	36.0	13.7	11.4	28.9	26.7	22.0	36.3	8.2	11.1	2.0	135.2
1970	9.5	21.9	17.2	35.0	43.4	15.6	13.1	28.3	27.4	22.5	38.5	10.3	26.7	2.0	155.7
1971	9.7	25.2	19.1	39.9	48.8	18.8	13.8	28.5	28.0	22.6	39.4	12.0	42.5	2.3	175.3
1972	9.7	25.3	19.9	43.0	57.8	22.6	15.6	27.9	28.1	22.5	40.2	13.1	58.6	3.5	193.9

*"Convenience": Liberia, Panama, etc.
Source: British Petroleum Statistical Review of the World Oil Industry, 1972, The British Petroleum Company, Ltd., London.

TABLE 9.4

Employment of Tankers in 1972
(estimated percentages of world's active ocean-going fleet
on main voyages)

Voyages to	United States	Caribbean	Middle East	North Africa	Others	Total
United States	4.0	3.0	2.5	0.5	1.5	11.5
Canada	—	1.0	1.0	—	0.5	2.5
Other Western Hemisphere countries	—	—	5.5	0.5	1.5	7.5
Western Europe, North and West Africa	—	1.5	44.5	4.0	4.5	54.5
East and South Africa and Asia	—	—	2.0	—	—	2.0
Japan	—	—	14.0	—	2.5	16.5
Other Eastern Hemisphere countries	—	—	4.5	—	—	4.5
USSR, Eastern Bloc, China	—	—	1.0	—	—	1.0
Total	4.0	5.5	75.0	5.0	10.5	100

Source: British Petroleum Statistical Review of the World Oil Industry,
1972, The British Petroleum Company, Ltd. , London.

TRANSPORTATION COSTS

Costs of transportation influence the selling prices of crudes and refined
oil products. For example, if transportation costs range from $6 to $8 a
ton (or roughly $0.80 to $1.07 a barrel) on shipments to Europe or the United
States from the Middle East, transportation costs can amount to approximately one-third of the final selling price. The fact that such a large percentage of the selling price can be attributed to transportation has a considerable effect on profits. Buyers of transportation services in Europe and
the United States will obviously seek sources of supply closer than the Middle East if at all possible.

It is difficult to determine accurately the cost of transportation from
one point to another, because the cost of shipping depends on the size of the

charter conditions—that is, long-term contract or spot (one-time) shipping—the type of product being shipped—that is, crude (dirty) or refined oil
(clean) product—and the existing shipping situation relative to supply and
demand for tankers.

Theoretically, the use of large tankers under long-term charter contracts should lead to lower transportation rates. But the best size of tankers to use, whether for hauling refined or crudes, often depends on the
receiving facilities of the consuming markets. However, even if receiving
terminals are too small for the size of tankers involved, large-size tankers
(say 200,000 dwt and over) can be sent to transit points located near such
consuming areas and then smaller tankers can be used to trans-ship to the
consuming point. This procedure also raises transport costs because of
extra transit, extra handling, and an intermediate storage expense, but it
can reduce the overall long-distance costs of shipment.

THE TANKSHIP MARKET

The tanker market is an example of a perfectly competitive market, that
is, one in which both buyers, or oil owners, and sellers, or tanker owners,
are "price takers" and "quantity setters." Each buyer and each seller regards price as given and responds to it by selecting the quantity in tonnage
that he wishes to buy or sell at that price.

Tankers may be chartered for a single voyage, known as spot shipping,
for a number of consecutive voyages, or for a long-term period. Since the
spot market is the most interesting and one which often affects long-term
charters, it is the one we shall discuss here.

When plotted on a graph, the quantity, transportation tonnage, and
response to price of buyers as a group is represented by a downward-sloping
market demand curve; the quantity response to price of sellers as a group is
represented by an upward-sloping market supply curve (Fig. 9.3). The
equilibrium price at any one time, E in Figure 9.3, is where the demand
curve intersects the supply curve, and is the one which "satisfied" both
buyer and seller evaluations of tonnage at a particular time. Buyer and
seller evaluations are constantly changing, so a new equilibrium price is
always emerging. Thus, the equilibrium price is the price at any one time
at which buyers are willing to "take" tankers and tanker owners are willing
to "give" tanker services. So E is the price that equates market demand
with market supply.

The tankship market, as pointed out by Z. S. Zanetos [1], is a highly
competitive and extremely volatile one in which the price, or tanker rate,
is determined by the interaction of supply and demand. Thus, tanker companies do not influence the tankship market to the extent that the market

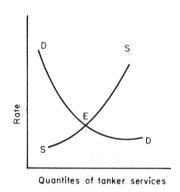

FIG. 9.3. The price of transportation services.

becomes one of an imperfect type, even though the major oil companies are the major users of tankers.

DEMAND FOR TANKSHIP SERVICES

Oil companies depend on their own tankers, and on ships of private tanker companies under long-term contract, or charter, to meet approximately 90% [1] of their transport needs. The remaining 10% are spot chartered.

In the spot market, many crude oil producers and refiners are active bidders. No one buyer of tanker services controls or appreciably influences market price. With many buyers bidding for tankship services, they actually become price takers. Their market behavior, or demand, can thus be represented by an inelastic demand curve that slopes from left to right and down, as indicated in Figure 9.4.

Thus the price (spot rate) elasticity of demand is highly inelastic. The extent of this inelasticity is dependent on three factors:

1. How long tankship services, for which no substitute except to a limited extent lower-priced pipelines, continue to exist. When substitutes for tankship services can be found, the highly inelastic curve may change to an elastic one.

2. How long the cost of transport by tankship remains only a small fraction of the total cost of producing final petroleum products. This question makes any price (rate) reduction less meaningful relative to increased demand for tankship services, since tankship services depend on market behavior of oil and oil products.

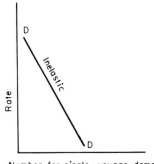

FIG. 9.4. Buyer curve for tankship curves.

3. And finally, whether the demand for final petroleum products will
    change from one of a highly inelastic nature to one of an elastic
    nature, especially over the short run.

Each of these three factors—lack of substitutes, small role in total
cost, and inelastic demand for output—tends to make the demand for the input tankship service inelastic, as shown in Figure 9.5.

Usually, in the short run, changes in transportation rates will induce
changes in other factors, such as marginal sales of oil and oil products, but
will not decrease or increase the quantity of transportation services demanded. In the long run, however, increases or decreases in rates will
cause decreases or increases, respectively, in quantities of transportation
services demanded.

More stable (less fluctuating) freight rates, to "normalize" f.o.b.
crude and refined oil product prices, are needed in the oil industry. Increases in transportation rates in the long run have a negative effect on oil
and oil product sales, thus causing increases in oil product prices.

DIFFICULTIES IN IDENTIFYING DEMAND FOR TANKSHIP SERVICES

It is difficult to compose a demand schedule, or to record the number of
transactions for tankship services completed at various rates. The number
of transactions by itself, though important, is not enough. Each transaction
has more than one dimension; for instance, there are agreements, or contracts, for one voyage, and agreements for several voyages over a period
of time. Also, the size and speed of tankers are not reflected in the number of transactions completed; these dimensions also have a bearing on

FIG. 9.5. Demand for tankship services.

capacity. Furthermore, some "lags" may exist: Rates prevailing for one period may not be the result of factors operating in the same period, but may be due to conditions existing in a previous period. Also, the average time duration of contracts, or agreements, can affect the number of completed transactions for tankship services. Actually, during periods of rising prices, the number of transactions increases, the average time duration of each contract increases, and the percentage of transportation capacity operating on time charters also increases. So it is difficult to determine what the demand for tankship services is at any one time.

SUPPLY OF TANKSHIP SERVICES

Seller concentration in the tankship market is small. According to Zanetos [1], there are more than 600 companies that own and operate tankships, and not one of them owns and controls more than 7% of all tankers afloat. This 7% includes the independents, shippers as well as the oil companies themselves, which as the principal users of tankers own and operate their own fleets. The oil companies own the largest fleets, but the independent tanker owners and operators are the ones usually on the supply side of the market.

Sellers in the spot market for tanker services, that is, tankship owners and oil company operators, must be "price takers." Their market behavior is represented by a single-market supply curve.

The shape of the single supply curve, as indicated in Figure 9.6, is almost vertical or inelastic. A supply curve of tankship services usually has this shape. When all tankers are in operation it is impossible, at least in the short run since it takes many months and usually years to build a tanker, for tankship owners to significantly expand capacity or increase the

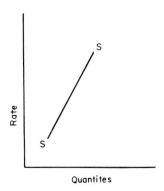

FIG. 9.6. Supply curve for tankship services.

number of tankers in their fleet to meet an increase in demand for tankship
services. A short-run increase in demand forces tanker rates up, but
tanker operators may lose profits if they are unable to meet this short-run
increase in demand for tankship services. A short-run increase in demand
forces tanker rates up, but tanker operators may lose profits if they are
unable to meet this short-run increase in demand because of a lack of
supply.

The high costs involved in converting ships from other uses, such as
the carrying of grain, almost eliminates the possibility of competition from
other tankship companies or the transferring of ships from other uses to
the hauling of crude oil and refined oil products. Thus, competition from
other shippers does not appreciably affect the shape of the supply curve.

When tankship rates go down, there is little that tankship operators
can do but accept lower rates for their services. When tankship rates fall
below the variable costs of marginal ship owners and operators, who have
the highest costs of operation, these marginal ship owners and operators
are often forced to cut their services temporarily.

The supply curve for tankship services can be altered materially, but
only in the long run. In the short run, from a few months to one year,
tankship operators are usually not able to increase the size of their fleet,
that is, their capital stock, and thus the output of their tankship services.
Sometimes, however, tankship operators can increase the size of their fleets
and thus their tankship services in the short run to a certain extent: They
may receive delivery on previously ordered ships, or they may decide to
continue using older ships that normally would have been scrapped. Over
the long run, or a year or more, since it takes a minimum of 15 to 20 months
to build a new ship, tankship operators can substantially increase the size
of their fleet and output of tankship services.

Actually, the tankship service market can be represented by three different supply curves. These are illustrated by Figure 9.7, which shows the effect on supply of tankship services for three different demand increases. In each case an inelastic demand curve for tankship services is illustrated by curve DD and an increase in demand by a curve D'D' to the right of DD. The supply curve for tankship services changes from an inelastic curve in Figure 9.7a to an elastic curve in Figure 9.7b and a highly elastic one in Figure 9.7c.

In Figure 9.7a, demand is increased substantially, so that the price of tankship services goes up from OP to OP' with an inelastic supply curve. In Figure 9.7b, supply reacts to the increase in demand in Figure 9.7a by becoming more elastic, but demand has dropped off somewhat due to high price in Figure 9.7a with limited supply; thus, the new price, OP' in Figure 9.7b, drops with a more elastic supply curve. In Figure 9.7c, the supply curve is highly elastic; or more tankship services are now available, but demand has dropped to the point now where the new price OP' is almost the same as the original price OP.

According to Zanetos, tankship owners have extremely elastic expectations of the future, tankship owners expect rate (price) trends once established to continue, whether the trend be upwards or downwards. Therefore, in periods of high demand or high spot rates, shipowners will respond by placing orders for more ships, and may even tend to overbuild. But the result is that each upswing in demand for more tankers is followed by a period in which rates are so depressed that shipowners can barely cover their variable costs of operation. So even under perfect competition, under which the tankship service market operates to a great degree, the life of a shipowner can be quite precarious. Thus, peculiarities in the construction of tankers and the length of time needed to bring new tankers into the tankship service market are the factors which affect the supply curve of tankship owners, and also the market rates which they charge buyers of tankship services.

TANKSHIP OPERATORS GO ANYWHERE

There is no standard route of travel for tankships. This factor favors tanker owners, and helps them overcome their natural disadvantages of limited supply and length of time needed to increase their supply to meet increased short-run demand. Geographically speaking, the tankship market is extremely dispersed.

From the standpoint of economics, for the market to be in equilibrium, that is, for the supply of tankers to equal the demand for tanker services, it is necessary (1) that prices (or rates) be set at an "appropriate" level, and (2) that ships be appropriately allocated among different geographic areas. This is the function of tankship brokers.

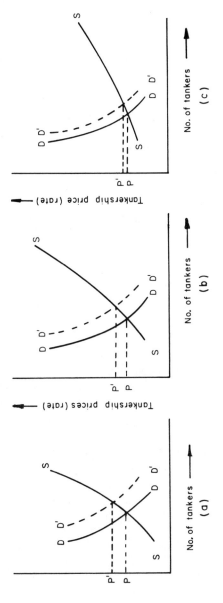

FIG. 9.7. Inelastic demand for tankship services.

Tankship brokers bring widely scattered tanker operators into a single competitive market. The prices (rates) established by them in the market represent an equating of supply and demand. When representing crude oil shippers, brokers attempt to get the highest possible prices; when they represent buyers of crude and refined oil products, brokers go for the lowest possible prices.

TANKER RATE STRUCTURES

Whenever either demand or supply shifts, there is a fluctuation in rates. The more shifts there are, the more inelastic the market demand and supply becomes. Because demand and supply for tankship services are extremely inelastic, there are wide fluctuations in price. Also, unusual events, such as the closing of the Suez Canal, can create wide fluctuations in rates. The Suez closing, for example, caused a sharp upward shift in demand, and an upward turn in price as well, for tankship services. The Korean War had the same effect of increasing demand over supply of tankship services, which in turn caused an increase in rates.

The basic determining factors of freight and charter hire rates are demand and supply of tanker tonnage. Long-term freight rates fluctuate much less violently than short-term rates, or spot rates. But both rates, short-term and long-term, are influenced by many uncertainties. For example, postwar currency restrictions have resulted in two "chartering markets," the sterling market and the dollar market. Each market has its own nominal voyage freight scales, which form a standard of reference for various voyages against which the market level at any one time can be readily measured and by which rates for various voyages can be compared.

Since 1969, the World Scale has been the basis for tanker voyage charter rates. The World Scale replaced the "International Tanker Nominal Freight Scale," called "Intascale," which in turn had replaced the London Market Tanker Nominal Freight Scale, known as the "London Scale," in May 1962.

The World Scale, designed and operated by a panel of London tanker brokers, serves as the standard of reference, or index, in tanker markets outside the United States for freight rates. The World Scale system makes it possible to measure the tanker market at any given time for various voyages, as well as providing a means of comparison of rates for various voyages. World Scale rates can be converted into dollar rates; all calculations of dollar rates are in tons of 2,240 lb (long tons). These rates are based on a round trip from a control port of loading in the loading range to a central port in the discharge range and back again to the central port in the same loading range. So freight rates, on the basis of World Scale, are calculated on a port-to-port basis and not to a range of ports, as formerly.

When oil companies or oil consumers do not furnish their own oil tanker services, the tanker market is usually determined in one of two ways: (1) for long-term chartering, the oil tanker company usually negotiates directly with the oil consumer or supplier; (2) for short-term, or spot chartering, usually involving single charters, most often the oil tanker company contacts a tanker broker who handles tankship services for either the oil consumer or the oil supplier.

In any long or short charter, those engaged in hiring carriers are known as charterers. They may be either buyers of oil or petroleum products on f.o.b. basis, or sellers of oil or petroleum products on a c.i.f. basis. The tanker operator, as the seller of transport services, may also be represented by a broker.

WORLD SCALE RATES

Here is how the World Scale system works. The example used involves a spot charter in the Middle East. Spot chartering is often used as a basis for setting long-term chartering rates.

EXAMPLE 9.1: World Scale Rates

Assume that an oil refinery in Rotterdam wishes to purchase some tanker spot services to bring 1.8 million bbl of crude oil (240,000 tons) from Ras Tanura, Saudi Arabia, the source of supply of the refinery. The tanker selected for this one-voyage haul is one of 250,000 dwt, with 10,000 dwt allowed for ship provisions, fuel, and spare parts. The round-trip journey will take 59 days, based on an average speed of 12 knots. The route is Ras Tanura to Quoin Island at the mouth of the Arabian Gulf, where all Gulf port shipment rates are standardized to Europe. From Quoin Island, the route is around the Cape of Good Hope and then on up to North Europe and Rotterdam, since the Suez Canal cannot at present handle 250,000 dwt ships.

The oil refinery wants to know what the transportation will cost. A tanker broker in London is approached who will quote a rate and make provisions for a 250,000-dwt ship for the designated scheduled time desired for loading at Ras Tanura, Saudi Arabia.

Current World Scale rate for a 250,000-dwt tanker, Arabian Gulf to Rotterdam (North Europe), is assumed to be 100 at this time. World Scale 100 represented in dollars, and rounded off, is $1,555 per hour per ton, as indicated in Figure 9.8. The tanker broker calculates the freight rate to the refinery by multiplying the cost per hour per ton by the total number of steaming hours to Rotterdam and return, and divided by the size of the tanker. In this case, the freight rate is $1,555 × $1,416 hr (59 days × 24

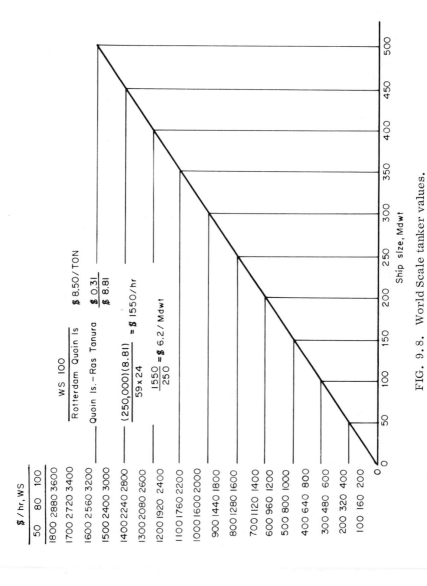

FIG. 9.8.   World Scale tanker values.

hr) + 250,000 dwt. Thus, the total freight rate to the refinery is $8.81/ton.
The breakdown of this rate is indicated in Figure 9.8: $0.31 from Ras Ta-
nura to Quoin Island, and $8.50 from Quoin Island to Rotterdam and return.

The oil refinery pays the tanker broker a total of $2,114,400 (240,000
dwt × $8.81) for carrier payment. The tanker broker takes his commission
out of the $2,114,400 and pays the tanker operator after shipment has been
delivered in Rotterdam.

Assuming that the breakeven point of the oil tanker is $1,250 per hour
per ton, and 1,416 hr are involved on the single voyage, his total costs will
be $1,770,000 (omitting the broker's commission for simplicity). He will
earn a profit of $344,400 ($2,114,400 - $1,770,000) for this single voyage.
The $1,250 is also the lowest cost per hour per ton under which the tanker
could operate. In other words, the lowest freight rate that the tanker broker
could quote the oil refinery, assuming the World Scale rate is still 100,
would be $7.375 per ton ($1,770,000 ÷ 240,000 tons). At this rate the tanker operator would only break even between costs of operation and revenues
received provided that a speed of 12 knots is maintained for 59 steaming
days.

To find the cost per hour per 1,000 dwt, the $1,550 cost per ton is
divided by 250 (thousandths) deadweight tonnage of the tanker, or $6.20 per
1,000 dwt, as the cost per hour. Thus, for every 1,000 tons of oil shipped
at WS 100 for hire of a 250,000-dwt tanker at Ras Tanura to deliver in
Europe, it costs the shipper $6.20 every hour. Therefore, any time that
can be saved enroute, that is, less than 59 days, will cut the cost of transporting the load of oil. Figure 9.8 indicates tanker values by graph and
formula. The formula used and plotted on the graph is the deadweight tonnage tanker times transportation cost per ton of oil divided by the number
of steaming days from source of oil to consumer destination and return,
and then multiplied by 24 hr in a day.

In Figure 9.8, tanker values in costs per hour per ton of ship for
World Scale rates of 100, 80, and 50 are listed vertically. Ship sizes in
1,000 dwt are listed horizontally. In addition, a 45° line is drawn as a
result of plotting values derived from the formula. For instance, when
calculating costs per hour per ton of ship for a 500,000-dwt tanker, by use
of the formula, a figure of $3,110 per hour per ton is derived; similarly,
when calculating costs per hour per ton for a 250,000-dwt ship, a figure of
$1,550 per hour per ton is obtained. The extension of a vertical line up-
ward from 500,000 and 250,000 dwt, respectively, and a horizontal line
from left to right of $3,110 and $1,550, respectively, produces intersection
points which, when connected to the zero axis, develops an appropriate 45°
line. This line can be used to obtain, graphically, costs per hour per ton
for any size tanker, in preference to use of the given formula, by merely
reading up from the desired deadweight tonnage to the 45° line and across
to the left and World Scale rates.

Any changes in source and/or destination point, including the size of tanker, will change the figures in the formula and the corresponding cost per hour per ton of the tanker. When the World Scale is less than 100, as 80 or 50, cost per hour per ton can be determined by merely multiplying the rate of 100 by 80 or 50. This will give the cost per hour per ton, at 80 and 50, respectively.

A World Scale rate of 70, 40, 110, 117.5, and so on, can always be represented in dollars for any size tanker by merely using the graph of Figure 9.8 to the nearest size. For example, spot charter World Scale rates for the week from October 28 to November 3, 1972, given in Table 9.5, are determined in dollars by use of the graph of Figure 9.8 and shown In Table 9.6.

Table 9.6 shows the smaller-size tankers have the lowest dollar rates, when converting World Scale rates to dollars. Actually, World Scale rates reflect the supply, or availability of tankers of various sizes in deadweight tonnage for the period involved, relative to demand. Over 100 indicates a shortage of that particular size tanker. For example, 117.5 was the World Scale for 20,000-dwt tankers in the week of October 28-November 3, indicating supply behind demand for that size tanker, while under 100, the World Scale of 65 for 130,000-dwt vessels in the same week is indicative of an ample supply of 130,000-dwt ships in relation to demand.

The World Scale provides the advantage of a single schedule which is suitable for use in all markets. In reality, the World Scale rate system is truly a tanker rate stabilization scheme on an international basis, indicating the availability of different-size tankers at the time World Scale rates are published. The chief aim of World Scale rates is to establish a fair and stable system of rates relative to size or capacity of oil tankers. Indirectly, also, the World Scale system points out the need for additional building of tankers, since it indicates scarcities of tankers in the face of increasing demands for these tankers, especially of the supertanker type.

EXAMPLE 9.2: World Scale Rates, with Rates of Return

Sometimes rates of return for shipping a certain type of crude oil are given in terms of percent of return at World Scale rates of 100, 80, and 50. For example, assume that we are contemplating continuing to ship Arabian Heavy at rates of return given as 6% at WS 50, 12% at WS 80, and 15% at WS 100 for different-size tankers. If we assume that the tanker being considered is a 270,000-dwt ship, we can then check the rate of return by converting WS rates to dollars. We consult Figure 9.8 as to what we can afford to pay in order to meet our return on investment targets of 6%, 12%, and 15%. If

TABLE 9.5

Spot Charter Tanker Rates, World Scale,
October 28 to November 3, 1972

Trade	Deadweight tonnage (thousands)	Cargo	Loading	World Scale rate
Arabian Gulf/West	140	Dirty	11/16	70
Arabian Gulf/East	130	Dirty	11/9	65
Red Sea/U.S. Gulf	34	Dirty	Prompt	85
Eastern Mediterranean/ Rostock	20	Dirty	Early Nov.	117.5
Arabian Gulf/Puerto Rico	16	Clean	12/1-12/20	125
Arabian Gulf/South Africa	24	Clean	11/12	125
Arabian Gulf/Bangkok	23	Clean	10/30	135.75
Arabian Gulf/U.K.	60-70			80
Mediterranean/U.K.	25			112.5
Nigeria/U.K.	50			90

*Aramco, Economics Department, News Digest, November 10, 1972.
Source: Aramco (Arabian American Oil Company), Economics Department, News Digest, November 10, 1972.

TABLE 9.6

World Scale Tanker Rates, in Dollars
October 28 to November 3, 1972

Deadweight tonnage (thousands)	World Scale rate	World Scale rate
140	70	$665 (70% of 905)
130	65	585 (65% of 900)
34	85	212.50 (85% of 250)
20	117.5	235 (117.5% of 200)
16	125	218.75 (125% of 175)
24	125	262.50 (125% of 210)
23	125	262.50 (135% of 210)
60-70	80	380 (80% of 475)
25	112.5	247.50 (112.5% of 220)
50	90	315 (90% of 350)

WS 100 is $1,490/hr for a 270,000-dwt tanker, a 15% rate of return when shipping Arabian Heavy must be obtained as follows: 59 steaming days at 24 hr, or a total of 1,416 hr. Then 1,416 × $1,490 = $2,109,840. This means that total investment is $14,065,600 for a 15% rate of return when the WS rate is 100. (Proof: $14,065,600 × 0.15 = $2,109,840.)

If WS is 80, the rate is 0.80 × $1,490, or $1,192/hr, so if the desired rate of return is 12%, total investment must be $14,065,600 since 12% is 1,416 times 1,192 = $1,687,872. (Proof: $14,065,600 × 0.12 = 1,687,872.)

Finally, if WS is 50, the rate is 0.50 × $1,490, or $745/hr; and if the desired rate of return is 6%, total investment must be $17,582,000. (Proof: 1,416 hr × $745 = $1,054,920, 6% rate of return; and $17,582,000 × 0.06 = $1,054,920.)

Savings accrued by pipelining might increase the rate of return, to say 8%, 14%, and 20% at WS 50, 80, and 100. In that case Arabian Heavy might be transported by pipeline instead of tanker, all other things being equal.

INFLUENCE OF BIG OIL COMPANIES ON TANKSHIP RATES

Big oil companies influence tankship rates, much as they do selling prices of crude and refined oil products. For instance, their merely postponing purchases of transport for a few months or so, because of ample stocks from previous commitments, will affect rates. Also, "owning" enough tonnage under long-term charters that their withdrawal from bidding for tanker services can be interpreted as a loss of confidence in the tanker service market can influence transportation rates, so much so that small charterers may withdraw from the market, thereby cutting available supplies of tanker services. The large oil companies can accentuate fluctuations in the tankship service markets, especially since the spot market is usually thin, consisting only of roughly 10 to 15% of total tankship capacity. Large oil companies can influence both increases and decreases in spot service rates by influencing supply cuts or supply increases through their movements.

Over a period of time, rates often can also be interpreted as a signal of a future rate decline of greater consequence. This can cause an increase in spot purchases at the expense of long-term charters as oil shippers become less likely to commit themselves to long-term charters for fear of paying higher prices than will be necessary in the figure. Also, if tanker operators are large enough, they can postpone the acceptance of orders for tankship services of all kinds. This can depress tanker service markets. In periods of rising tanker rates, the opposite holds true, as time charters increase at the expense of spot services.

BASIC TERMINAL FACTORS

The first basic terminal factor is tankage for both crude oil and refined oil products. Associated with tankage, of course, is crude production and refinery processing. For example, a 6 million bbl tankage capacity might be provided for an oil production of 1.5 million bbl/day. The procedure would be as follows.

Incoming ships enter the cargo queue and await their turn for a berth to load. When a tanker is next in the queue for cargo loading, its capacity is matched with available inventory already in the tanks at the terminal. The crude inventory, plus an assumed minimum of 10 hr production run capacity, should be sufficient for the tanker's order of shipment. And when tankage is filled to capacity, perhaps because of an extensive port closure due to inclement weather, crude production must be halted and tabbed as "lost production."

The second basic terminal factor is the number of berths required. The number of berths should equal normal demand and also be sufficient to keep up with tankage capacity and increased production in the oil fields. Berths are usually assigned to ships on a first in, first out basis. "Berth time" must include times for mooring, deballasting, documenting, and unmooring after loading, as well as the actual time needed for loading of oil and oil products. In other words, total berth occupancy time equals cargo loading time + the four constants of mooring, deballasting, documenting, and unmooring.

The third basic terminal factor is loading system capacity in thousand barrels or tons per hour. Loading time of a tanker depends on the loading system capacity of a terminal, since loading time equals ship size divided by the capacity of the loading systems assigned.

The three basic terminal factors are interrelated because of the interaction among all of the factors from oil field production to loading system capacity. This concept can best be illustrated by following a typical crude oil production to terminal facilities.

EXAMPLE 9.3: Efficient Use of Terminal Facilities

Assume 6,000,000 bbl of tankage, 1,500,000 bbl of crude per day in oil field production capacity, 2 loading berths, and a 200,000 bbl/hr loading system capacity. Figure 9.9 illustrates these figures.

Assume further that there are two tankers arriving in the terminal for oil cargo at approximately the same time. One is a 300,000 dwt tanker

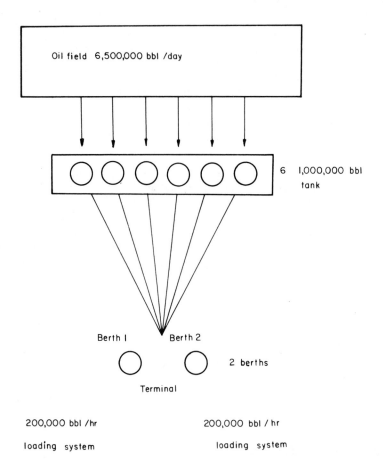

FIG. 9.9. Interrelations among basic terminal factors.

which carries 2,250,000 bbl, and the other is a 500,000-dwt tanker which can haul 3,750,000 bbl. There is no cargo queue, so the 300,000-dwt tanker is assigned to Berth 1 and the 500,000-dwt tanker to Berth 2.

Now further suppose that six 1,000,000-bbl tanks have been loaded before these two tankers arrive in port, thus there is no problem relative to immediate loading. Both tankers can be loaded at the same time when the loading system capacity is 200,000 bbl/hr.

Loading time for the 300,000-dwt tanker will be 11.25 hr (2,250,000 ÷ 200,000). Loading time for the 500,000-dwt tanker will be 18.75 hr (3,750,000 tanker capacity ÷ 200,000 loading system capacity).

Thus, in 18.75 hr, upon completion of all loadings, all 6,000,000 bbl would be drained from the 6,000,000-bbl tank capacity. But the tanks would not actually be empty. The oil fields are producing 1,500,000 bbl/day, or 62,500 bbl (1,500,000 ÷ 24 hr) per hour. So in 11.25 hr, upon completion of the loading of the 300,000-dwt tanker, some 703,125 bbl (or 11.25 hr × 62,500 bbl/hr) would have entered the tanks, and by 18.75 hr, or after the 500,000-dwt tanker had been loaded, at least 1,171,875 bbl (62,500 bbl/ hr × 18.75 hr) would have entered the tanks. At the end of the 19th hr there will be 1,187,500 bbl in the tanks. The calculations are shown in Table 9.7.

The ratio of loading system capacity to oil field production as shown in Table 9.7 is 3.2 to 1 (or 200,000 bbl:62,500 bbl) when only one tanker is being loaded; for two tankers it is 6.4 to 1 (400,000 bbl:62,500 bbl). It takes 96 hr, or 4 days (6,000,000 bbl ÷ hours), to fill the tanks to capacity with the current oil field rate of production of 1,500,000 bbl/day.

In such circumstances, it would appear that the loading of smaller vessels, when oil field capacity is only 1,500,000 bbl/day, would be the best procedure. However, after 18.75 hr, two smaller tankers of sizes 100,000 dwt (or 750,000 bbl capacity) might each start loading, since the tanks would have over 1,187,500 bbl of oil in tank storage. Table 9.8 indicates the results of the loading of the two 100,000-dwt tankers.

In summary, it is obvious from Table 9.8 that the planning of tanker loadings at the terminal was poor. This points up the importance of planning for tankers. Efficient scheduling of tankers by size, taking into account not only the basic terminal factors of tankage capacity, number of berths, and loading system capacity, but also production in the oil fields, is necessary for efficient operation.

This example points out that with an oil field production of 1,500,000 bbl/day, the scheduling of smaller vessels, possibly of 100,000 dwt and under, should be considered, with additional loading facilities, possibly a SPM (Special Point Mooring) system, used when large-size tankers are being loaded at the same time.

TABLE 9.7

Summary of Loading with a 300,000-dwt Tanker
and a 500,000-dwt Tanker

At end of	Into tank per hour, in bbl	Out of tanks, bbl/hr			Balance, in tanks, bbl
		Berth 1	Berth 2	Total	
					6,000,000
1 hr	62,500	200,000	200,000	400,000	5,662,500
2 hr	62,500	200,000	200,000	400,000	5,325,000
3 hr	62,500	200,000	200,000	400,000	4,987,500
4 hr	62,500	200,000	200,000	400,000	4,650,000
5 hr	62,500	200,000	200,000	400,000	4,312,500
6 hr	62,500	200,000	200,000	400,000	3,975,000
7 hr	62,500	200,000	200,000	400,000	3,637,500
8 hr	62,500	200,000	200,000	400,000	3,300,000
9 hr	62,500	200,000	200,000	400,000	2,962,500
10 hr	62,500	200,000	200,000	400,000	2,625,000
11 hr	62,500	200,000	200,000	400,000	2,287,500
12 hr	62,500	50,000[a]	200,000	250,000	2,100,000
13 hr	62,500	200,000	200,000	200,000	1,962,500
14 hr	62,500	200,000	200,000	200,000	1,825,000
15 hr	62,500	200,000	200,000	200,000	1,687,500
16 hr	62,500	200,000	200,000	200,000	1,550,000
17 hr	62,500	200,000	200,000	200,000	1,412,500
18 hr	62,500	200,000	200,000	200,000	1,275,000
19 hr	62,500	200,000	150,000[b]	150,000	1,187,500[c]
Total, end of 19th hr	1,187,500	2,250,000	5,750,000	6,000,000	

[a]Only 50,000 bbl loaded at Berth 1 in 12th hr, as the 300,000-dwt tanker concludes its full load of 2,250,000 bbl (1/4 × 200,000 bbl).
[b]Only 150,000 bbl loaded at Berth 2 in the 19th hour as the 500,000-dwt tanker concludes its full load of 3,750,000 bbl (3/4 × 200,000 bbl).
[c]At 0.75 of the 19th hr, the 500,000-dwt tanker should be fully loaded, as the balance of the needed 150,000 bbl to make up the total 3,750,000-bbl load is completed in 0.75 of the 19th hr. At that point, with a rate of 1,500,000 bbl/day, or 62,500 bbl/hr, the tanks should also have a balance of 1,171,875 bbl. Fifteen minutes later (0.15 hr) or the end of the 19th hr, the tanks will be up to 1,187,500 bbl.

TABLE 9.8

Summary, Loading Two Tankers, 100,000 dwt Each,
After Loading 300,000 and 500,000 dwt Tankers

| | Into tank per | Out of tanks, bbl/hr | | | Balance, in |
At end of	hour, in bbl	Berth 1	Berth 2	Total	tanks, bbl
19 hr					1,187,500
20 hr					
(mooring)	62,500	0	0	0	1,250,000
21 hr	62,500	200,000	200,000	400,000	912,500
22 hr	62,500	200,000	200,000	400,000	575,000
23 hr	62,500	200,000		400,000	237,500
24 hr	62,500	150,000	150,000	300,000	0
Total at end					
of 24 hr	312,500	750,000	750,000	1,500,000	0
(1 day)					
Add previous					
figures	1,187,500	2,250,000	3,750,000	6,000,000	
(Table 9.7)					
Total for	1,500,000	3,000,000	4,500,000	7,500,000	
24 hr					

In this case, continuance of operation at the terminal was the problem, otherwise a shutdown in oil field production was imminent. Only if oil field capacity were increased, say to 3 or 4 million bbl/day, with corresponding additional tank capacity and an increase in number of berths at the terminal, could the oil company considered here afford to accommodate large-size tankers on a regular basis without running the risk of cutbacks or shutdowns in production in the oil fields.

As given in this example, the present loading system capacity of 400,000 bbl/hr (200,000 bbl/hr for each berth), with two berths, is heavily in excess of tank capacity of 6,000,000 bbl and oil production of 1,500,000 bbl/day (62,500 bbl/hr).

TANKER PLANNING WITH TERMINAL FACILITIES

Terminal facilities for efficient loading of tankers, especially the larger ones and the supertankers, are needed in major crude-oil shipping ports. Planning ship arrivals and loading these ships within a minimum amount of time, since awaiting time involves money expenditures, is the main objective of an export terminal of an oil company. It is most important to keep the tankers moving. The importance of this can best be explained by example.

EXAMPLE 9.4: Matching Terminal Facilities to Liftings

Assume the following facts relative to the programming of ships through basic terminal facilities for a 160,000-dwt maximum at the terminal.

Liftings are 1,000,000 bbl/day, or 133,333 tons.

Production in the oil field is 1,500,000 bbl/day.

Tankage capacity is 6,000,000 bbl.

Number of berths at the terminal is 1.

Number of 100,000 bbl/hr loading systems is 1.

Loading of ships is from 40,000 to 100,000 bbl/hr, depending on the size of the tanker, but straight liftings take a maximum of 10 hr.

With the above facts in mind, assume the following steps on arrival and departure from the terminal. (In actual practice, this computation undoubtedly be performed with the aid of a computer, with which a more precise program could be put together.) Also, assume these steps with "times" in hours for each ship to be loaded.

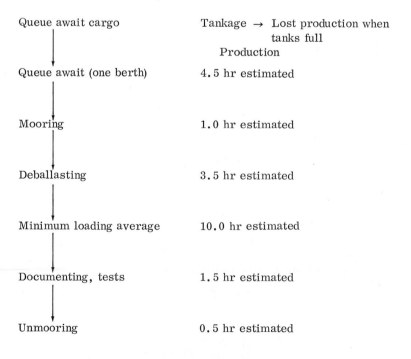

Queue await cargo	Tankage → Lost production when tanks full
	Production
Queue await (one berth)	4.5 hr estimated
Mooring	1.0 hr estimated
Deballasting	3.5 hr estimated
Minimum loading average	10.0 hr estimated
Documenting, tests	1.5 hr estimated
Unmooring	0.5 hr estimated

Since no computer is available, the next step after receiving notification of order is to program a list of tanker arrivals by date for cargo loading, bearing in mind that a minimum amount of awaiting cargo time per tanker is most desirable.

Table 9.9 shows how 45 ships might be programmed for a typical January, using the following data. (Usually this programming is done by computer.)

Preloading time	4.5 hr per tanker
Mooring 1.0 hr	
Deballasting 3.5 hr	
Average loading time	10.0 hr per tanker
Postloading time	2.0 hr per tanker
Documenting 1.5 hr	
Unmooring 0.5 hr	
Total average cargo hours	
per ship	16.5 hr

Efficient programming of tankers reduces average awaiting cargo times in a terminal, and minimizes the required investments for production capacity and/or tankage.

Since in this case production in the oil fields is 1,500,000 bbl/day and liftings at the terminal are 1,000,000 bbl/day, an excess of 500,000 bbl/day will build up each day. A single point mooring (SPM) investment in the sea of, say $1 million, might be a possibility for the loading of two 200,000-dwt tankers every 3 days, which would take 3,000,000 bbl (100,000 ton × 7.5 bbl × 2) and still leave some capacity available for berth loading in case of emergencies.

The results of programming in Table 9.9 are given in Table 9.10, showing a total of 45 different-sized ships loaded during January with an average loading time of 10 hr, using one berth, and a loading system capacity of 100,000 bbl/hr. Liftings per 10 hr totaled 1,000,000 bbl of crude, or 133,333 dwt, and the loading system operated from 40,000 to 100,000 bbl/hr per hour, depending on ship size.

If we assume that total average port time per tanker was 20.5 hr, average awaiting cargo time per tanker would have been 4.0 hr, since there are 16.5 cargo hours per ship.

Summarizing Table 9.10 for the month of January,

Total hours in a 31-day month	744
Total pre- and postloading time (30 × 6.5 hr)	195
Total loading time (30 × 10 hr)	300
Total awaiting time (30 × 4 hr)	120
Total hours accounted for on 45 ships	
(16.5 hr × 45 ships)	732.5

TABLE 9.9

Programming Tanker Distribution at Terminal

Date	Time, hr	Ship size, dwt	Liftings, tons
Jan. 1	4.5		
	10.0	105,000	100,000
	2.0		
	4.5		
2	10.0 $\begin{cases} 3 \text{ hr} \\ 7 \text{ hr} \end{cases}$	105,000	33,333 70,000
	2.0		
	4.5		
	10.0	65,000	63,333
3	2.0 $\begin{cases} 0.5 \text{ hr} \\ 1.5 \text{ hr} \end{cases}$		
	4.5		
	10.0	125,000	123,333
	2.0		
	4.5		
4	10.0 $\begin{cases} 1.5 \text{ hr} \\ 8.5 \text{ hr} \end{cases}$	70,000	10,000 56,667
	2.0		
	4.5		
5	10.0 $\begin{cases} 9.0 \text{ hr} \\ 1.0 \text{ hr} \end{cases}$	85,000	76,667 7,667
	2.0		
	4.5		
	10.0	130,000	125,667
	2.0		
	4.5		
6	10.0	90,000	88,333
	2.0		
	4.5		

TABLE 9.9 (continued)

Date	Time, hr		Ship size, dwt	Liftings, tons
Jan. 7	10.0	$\begin{cases}7.5\ \text{hr} \\ 2.5\ \text{hr}\end{cases}$	65,000	45,000 15,000
	2.0			
	4.5			
	10.0		120,000	118,333
	2.0			
8	4.5	$\begin{cases}3.0\ \text{hr} \\ 1.5\ \text{hr}\end{cases}$		
	10.0		58,000	55,333
	2.0			
	4.5			
9	10.0	$\begin{cases}6.0\ \text{hr} \\ 4.0\ \text{hr}\end{cases}$	130,000	78,000 50,000
	2.0			
	4.5			
	10.0		85,000	81,333
	2.0			
10	4.5	$\begin{cases}1.5\ \text{hr} \\ 3.0\ \text{hr}\end{cases}$		
	10.0		65,000	61,333
	2.0			
	4.5			
11	10.0	$\begin{cases}4.5\ \text{hr} \\ 5.5\ \text{hr}\end{cases}$	155,000	72,000 80,000
	2.0			
	4.5			
	10.0		55,000	53,333
12	2.0			
	4.5			
	10.0		115,000	114,000
	2.0			
	4.5			

TABLE 9.9 (continued)

Date	Time, hr	Ship size, dwt	Liftings, tons
Jan. 13	10.0 $\begin{cases} \text{3 hr} \\ \text{7 hr} \end{cases}$	70,000	19,333 46,667
	2.0		
	4.5		
	10.0	90,000	87,000
14	2.0 $\begin{cases} \text{0.5 hr} \\ \text{1.5 hr} \end{cases}$		
	4.5		
	10.0	120,000	118,333
	2.0		
	4.5		
15	10.0 $\begin{cases} \text{1.5 hr} \\ \text{8.5 hr} \end{cases}$	100,000	15,000 85,000
	2.0		
	4.5		
16	10.0 $\begin{cases} \text{9 hr} \\ \text{1 hr} \end{cases}$	55,000	46,000 5,333
	2.0		
	4.5		
	10.0	130,000	128,000
	2.0		
17	4.5		
	10.0	60,000	58,333
	2.0		
	4.5		
18	10.0 $\begin{cases} \text{7.5 hr} \\ \text{2.5 hr} \end{cases}$	100,000	75,000 25,000
	2.0		
	4.5		
	10.0	110,000	108,333
	2.0		

TABLE 9.9 (continued)

Date	Time, hr	Ship size dwt	Liftings, tons
Jan. 19	4.5 $\begin{cases} 3.0 \text{ hr} \\ 1.5 \text{ hr} \end{cases}$		
	10.0	105,000	100,000
	2.0		
	4.5		
20	10.0 $\begin{cases} 6.0 \text{ hr} \\ 4.0 \text{ hr} \end{cases}$	60,000	33,333 22,000
	2.0		
	4.5		
	10.0	115,000	111,333
	2.0		
21	4.5 $\begin{cases} 1.5 \text{ hr} \\ 3.0 \text{ hr} \end{cases}$		
	10.0	80,000	76,000
	2.0		
	4.5		
22	10.0 $\begin{cases} 4.5 \text{ hr} \\ 5.5 \text{ hr} \end{cases}$	130,000	57,333 76,000
	2.0		
	4.5		
	10.0	60,000	57,333
23	2.0		
	4.5		
	10.0	80,000	76,000
	2.0		
	4.5		
24	10.0 $\begin{cases} 3.0 \text{ hr} \\ 7.0 \text{ hr} \end{cases}$	100,000	57,333 40,000
	2.0		
	4.5		
	10.0	95,000	93,333

TABLE 9.9 (continued)

Date	Time, hr	Ship size, dwt	Liftings, tons
Jan. 25	2.0 $\begin{cases} 0.5 \text{ hr} \\ 1.5 \text{ hr} \end{cases}$		
	4.5		
	10.0	110,000	105,333
	2.0		
	4.5		
26	10.0 $\begin{cases} 1.5 \text{ hr} \\ 8.5 \text{ hr} \end{cases}$	100,000	28,000 68,333
	2.0		
	4.5		
27	10.0 $\begin{cases} 9.0 \text{ hr} \\ 1.0 \text{ hr} \end{cases}$	75,000	65,000 7,333
	2.0		
	4.5		
	10.0	130,000	126,000
	2.0		
28	4.5		
	10.0	80,000	76,300
	2.0		
	4.5		
29	10.0 $\begin{cases} 7.5 \text{ hr} \\ 2.5 \text{ hr} \end{cases}$	75,000	57,033 15,900
	2.0		
	4.5		
	10.0	120,000	117,433
	2.0		
30	4.5 $\begin{cases} 3.0 \text{ hr} \\ 1.5 \text{ hr} \end{cases}$		
	10.0	70,000	67,250
	2.0		
	4.5		

TABLE 9.9 (continued)

Date	Time, hr	Ship size, dwt	Liftings, tons
Jan. 31	10.0 $\begin{cases} 6.0\ \text{hr} \\ 4.0\ \text{hr} \end{cases}$	115,000	66,083 45,920
	2.0		
	4.5		
	10.0	90,000	87,413
	2.0		
Feb. 1	4.5 $\begin{cases} 1.5\ \text{hr} \\ 3.0\ \text{hr} \end{cases}$		

TABLE 9.10

Breakdown by Size of Ships Loaded in January

Ship size	Number loaded
40,000 to 50,000 dwt	0
50,000 to 60,000 dwt	3
60,000 to 75,000 dwt	10
75,000 to 100,000 dwt	10
100,000 to 150,000 dwt	21
150,000 to 200,000 dwt	1

Although this example is oversimplified, it does point out the important terminal considerations involved in planning terminal loadings, such as number of barrel liftings, oil production in the field, tanker capacity, number of berths involved, and loading system capacity.

Obviously, any change in one of the terminal factors, such as an added berth or an improved loading system, should bring increases in the other factors—otherwise there might be excess idle capacity. Also, any increase in oil production must include a corresponding increase in tank capacity lifting berths, and possibly loading system capacity. Thus, some important ratios which are worthwhile to watch are (1) tank capacity to oil production and to liftings, (2) production capacity to liftings (this ratio decreases as liftings increase), since this ratio affects awaiting times, (3) number of berths to liftings, (4) awaiting times to number of berths, and (5) awaiting times to lifting level.

In an actual terminal situation, optimum average awaiting time per ship could be determined by balancing the value of ship time spent awaiting cargoes against corresponding investments in terminal facilities. Thus, if average awaiting time per ship is 12 hr at an average cost of $4,000/hr to a shipper and 400 ships are loaded per year, total added cost to the shipper is 400 × 12 × $4,000 or $19,200,000. If an investment for added terminal facilities involves, say $14,000,000, a savings of $5,200,000 per year can be made by cutting down on the average awaiting time.

An optimum combination of production capacity and tank capacity can also be determined. Investment required for additional production capacity is considered relative to the increase in investment required for equivalent tankage.

EXAMPLE 9.5: Investment Relative to Increase in Productivity and
              Decrease in Costs

An investment for raising oil production from 1 million bbl/day to 2 million bbl/day is assumed to be $30 million, and the investment for increasing tank capacity from 6 million bbl to 12 million bbl, or a ratio of 6 to 1 tankage: production, is regarded as $4/bbl. This represents $24 million required in order to achieve at least the same assumed average awaiting time of, say 5 hr per ship.

Then, if we assume that crude liftings are 2 million bbl/day with 12 million bbl of tankage assigned, oil production capacity is 3.4 million bbl/day, with a production-liftings ratio of 1.7 to 1. The ratio of tankage capacity to production is then 3.5 to 1 (12,000,000 ÷ 3,400,000).

If there are 6 million bbl/day in liftings with 12 million bbl of tankage assigned, 10.2 million bbl/day of production capacity is required to maintain an average of 5.0 hr awaiting time. But 12 million bbl tankage capacity will not be sufficient. To maintain a 6 to 1 ratio, an increase to 20.4 million bbl tankage capacity is necessary (3,400,000 × 6). This will also raise tankage investment to at least $61.2 million based on $3/bbl.

EXAMPLE 9.6: Changes in Requirements Relative to Production,
              Tankage, Liftings, and Terminal Facilities

This example illustrates particularly how important it is to provide for and coordinate such requirements as tank capacities, oil field production, tanker liftings, and terminal facilities (loading system capacities and number of tanker berths), particularly when production, liftings, and loading system capacities are increased. Furthermore, awaiting times are costly, and so any savings in hours per ship in port through the provision and coordination of adequate facilities improves port performance and saves money for the oil shipper.

Given: Assume that Port X in the Gulf of Siesta near the Kolbi oil fields, with terminal facilities for handling up to 120,000-dwt tankers, has been averaging an estimated total awaiting time per ship of 45 hr. The production-to-liftings ratio has been either 1.2 to 1 or 1.4 to 1, or 1,134,000 bbl/day (151,200 tons/day) production to 945,000 bbl/day (126,000 tons/day) liftings on a 1.2 ratio, and 1,323,000 bbl/day (176,400 tons/day) to 945,000 bbl/day (126,000 tons/day) on a 1.4 ratio. Tank capacity is 1,800,000 bbl (240,000 tons); the loading system capacity is 40,000 bbl/hr (4,000 tons/hr) or 960,000 bbl/day.

With one berth, loading is as follows.

40,000 bbl/hr, or 5,333 tons/hr. This is equal to 960,000 bbl/day, or 128,000 tons/day.

Tank capacity is 1,800,000 bbl (240,000 tons).

Production-to-liftings ratios are as follows: At 1.2 to 1, the ratio is 1,134,000 bbl/day (151,200 tons/day) for production in oil fields to 945,000 bbl/day (126,000 tons/day) for liftings at terminal. At 1.4 to 1, the ratio is 1,323,000 bbl/day (176,400 tons/day) for production in oil fields to 945,000 bbl/day (126,000 tons/day) for liftings at terminal.

On this basis, the maximum-size tanker, 120,000 dwt that can be handled at this terminal can be loaded in 22.5 hr (120,000 ÷ 5,333 tons/hr).

With the addition of, for example, an increase in loading system capacity to 80,000 bbl/hr (10,666 tons/hr), or 1,920,000 bbl/day (255,555 tons/day) and another berth, giving two tanker berths total, two 120,000-dwt tankers can be loaded in 22.5 hr, or one tanker in 11.25 hr.

Wanted: With the loading capacity now 255,555 tons/day (1,920,000 bbl/day), tank capacity and the production-to-liftings ratio per hour need to be increased to take full advantage of the extra loading capacity and extra tanker berth now available. The question here is how much increase.

Solution

One way is as follows. With the unknown quantity in this case, X, being new tank capacity, "old" tank capacity is multiplied by the new loading capacity figure. Then, dividing by the "old" loading capacity figure times X gives the amount of "new" tank capacity.

240,000 tons × 255,555 tons + 128,000 tons × X

$$= 480,000 \text{ tons, or } 3,600,000 \text{ bbl as new tank capacity}$$

On "new" production to liftings based on the same ratios of 1.2 to 1 and 1.4 to 1, the figures are as follows, all in tons (X now represents "new" production):

For the 1.4 to 1 ratio,

("old" production at 1.4) × ("new" tank capacity)

÷ ("old" tank capacity)(X)

Using this formula, we have

$$(176,000)(480,000) \div (240,000)(X) = 352,800 \text{ tons } (2,646,000 \text{ bbl/day})$$
for oil field production

Liftings would then be 252,000 tons/day, or 1,890,000 bbl/day (2,646,000 ÷ 1.4).

For the 1.2 to 1 ratio of production to liftings, it is

$$(352,800)(1.2) \div (1.4)(X) = 302,400 \text{ tons/day, or } 2,268,000 \text{ bbl/day}$$

Then 302,400 tons/day divided by 1.2 gives 252,000 tons/day liftings, which is also 1,890,000 bbl/day.

Conclusion:

New production to liftings ratio at 1.4 to 1 is

2,646,000 bbl/day:1,890,000 bbl/day

New production to liftings ratio at 1.2 to 1 is

2,268,000 bbl/day:1,890,000 bbl/day

Loading capacity is now 1,920,000 bbl/day

New tank capacity is now 3,600,000 bbl.

120,000-dwt tankers can now be loaded in 11.25 hr instead of 22.5 hr, if only one berth were used. A savings of 11.25 hr in awaiting time per ship at, say, $1,000/hr, is a total savings per each tanker in port of $11,250. With several hundred tankers being loaded each year, this savings can be substantial.

TANKER SCHEDULING

Actually, scheduling of tankers, as such, precedes programming or planning of tankers. Scheduling involves the determination of the number of different-sized tankers which will be needed. Scheduling tankers is not an easy job, what with changing relationships between supply and demand of tankers, and the different-sized tankers required in connection with terminal facilities. Many terminal facilities, for instance, are not equipped to take larger-size tankers, including the new supertankers of today.

EXAMPLE 9.7: Scheduling with Estimated Total Tonnage Requirements

The number of tankers of a given sizes can usually be determined on the basis of the number of ton-days needed. For example, if a refinery requires 80 million tons of crude a year (600 million bbl) and has port facilities for tankers up to 270,000 dwt, it should be able to tell how many tankers of this size would be required on a time charter, one year, contract.

Assuming the net weight of a 270,000-dwt tanker is 248,400 tons, after allowing for fuel, crew provisions, and spare part inventory, and that such tanker will be in operation 323 days a year, Arabian Gulf source, and since 42 days are estimated as being normally required for docking repairs, surveys, and so on, average speed is estimated at 12 knots, the number of 270,000-dwt tankers needed can be determined. With this information, and assuming the travel time from the Arabian Gulf source of supply of crude to the refinery to be 19.5 days, the number of 270,000-dwt ships needed per year can be roughly determined as follows.

1.
$$\begin{array}{ccc} \text{Requirements} & \text{no. of days required} & \text{total number of} \\ \text{per year} \times & \text{by tanker for round} = & \text{ton-days} \\ & \text{trip (19.5} \times 2) & \text{needed} \end{array}$$

$$80,000,000 \text{ tons} \times \quad 39 \text{ days} \quad = 3,120,000,000 \text{ ton-days}$$

2. Number of ton-days of one 270,000-dwt tanker per year

$$\begin{array}{ccc} \text{Running time per year} & \text{net weight (dwt) per} & \text{ton-days per year per} \\ \text{for one 270,000-dwt} \times & \text{270,000-dwt} = & \text{each 270,000-dwt} \\ \text{tanker} & \text{tanker} & \text{tanker} \end{array}$$

$$323 \text{ days} \quad \times \quad 248,400 \text{ tons} \quad = 80,233,200 \text{ ton-days}$$

3. Then, total number of ton-days needed ÷ ton-days per year + 279,000
   per tanker

$$= \text{no. of ships required.}$$

$$3,120,000,000 \text{ ton-days} \div 80,233,200 \text{ ton-days} = 40 \text{ ships}$$

So, average cargo ± 248,400 tons of crude oil times 323 days =
80,233,200 ton-days. On this basis, forty 270,000-dwt tankers are required
for delivery of 80,000,000 tons of crude oil from the Arabian Gulf course.

Tonnage requirements from other sources of supply can be calculated
in the same way by merely changing the number of travel days involved between source and customer. Also, tonnage requirements can be determined
similarly when larger or smaller tankers are being considered.

Separate assessments of the number of each type of tanker required
must be made well ahead of need. This is especially true when dealing with
supertankers.

EXAMPLE 9.8: Matching Terminal Facilities to Liftings

In coordinating tankage requirements, loading facilities, and production of
oil in the oil fields, we want to load a supertanker of 700,000 dwt (5,250,000
bbl) capacity in 24 hr. (There are 7.5 bbl to the ton.) We are producing
2 million bbl/day, so we need tank capacity of at least 6 million bbl. This
includes a safety factor against any temporary shortage of shipping.

Actually, six 1-million bbl tanks for one loading of a 700,000-dwt
tanker requires an order of 5,250,000 bbl. Excessive tank capacity is not
desirable because of heavy tank investment costs. Estimated tank investment costs can be as high as $3 to $5/bbl. Thus, if we have a 6-million bbl
capacity, our total investment could be anywhere from $18 to $30 million.
It is wise, however, to have some excess capacity in order to have some
excess capacity in order to accommodate late ship arrivals due to inclement
weather.

Now, when production in the oil fields is boosted, tank requirements
must also change. If we then assume an oil field increase in production of
say, 1 million bbl/day, tank requirements and possibly loading facilities
must also increase in proportion, in order to guarantee a continuance in
supply of tankers and to safeguard against cutbacks in oil production.

If the 700,000-dwt tanker takes 52 days, including one day for loading
and one day for unloading, for its voyage to Rotterdam, The Netherlands,
from Ras Tanura, Saudi Arabia, and return, and seven voyages are made
annually (365 ÷ 52 = 7), the cost per hour per ton at a World Scale rate of
100 is $3,250 ($5.79/ton, or $3,250 × 1,240 hr ÷ 700,000 tons). Then a
savings in loading time of just 3/10 of each hour of loading, or 7.2 hr on a
24-hr day loading (50.4 hr total on seven voyages) can means an annual savings of $1,146,600 (50.4 × $3,250 × 7 voyages) for a tanker of 700,000 dwt
capacity.

If we further assume that we have increased our oil field production to 3 million bbl/day (an increase of 1 million bbl/day or 365 million bbl/year), a total of 209 (1,095,000,000 ÷ 5,250,000) 700,000-dwt tankers would be needed to transport 1,095,000,000 bbl of oil annually (3 million bbl/day × 365 days). Total annual savings could then amount to $163,800 × 209 ships, or $34,234,200 based on the maintenance of at least a 3/10 of an hour saving in loading time, as tankage capacity and the loading system capacity are increased in proportion with an increase in oil field production.

The overall objective at the terminal in terms of tanker scheduling and programming is to load the ships as quickly as possible, with a minimum of delay. Orders for oil, or for oil products, should be available or at least partly ready when tankers "berth in" for loading at the terminal. Furthermore, by eliminating the hiring of even one spot charter per year, which may cost $2 million for a single voyage, by savings possibly 1,000 days on time charters, the planning for terminal operations and the efficient scheduling of tankers in and out of the terminal will certainly have been worthwhile.

Now let us apply some of the financial aspects of oil management discussed in Chapter 4 to a proposed investment in additional terminal facilities. Cash flow, present values, and annual cost of the project become important factors in determining the economic feasibility of the investment and in aiding us to make the correct decision relative to the investment.

EXAMPLE 9.9: Investment in Additional Terminal Facilities to Cut Costly Awaiting Time

Given: Number of tankers in X terminal per year as 300; average size tanker loaded is 240,000 dwt. Rates are WS 100 or $1,490/hr; WS 80 or $1,192/hr (80% of $1,490); and WS 50 or $745/hr (50% of $1,490).

We are concerned about the long and costly awaiting times of ships in port. Therefore, we are considering an investment in a SPM (Special Point Mooring) loading facility. We discover that if we install a SPM at $1 million total investment, we will save 1 hr per ship coming into the terminal for crude loading. Current tank capacity is assumed to be 2 million bbl. In addition, the following costs for the SPM terminal must be considered:

Estimated life of the investment is 20 years, with the government tax being 55%. An amount is then calculated, as 0.55 × 1/18 × $1,000,000 or $30,555 each year as a cash inflow. (Although depreciation allowance is for 20 years, we will not actually begin using the SPM until the third year.) Assuming operating costs of the SPM to be the same each year at $100,000, and with the government assuming a 55% portion of the running costs, operating costs per year for the SPM are then $0.45 \times \$100,000 = \$45,000$ as an annual cash outflow (maintenance costs after taxes).

We also assume that it will take the first two years for designing and engineering, building, and installation of the SPM, with the third year then pegged for the beginning of operations. We estimate that 1 hr per ship at WS 100, or $1,490/hr per ship, will be saved with this investment. This will amount to savings of $447,000/year, or 300 ships × $1,490/hr per ship.

Wanted: To determine: (1) cash flow, (2) discounted cash flow or present value of the investment, and (3) annual cost.

Table 9.11 gives the cash flow. Table 9.12 indicates the discounted cash flow (present values), using 15% as the discount factor, which then indicates present values declining each year with the smallest present value at 20 years. Table 9.13 gives the annual cost at 15% and Figure 9.10 shows optimum tankage at the WS 100 rate.

Solution

Tank costs (total cost of capital investment in tanks per year):

1. For a 1 million bbl tank investment of $5 million—$5,000,000 × 0.129 annual equivalent gives us $645,000 for cost of capital per year. (0.129 annual equivalent is used for all capital investments involving a 15% discount factor, where depreciation is 20 years and tax is 55%. This figure is derived from Table 9.13 by dividing 6.129, the 15% factor, by $488,890 total of depreciation and tax computation). Next, $645,000 + $55,000 (average annual maintenance and repair costs of tanks) = $700,000 total costs of a 1 million bbl tank per year.

2. For a 2 million bbl tank capacity, the cost per year is $1,400,000 ($700,000 × 2).

3. For a 3 million bbl tank capacity, the cost per year is $2,100,000 ($700,000 × 3).

4. For a 4 million bbl tank capacity, the cost per year is $2,800,000 ($700,000 × 4).

5. For a 5 million bbl tank capacity, the cost per year is $3,500,000 ($700,000 × 5).

Total cost per year is awaiting times, in hours, multiplied by $1,490 (WS 100 rate) × 300 ships. Table 9.14 shows the calculation of awaiting times.

Table 9.14 points out that awaiting times and total costs per year increase with decreased tank capacity. These costs must be balanced against tank investment amounts to determine the feasibility of investing.

Figure 9.10 illustrates the optimum tankage when the liftings are 1 million bbl/day of crude. As Figure 9.10 shows, the optimum tankage is 2,800,000 bbl. At 2,800,000 bbl, capital cost plus maintenance and repair costs averaging $55,000 is $1,960,000.

Table 9.14, illustrated by Figure 9.10, shows that when tankage capacity is reduced, awaiting times of tankers increases and thus costs per ship are increased. Only with comparable tank capacity, or balancing of tank costs with costs per year of awaiting times, can we arrive at an optimum point where ship savings overcome tank costs. This figure is reached at a tank capacity of 2,800,000 bbl when the total cost per year of average ship awaiting time of 4.4 hr ($1,960,000 ÷ 300 × $1,490 = 4.4 hr) times 300 ships times the WS 100 rate of $1,490 is $1,960,000. It is also the cost of 2,800,000 bbl of tank capacity ($700,000 × 2.8 = $1,960,000).

ESTIMATING TANKER TONNAGE REQUIREMENTS BY USERS

The oil industry needs to maintain a balance between supply of available tankers and demand for their carrying capacities. Shippers and users of oil transportation services need to estimate their tonnage requirements and their "security tonnage" (stores, fuels, spare parts, etc.) by determining in advance how much in charters are required to cover refinery needs. This applies to time charters and to charters for consecutive "hauls" of oil. Less dependency needs to be placed upon the freight market, with its wide fluctuations in rates. This also means that there is less need for charters for single voyages (spot carriers), which represent last-minute corrections to shipping programs. Spot arrangements in shipping oil can be subject to fluctuations in rates in the market.

Forward planning is required to ensure that tanker tonnage requirements are available at the right time and at the right place. Because of changing relationships between demand and supply, and among different types of tankers required in connection with terminal facilities, forward planning of tanker tonnage requirements is not easy to do.

Total tonnage requirements are expressed in terms of a standard tanker. For instance, tonnage requirements might be for a T-2 tanker of 16,000 dwt tons, that has a speed of 14 knots. On such a tanker as a T-2, it might be assumed that it will be in operation 330 days out of a year, with 35 days per year required for docking, repairs, survey, and so on. On the basis of this information, a refinery might wish to determine what will be the probable source of supply of T-2 tankers in line with its tonnage requirements at the refinery plant. The amount of crude oil required is then multiplied by the number of days required by a T-2 tanker for a round-trip voyage.

TABLE 9.11

Cash Flow

	1 year	2 years	3 years	4 years	5 years	6 years
When WS = 100						
Investment $1,000,000 +0.55 × 1/20 × 1,000,000	-500,000	-500,000	+30,555	+30,555	+30,555	+30,555
Operating cost 0.45 × $100,000			-45,000	-45,000	-45,000	-45,000
Subtotals	-500,000	-500,000	-14,445	-14,445	-14,445	-14,445
Savings per ship at $1,490/ hr (WS 100 rate)			+447,000	+447,000	+447,000	+447,000
Net cash flow when WS = 100	-500,000	-500,000	+432,555	+432,555	+432,555	+432,555
When WS = 80	-500,000	-500,000	-14,445	-14,445	-14,445	-14,445
Savings at $192 per ship			+357,600	+357,600	+357,600	+357,600
Net cash flow when WS = 80	-500,000	-500,000	+343,155	+343,155	+343,155	+343,155
When WS = 50 subtotals	-500,000	-500,000	-14,445	-14,445	-14,445	-14,455
Savings at $745 per ship			+223,500	+223,500	+223,500	+223,500
Net cash flow when WS = 50	-500,000	-500,000	+209,055	+209,055	+209,055	+209,055

TABLE 9.11 (continued)

	7 years	8 years	9 years	10 years	11 years	12 years
When WS = 100						
Investment $1,000,000						
+0.55 × 1/20 × 1,000,000	+30,555	+30,555	+30,555	+30,555	+30,555	+30,555
Operating cost 0.45 × $100,000	−45,000	−45,000	−45,000	−45,000	−45,000	−45,000
Subtotals	−14,445	−14,445	−14,445	−14,445	−14,445	−14,445
Savings per ship at $1,490/ hr (WS 100 rate)	+447,000	+447,000	+447,000	+447,000	+447,000	+447,000
Net cash flow when WS = 100	+432,555	+432,555	+432,555	+432,555	+432,555	+432,555
When WS = 80						
subtotals	−14,445	−14,445	−14,445	−14,445	−14,445	−14,445
Savings at $192 per ship	+357,600	+357,600	+357,600	+357,600	+357,600	+357,600
Net cash flow when WS = 80	+343,155	+343,155	+343,155	+343,155	+343,155	+343,155
When WS = 50						
subtotals	−14,445	−14,445	−14,445	−14,445	−14,445	−14,445
Savings at $745 per ship	+223,500	+223,500	+223,500	+223,500	+223,500	+223,500
Net cash flow when WS = 50	+209,055	+209,055	+209,055	+209,055	+209,055	+209,055

TABLE 9.11 (continued)

	13 years	14 years	15 years	16 years	17 years	18 years
When WS = 100						
Investment $1,000,000						
+0.55 × 1/20 × 1,000,000	+30,555	+30,555	+30,555	+30,555	+30,555	+30,555
Operating cost 0.45 × $100,000	−45,000	−45,000	−45,000	−45,000	−45,000	−45,000
Subtotals	−14,445	−14,445	−14,445	−14,445	−14,445	−14,445
Savings per ship at $1,490/hr (WS 100 rate)	+447,000	+447,000	+447,000	+447,000	+447,000	+447,000
Net cash flow when WS = 100	+432,555	+432,555	+432,555	+432,555	+432,555	+432,555
When WS = 80	−14,445	−14,445	−14,445	−14,445	−14,445	−14,445
Savings at $192 per ship	+357,600	+357,600	+357,600	+357,600	+357,600	+357,600
Net cash flow when WS = 80	+343,155	+343,155	+343,155	+343,155	+343,155	+343,155
When WS = 50	−14,445	−14,445	−14,445	−14,445	−14,445	−14,445
subtotals						
Savings at $745 per ship	+223,500	+223,500	+223,500	+223,500	+223,500	+223,500
Net cash flow when WS = 50	+209,000	+209,055	+209,055	+209,055	+209,055	+209,055

TABLE 9.11 (continued)

	19 years	20 years
When WS = 100		
Investment $1,000,000		
+0.55 × 1/20 × 1,000,000	+30,555	+30,555
Operating cost 0.45 × $100,000	-45,000	-45,000
Subtotals	-14,445	-14,445
Savings per ship at $1,490/ hr (WS 100 rate)	+447,000	+447,000
Net cash flow when WS = 100	+432,555	+432,555
When WS = 80	-14,445	-14,445
Savings at $192 per ship	+357,600	+357,600
Net cash flow when WS = 80	+343,155	+343,155
When WS = 50 subtotals	-14,445	-14,445
Savings at $745 per ship	+223,500	+223,500
Net cash flow when WS = 50	+209,055	+209,055

TABLE 9.12

Present Values

Year	At 0%, $m	At 15%, $m cash flow discount factor		At 30%, $m	
1	-500	-500 × 0.870	= -435.0	-500 × 0.769	= -384.500
2	-500	-500 × 0.756	= -378.0	-500 × 0.592	= -296.000
3	+432.5	+432.5 × 0.658	= +284.6	+432.5 × 0.455	= +196.7875
4	+429.5	+429.5 × 0.527	= +247.4	+432.5 × 0.350	= +151.3750
5	+429.5	+429.5 × 0.497	= +214.95	+432.5 × 0.269	= +116.3425
6	+429.5	+429.5 × 0.432	= +186.8	+432.5 × 0.207	= +89.5275
7	+429.5	+429.5 × 0.376	= +162.6	+432.5 × 0.159	= +68.7675
8	+429.5	+429.5 × 0.327	= +141.4	+432.5 × 0.123	= +53.1975
9	+429.5	+429.5 × 0.284	= +122.8	+432.5 × 0.094	= +40.6550
10	+429.5	+429.5 × 0.247	= +106.8	+432.5 × 0.073	= +31.5725
11	+429.5	+429.5 × 0.215	= +93.0	+432.5 × 0.056	= +24.2200
12	+429.5	+429.5 × 0.187	= +80.9	+432.5 × 0.043	= +18.5975
13	+429.5	+429.5 × 0.163	= +70.5	+432.5 × 0.033	= +14.2725
14	+429.5	+429.5 × 0.141	= +61.0	+432.5 × 0.025	= +10.8125
15	+429.5	+429.5 × 0.123	= +53.2	+432.5 × 0.020	= +8.6800
16	+429.5	+429.5 × 0.107	= +46.3	+432.5 × 0.015	= +6.4875
17	+429.5	+429.5 × 0.093	= +40.2	+432.5 × 0.012	= +5.1900
18	+429.5	+429.5 × 0.081	= +35.0	+432.5 × 0.009	= +3.8925
19	+429.5	+429.5 × 0.070	= +30.3	+432.5 × 0.007	= +3.0275
20	+429.5	+429.5 × 0.061	= +26.4	+432.5 × 0.005	= +2.1625

COMBINED CARRIERS

Today the combined carrier, which carries oil as well as other commodities, is coming into prominence, representing about 10% of the world's crude oil carrying capacity. It was reported that, by 1975, 90% of this rapidly growing combined carrier fleet, according to London tanker brokers, was expected to have a total capacity of almost 45 million dwt [2], but executives of oil companies using these carriers actually reported a total capacity of 50 million dwt in 1975.

But the primary function of all sizes of combined carriers is the transport of crude oil over long ocean routes. The combined carrier was designed and developed to haul oil, or dry cargo, over two-way and triangular routes. That is, it was built to operate as an oil tanker, with the added ability to take on ore or coal when that would be economically advantageous,

TABLE 9.13

Annual Cost of $1 Million Investment, 20 Years Depreciation
and 55% Tax Rate at 15% Interest

Year*	Depreciation and tax on total investment per year	×	15% factor	=	What we get back
3	+30,555		0.870		26,582.850
4	+30,555		0.756		23,099.580
5	+30,555		0.658		20,105.190
6	+30,555		0.572		17,477.460
7	+30,555		0.497		15,185.835
8	+30,555		0.432		13,199.760
9	+30,555		0.376		11,488.680
10	+30,555		0.327		9.991,485
11	+30,555		0.284		8,677.620
12	+30,555		0.247		7,547.085
13	+30,555		0.215		6,569.325
14	+30,555		0.187		5,713.785
15	+30,555		0.163		4,980.465
16	+30,555		0.141		4.308.255
17	+30,555		0.123		3,758.265
18	+30,555		0.107		3,269.385
19	+30,555		0.093		2,841.615
20	+30,555		0.081		2,474.955
Total for 20 years	488,890		6.129		187,271.595

*SPM not in use during first 2 years.

on one or more legs of its round-trip voyages in order to minimize the cost of ballast. But the combined carrier was not designed to switch from one product market to another in accordance with changes in freight rates in different product markets. Trading patterns of the individual combined carrier companies are generally determined by their commitment to oil charters.

H. P. Drewry Shipping Consultants estimate that a dual-purpose ship, such as the combined carrier, must earn at least 20% more than a conventional tanker or bulk carrier, a 60% sailing time loaded to 40 percent in ballast, as compared to the 50-50 schedule on which a conventional carrier operates, if the additional investment in combined carriers is to be justified. The best recorded operating schedule achieved by an ore-oil carrier

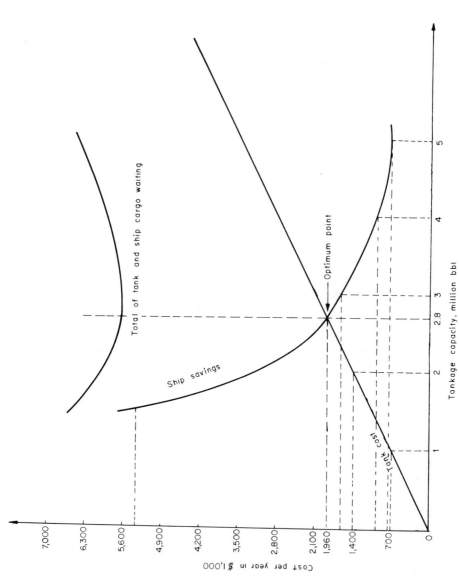

FIG. 9.10. Optimum tankage, at WS 100, for a 270,000-dwt tanker, with 1 million bbl/day liftings.

TABLE 9.14

Costs of Awaiting Times
(300 ships loaded, 1 million bbl/day liftings
for the 270,000-dwt tanker average)

	Awaiting times, hr	Total cost per year at $1,490/hr
With 5 million bbl capacity ($750,000 cost per year ÷ 300 = 2500 ÷ 1,490)	1.7	$750,000
With 4 million bbl capacity ($1,050,000 ÷ 300 = $3,500 and $3,500 ÷ 1,490)	2.3	1,050,000
With 3 million bbl capacity ($1,600,000 ÷ 300 = $5,333 and $5,333 ÷ 1,490)	3.6	1,600,000
With 2 million bbl capacity ($3,150,000 ÷ 300 = $10,500 and $10,500 ÷ 1,490)	7.0	3,150,000
With 1.5 million bbl capacity ($5,450,000) ÷ 300 = $18,167 and $18,167 ÷ 1,490)	12.2	5,450,000

in the 100,000-120,000 dwt class, according to Drewry, was a 67.33% ratio, or 246 days at sea loaded, 82 days in ballast, and 37 days in port (246:82 + 37). This, however, was the result of a study in which the average for all ships was under 58% for ore/oil and 56% for ore/bulk/oil carriers [3].

The costs of building a dual-purpose ship are estimated by Drewry to be 10 to 15% more than that of an equivalent-size single-purpose carrier. For example, a 160,000-dwt ore/oil carrier costs over $25 million, or nearly $160/ton. Also, a dual-purpose ship committed on charter or long-term contract of affreightment to only one of its two potential markets can be expected to be used and earn profits equal to a single-purpose ship, but the operating costs of a dual-purpose ship will usually be higher. Thus, ballast time of the dual-purpose ship must be reduced substantially and replaced with loaded time so earnings can more than compensate for these higher costs.

The trend today, surprisingly, is for the dual-purpose carrier to trade exclusively in oil. Drewry feels that this is due to trends toward higher freight rates in the oil market and to the tanker rate structure itself. Not only that, the major oil companies, or largest shippers of oil and purchasers of tankship service, appear to prefer the combined carrier, as these oil companies charter them at rates equivalent to or better than the rates being paid for conventional tankers of equivalent size. Even some of the combined carriers are being operated in the spot market, although less profitably, exclusively for oil voyages. Thus, the trend is to use dual-purpose carriers exclusively for oil carrying, and on long-term charters rather than spot purchasing.

ECONOMICS OF DESIGN OF TANKERS

Today's deadweight tonnage continues to increase, and is also greater in proportion by reason of improvements in speed and turnaround of the latest types of tankers. In designing modern tankers, three major problems had to be overcome:

1. Provision was needed to reduce to a minimum loss of stability from movement of a liquid cargo and also to prevent the cargo, or water ballast, from surging so violently as to damage the ship in a storm.

2. Allowance was needed for changes in volume of a liquid cargo during extreme variations in sea and air temperatures on voyages from hot to cold and from cold to hot climates.

3. Construction of the tanker had to allow for safe ballasting, as many of the voyages had a high proportion of "ballast passages."

Other factors also had to be considered in design and layout, such as ventilation of cargo spaces, safety and fire precautions for ship and cargo, and so on.

Anyone entrusted with the job of designing the most economical tanker is faced with many problems which have to be considered carefully, especially since there may be conflicting considerations involved in resolving these problems. For instance:

1. The deadweight capacity, or carrying capacity of a vessel is the total carrying weight of the cargo, bunkers, provisions for the crew, fresh water, stores, and spare parts, expressed in tons of 2,240 lb (long tons), which a vessel can lift when loaded in salt water to her maximum draught. The draught of a vessel is the vertical distance between the water line and the keel, or bottom, of the ship.

2. Dimensions of supertankers are closely related to required speed. A high speed necessarily implies a fine hull, which affects cubic cargo capacity. A ship's length can be increased, which means a higher displacement in the sea, but also a corresponding increase in fuel consumption and a larger bunker capacity. Therefore, when designing and building large-size tankers, investment costs increase with increases in vessel dimensions.

3. Draught. Dimensions, in particular the draught, are also governed by the number of loading and discharging terminals around the world which can handle deeply laden tankers. The larger the tanker, the smaller the number of terminals where it can be accommodated without difficulty. In other words, the flexibility of operation of supertankers diminishes somewhat as the size of piping increases. The future, however, should bring an increase in the number of terminals that can accommodate supertankers, as more countries are presently building, or have on the drawing boards, improved and enlarged terminal facilities designed to cope with the steadily increasing number of supertankers.

4. Speed. High speeds are attractive to oil producers and consumers from the viewpoint of cargo-carrying capacity per year. But high speeds are expensive from the viewpoint of fuel consumption. Fuel consumption varies, however, as the cube of the speed, and does not increase in direct proportion to the speed. For example, fuel consumption for tankers of approximately the same deadweight tonnage at varying speeds was reported in 1963 as shown in Table 9.15, indicating the effect on daily fuel consumption.

Higher costs of machinery and extra fuel consumption must be economically balanced with enough compensation from extra cargo-carrying

TABLE 9.15

Fuel Consumption of Tankers

Ship	Capacity, dwt	Speed, knots	Fuel consumption per day, tons
T.T. Golden Eagle	32,256	15	72
T.T. Opportunity	32,517	16	85
T.T. Thorshall	42,400	16	80
T.T. Pentelikon	41,831	16.5	92
T.T. Arietta S. Livanos	41,200	17	100
New (under construction)	80,000	15	115
New (under construction)	80,000	15 1/2	125

Source: J. Bes, Chartering and Shipping Terms (Amsterdam, Holland: Drukkery Press, 1962), pp. 220-250.

capacity per year to make sea transportation services offered worthwhile
to tanker operations.

5. Propelling Machinery. The larger the tanker, the more expensive
the propelling machinery becomes. High-cost propelling machinery is justified when enough cargo can be obtained each year at profitable rates.

CHARTERING

For a shipper of oil who is chartering a tanker, or fleet of tankers, the
above aspects of deadweight tonnage of cargo, space, draught, speed, and
daily consumption of diesel or fuel oil are very important. On a time charter basis, the daily fuel consumption is of special interest to the charterer,
since tankers taken up on long-term time charter are operated by oil companies in exactly the same way as their own tankers. All running expenses,
including crew wages, repairs, insurance on tankers and on cargo, docking
charges, including demurage, survey, and so on, are paid by the charterer.
In addition, the charterer must maintain tankers in efficient condition while
under contract, and is expected to deliver the tanker to its owner in the
same condition as it was received, although a "normal" allowance is made
for ordinary wear and tear (depreciation). Charterers can also sublet vessels
under contract to them to other parties. This does not, however, relieve
the charterer from his original responsibilities to the owner of the tanker.

SUMMARY ON TANKERS

Increased production capacities and greater tankage volumes are required
to provide efficient cargoes for the larger tankers of today. One way of
satisfying these requirements is to program the arrivals of the larger tankers, as the smaller tankers have been programmed. Planning beforehand
for arrivals of the larger tankers will reduce average awaiting times in the
terminal for these expensive vessels. Also, by efficient programming of
these tankers, less investment will be needed in tankage.

Loading systems with capacities in the 200,000 bbl/hr range are needed to handle large-size tankers of the 300,000, 500,000, 750,000 dwt, and
possibly larger tankers types. Greater maximum capacities of loading systems will also provide more effective usage of other facilities of a terminal,
such as berth capacity. With proper planning, and improved loading system capacity, the number of berths and tankage requirements for larger-
size tankers can actually be reduced. This means a saving of investment in
port facilities.

When computers are used in planning and programming of tanker loading, input data used for determining an efficient program of ship loading

must be complete. Computer input data should include predicted ship arrivals by size and any crude lifting patterns of the past, port selection, targeted production capacities, tankage, berths, loading systems, any tanker's loading rate restrictions, and any weather and repair information that can affect output, or loadings.

The output should include the awaiting times in hours for cargoes, berths, and loading systems. Also, the percentage of occupancy in berths and loading systems, and total port time and total time at the berth should be included in output data supplied by the computer.

OIL PIPELINES

The other means of transportation of oil discussed in this book is pipelines, the third largest type of freight carrier in the world, with load volumes exceeded only by sea shipping and railroads. As the giant industry of petroleum spans the earth today, the oil pipelines become the steel tracks to internationalism, a fact borne out by the growing evidence that World War II was, above all else, a war for oil. With good reason, international oil and oil pipelines are foremost topics of diplomatic conversations today, just as they were of military conversations throughout World War II.

Since 1929, there has been a marked trend toward increased use of pipelines for movement of oil, due mostly to a greater use of steel pipe with higher tensile strength and thinner walls, either seamless or electrically welded and available from suppliers in 40-foot lengths. Because of its larger inside diameter, this type of pipe provides an appreciable increase in capacity relative to its weight. Moreover, it costs less to ship from the mills, is cheaper to lay, and is more resistant to corrosion and leakage than other pipe.

These developments affecting construction were closely related to important advances in the operation of pipelines. With pipe of greater tensile strength, for example, it became economically feasible to speed up pipeline flow, and thus more barrels per hour were moved by increasing the pressure within the lines. Electric-powered pumps have been used, because with their use automatically controlled central pipeline stations and remote-controlled line stations can be used.

The combination of higher and more closely controlled line pressures contributed significantly to an increase in the efficiency of "batching" operations. Batching, it is true, is a practice almost as old as the pipelines themselves. But with greater line pressures it has been possible to ship various types of crude and refined products through the lines, one after the other, with a minimum of contamination or mixing. This can be done, however, only when the structure of the refining sector is kept at carefully predetermined levels.

TAPLINE AND TANKERS

Tapline is a 30-inch-diameter pipeline, 1,100 miles in length. It begins in the Aramco oil fields at Quaysomah, Saudi Arabia, and extends across Jordan and Syria to its Mediterranean terminal at Sidon, Lebanon, with a later extension supplying a second terminal at Tripoli, Lebanon. This pipeline, constructed in 1949-1950 at a cost of $230 million, had a capacity of 300,000 bbl/day, or 15 million tons/year. By constructing four new pumping stations located between five existing pumping stations, the capacity was increased by 1958 to 469,000 bbl/day or approximately 23.5 million tons/year.

The justification for the building of Tapline was to reduce the cost of transportation per ton-mile in competition with normal transport by tanker. At the time of the construction of Tapline in 1949, the Suez Canal was open and there were no supertankers.

EXAMPLE 9.10: Tapline

Let us compare the use of a T-2 tanker, deadweight capacity 16,000 tons, speed 14 knots, on a round-trip voyage with and without the use of Tapline. The comparison is shown in Table 9.16.

TABLE 9.16

Round-Trip Voyage With and Without Use of Tapline

	Sidon-Rotterdam-Sidon (with use of Tapline)	Ras Tanura-Rotterdam-Ras Tanura (without use of Tapline)
Steaming days	20 days	38 days
Extra days for Suez	—	2
Loading	2	2
Discharge	2	2
	24 days	44 days
Average cargo	15,300 tons	15,300 tons
Number of round trip voyages per year	14	7.5
Amount of cargo carried per year	214,000 tons	115,000 tons
Suez Canal dues (tolls)	—	£33,000

Source: J. Bes, Chartering and Shipping Terms (Amsterdam, Holland: Drukkery Press, 1962).

As Table 9.16 shows, the use of Tapline for movement of oil and oil products to Europe is preferred. The round-trip voyage is cut 20 days, the number of round-trip voyages per year is doubled, thus total oil cargo moved is considerably increased by using Tapline services. In addition, an estimated expense of £33,000 is saved in Suez Canal fees by use of Tapline. Thus, Tapline is more economical than use of oil tankers traveling through the Suez Canal.

The evolution of the supertanker since 1955 has radically changed the original cost relationship. Supertanker savings in cost of transportation, as compared with the T-2 tanker, are approximately 48% for a 45,000-dwt tanker and approximately 57% for a 70,000-dwt tanker [3]. Consequently, movement of crude oil by Tapline is no longer competitive with transport by tanker. The fixed capital charges largely determine the cost per ton-mile, which are lowest when spread over maximum capacity. Since 1957, the pipeline has never been operated to full capacity.

Another factor which has affected the movement of crude oil by Tapline to Sidon, Jordan, has been the payment of higher royalties to the three governments of Jordan, Syria, and Lebanon. (Saudi Arabia, the fourth country involved, now owns a 60% interest in the line.) The original transit agreement with Tapline provided for flat annual payments. Since 1956, the payments of transit royalties have increased as tanker rates have declined; this had also hurt Tapline operations. Finally, as tanker rates have been kept lower than pipeline fees, and the use of supertankers to carry volume has increased, it has become cheaper to transport by tanker. Thus, in 1970, the amount of crude being shipped through Tapline decreased to 100,000 bbl/day, and in 1973, Tapline shut down completely, and as of this writing, Tapline is still not operating at all.

NOTES

1. Zanetos, Zenon S., The Theory of Oil Tankship Rates (Cambridge, Mass.: M.I.T. Press, 1966).

2. Petroleum Press Service, "The Role of Combined Carriers," February 1972, pp. 61-64.

3. Bes, J., Chartering and Shipping Terms (Amsterdam, Holland: Drukkery Press, 1962), pp. 220-250.

Chapter 10

CONCLUSIONS

It is now time to sit back and review the materials presented in this book. In particular, we want to inquire as to what light these materials throw on the overall structure of the oil industry in real life.

Let us begin by recalling briefly the nature of the materials that we have read and examined. We began our excursion into the economics of oil engineering by observing the vast amount of supplies, both still in the ground and being produced on a worldwide basis. We found supplies to be unbelievably large, the figures far beyond the limits of the layman's imagination. And the demand around the world for these supplies easily matched the figures on supplies. We may have asked ourselves what kind of order could be discerned in the apparent confusion of satisfying requirements of demanders of oil and oil products by suppliers when consumption actually equals production and processing. The answer is through efficient planning and scheduling of orders by suppliers and by increasing discoveries of crude oil reserves in the ground through continuous explorations.

Three forces ultimately direct the continuous searching and buildup of oil reserves: (1) the continued willingness and ability of consumers to buy the various oil products offered to them; (2) a realistic price, with appropriate price increases, which give incentives to those to find more needed oil reservoirs, plus requirements to meet additional increasing costs of finding oil; (3) the technological capabilities and improved exploration/ development techniques and instruments.

We learned something about the raw material, crude oil, and the finished product, or refined oil product. For one thing, crude oil, like many other raw materials which must be processed before consumption can take place, is usually not the same in composition, or in characteristics, at each of its source of supply. Thus undifferentiated crude oil is the exception rather than the rule. This sometimes complicates refining.

Furthermore, the only demand for crude oil is a derived one. The demand is for oil products. Thus demand for crude depends on demand for oil products. When demand for oil products drops, demand for crude will drop. Refining of crude oil is a complex series of operations with production in a continuous flow through a maze of different piping of various sizes, and with the use of different-size vessels and tanks, featuring temperature and pressure control of the highest order on a strict time basis. The number of oil products produced and their uses is enormous, and new uses for oil products continue to be found, thus increasing demand for crude oil.

To keep up with demand and to exploit supplies, to process them, and to get the finished products to market costs money. Thus, large amounts of investment funds for exploring, developing, and producing in the oil fields, for building sufficient capacities in refineries and processing of the crudes are needed continuously. Since risks of investment are great, particularly in searching for crudes, we have found that a knowledge of financial aspects of the industry are exceedingly important. Such concepts as net cash flows, returns on investment, payback periods, and so on, as well as depreciation charges and depletion allowances, need to be understood and applied practically to projects and assets.

We have observed engineering economics in the oil fields in reference to economic balance in yield by reference to oil well drilling, spacing of wells, economic size of pipe diameter, and so on, taking into consideration total costs vs. values to be received. We have found that secondary recovery is highly important today, especially when primary recovery is low and discoveries of new reserves drop off.

We learned that economics of oil engineering is prevalent in refineries, as well as in the oil field, especially relative to economics of design of equipment of yield and recovery in process, and of economies of scale.

In Chapter 8, we discovered how complex the pricing of crude and refined oil products can be. Posted prices, which are administered prices set by producing countries and companies, determine selling prices in the industry. Most oil supplies rely on tankers and pipelines for long-distance movement. In most cases, it is more economical to ship by tanker than by pipeline. The oil industry is thus a tremendous industry to comprehend. For instance, consider the following.

ON OIL FIELDS

The costs of developing discovered reserves is a prime factor that prevents unprofitable flooding of the market with oil supplies. Uncertainties of exploration actually determine levels of output or supply in the industry. Thus the crude oil industry is inherently self-adjusting, contrary to most common beliefs.

The uncertainties of exploration also act as a barrier to funds for exploration and development. The more exploration, the higher is the finding cost per unit of oil that is found. Also, the more a given discovery is developed, usually the higher is the cost of additional development.

Furthermore, from the time a well begins to operate, the greater the production, the higher the cost of additional production, as marginal costs rise. Also, when production declines (as production decline curves illustrated), costs per unit go up. When additional costs (marginal costs) of another level of barrel output go above current prices per barrel and remain so, production will cease in the long run; also, when incremental development costs rise above anticipated prices for the oil, development slows and eventually stops. In general, when the cost per barrel for finding oil goes above any long-term anticipation of profit, exploration ceases. Decisions on investment for exploring and developing oil reserves, as well as production, are made on the basis of the life of the oil field.

ON COST OF DEVELOPMENT

The costs of drilling and equipping numerous wells with pipelines and tankage are astronomical. Furthermore, in calculating the cost of finding oil, producing wells are charged with exploration costs.

In deciding whether to go ahead and develop for production, or whether not to do so, it is thus well to evaluate the project in terms of the risk factor relative to investment. The ratio of estimated revenues to be received to risk capital is an important indicator to consider, as well as the ratio of revenues to total exploration and development. In both cases, revenues must be considerably greater than risk capital and total exploration and development costs.

For example, assume that estimated risk capital is $25 million, exploration costs are $5 million, estimated ultimate recoveries are 120 million bbl of oil, and estimated average revenue is $2/bbl of crude, after taxes. Payments are estimated at royalty one-eighth, thus working interest on ultimate recovery is 105 million bbl. Then 105 million bbl × $2 = $210 million estimated total revenue.

With risk capital investment of $25 million, the ratio of estimated total revenue to risk capital is $210 million:$25 million or about 9:1. The risk factor is thus low. With exploration costs of $5 million, the ratio of estimated total revenue to exploration costs is $210 million, to $5 million, or 42:1. The ratio of estimated revenues to exploration costs is high.

Thus, if the project rates high in accessibility, all other things being equal, development and drilling should proceed, beginning with the drilling of a test well.

Sharp pencil calculations on payback time, annual rate of production, ultimate net profit after income taxes, and so on, are usually postponed until after oil is found through test well drilling. It is obvious that oil must be found at a cost per barrel less than the posted price (see Chapter 8). Actually, the cost of finding oil should be much lower than the posted price so that the oil company can continue to search, develop, and produce oil that has been discovered, and still sell the oil at a profit. By knowing what it costs to find oil per barrel, a comparison may be made with any so-called average cost per barrel of finding oil for the oil industry. In reality, because the cost of finding oil varies widely from year to year with the degree of success in finding new sources of oil, an average of several years is more truly a representable cost of finding oil in any given area. After establishing the cost of finding oil, the next problem is to establish the quantity of recoverable oil against which to apply these costs. The ultimate recovery figure is always subject to revision, which is often upward, in the course of subsequent development.

In the United States, after exploration and test drilling is completed, a "normal" well takes, according to U.S. experts, between 110 and 120 days to complete at a cost of approximately $300,000, with about 275 tons of oil tubular goods used. Current top allowable production rate is 550 bbl per well, and a permissible gas-oil ratio is 1,500 cubic feet per barrel of crude oil. Wells are drilled approximately on a 40-acre spacing plan. Although completed wells with high productivity yield 500 to 550 bbl/hr, the productivity index of the well really depends on the natural pressure of the well, which can range from a 2 bbl per pound per square inch down to little or no pressure at all.

ON REFINERIES

At the refinery level, costs are generally regarded as common costs. That is, the cost of industrial refined products as well as their respective profitabilities are regarded as arbitrary. Theoretically, it is possible to devise a unique method of allocating total refinery costs against various refined oil products produced on the basis of relative marginal costs of varying yields of each product. This is not, however, necessarily practical.

For instance, refined oil products derived from crude in the United States are not strictly joint products in the sense that an increase in the output of one product necessarily involves an increase in the output of the other product in invariable proportions. On the contrary, a modern refinery often increases the output of one of its joint refined products at the expense of another joint product. For example, by cracking distillates and heavy residual fuel oils, and then polymerizing refinery gases, the cracking process can continue until only gasoline and carbon emerge.

But joint product and production makes it difficult to assess the cost of any one product produced. The refiner knows what his total costs of operation are, and it is possible to single out certain finishing processes, such as fractionation of gasolines and chemical treatment of lubes, but it is almost impossible to determine how much running, or operating, expenses and overheads should be allocated to any one product.

Recently, the thinking has been that each product is a by-product of all the others, or all products are considered as co-products. This kind of thinking reflects a shift of importance from one product to another, which goes on continuously among gas oil, diesel oil, distillate fuels, and so on. As it is impossible to calculate, except for refined finishing processes, the production cost of industrial refined products, selling prices cannot be set based on costs. Selling prices are determined, therefore, on the basis of a method which runs parallel to the method of costing, which is that each product sells at the price its market will bear.

There is some price discrimination under this method of setting prices for oil products. With the relation of fixed costs to variable costs always high, this method of price setting creates the temptation to charge high prices for refined oil products which are "secure" in their markets at the time of setting prices, and to cut prices in cases where there is danger of competition and loss of markets or sales because of competition.

Within limits, it is possible to measure also the opportunity costs of industrial oil refined products, that is, what products are produced and at what costs these products are taking the place of what other products at the same costs. Where this can be done, as for example between home heating oils and gasolines, heating oils ought not to be sold for less than their worth, after making allowances for additional processing costs, as substitutes for crude oil in the production of additional gasoline.

The redeeming constituents in the economics of petroleum products, however, are as follows.

1. A whole range of finished oil products can be made from a given crude, which can be sold at different prices.

2. There is the possibility of varying the percentage yield (product mix) and blending of these various products. Being able to vary the amount of the several products originating from a given crude, thus avoiding gluts in some product markets which may bring down the price of these products, operators of refineries are able to develop such crudes that yield the most refined oil products in demand at the time. On a limited scale, this makes for a fairly high elasticity of supply.

Income derived from refining depends on the yield pattern, or product mix, of a refinery. Generally speaking, the higher the yields of lighter and higher-priced refined products, such as gasoline and kerosene, the larger is the income from refining of crude oils.

The relation of prices to yields can only be understood fully when one considers that any one refined oil product can be made only if other refined oil products are derived simultaneously. Joint products, which cannot be produced one without the other, are not peculiar to the oil industry; for example, this is also true in agriculture, as in the case of wool/mutton. In the petroleum industry, the interplay of several products is of special significance because of the extensive variety of possible products.

Manufacturing "schemes" for refineries should be eliminated where marketing justification for these schemes is not present. For example, setting up a refinery aimed at increasing output of gasolines and middle distillates extracted from crudes at a high cost, involving investment in substantial secondary processing operations such as catalytic cracking, hydrocracking, and so on, can only be justified on the basis of guaranteed markets for such highly valued products as premium gasolines, jet fuels, and so on.

In many cases, however, refineries are built with more complex manufacturing schemes because of refinery desires to manufacture products satisfying specifications of buyers, together with the advantage of manufacturing more elaborate products so as to participate in many more markets of oil products, even if in limited quantities. In other words, the refinery follows a policy of diversification rather than one of simplification. For example, if a market for a substantial amount, in barrels, of high-octane gasoline was available, a catalytic reforming unit would be indispensable. And a large investment in a catalytic reforming unit would probably be economically feasible, since the potential revenues for the oil product involved would more than justify the expenditure for such a unit.

Properties of cuts are important, for there are natural yields from crudes obtained by simple distillation, such as orienting refinery production toward good-quality naphtha cut at 338° F. Some flexibility may also be achieved by refining variable proportions of each crude without any substantial secondary processing. For example, the amount of yield of middle distillate is highly dependent on the type of crude used; yields up to 32% may be obtained with Arabian Light and only a 15.4% yield of middle distillates with Arabian Heavy.

To illustrate the complexities involved in operating a refinery efficiently to get the most out of the different types of crude used, although in a single distillation range differ, only slightly for light and middle cuts, the sulfur content and viscosity of heavy cuts in a single range are substantially increased when going from Arabian Light to Arabian Heavy.

Furthermore, sulfur contents of total straight-run naphtha do not need any desulfurization treatment, but heavy naphthas must be desulfurized before they are reformed to produce a high-octane gasoline base.

Petroleum refinery processes may become quite complex in large refineries, where the crude feed and the oil product streams are often complex mixtures of compounds and where strict material and energy balances are not often possible. Crude distillation, the first separation process, is performed in a series of large distillation columns. One column may separate the most volatile heavy compounds from the crude oil; another column may separate the oil into several fractions; and a third column, operated under a vacuum, may separate the least volatile heavy components, obtained from the bottom of the second column, into several fractions.

Auxiliary equipment includes heat exchangers, furnaces for heating the crude feed, and steam strippers for purifying oil product streams. All oil product streams leaving the crude distillation unit are complex mixtures characterized by their boiling range.

Following crude distillation is a secondary process, as is catalytic cracking, a major refinery process for converting a heavy-fraction gas oil into high-octane gasoline. This process splits (cracks) large molecules into smaller ones. Because the crude feed is a complex mixture, many reactions occur. The major products from catalytic cracking are high-octane gasoline, heating oil, and gases such as butane, propylene, and butylene, which are used in various other refinery processes. Large catalytic cracking units will process from 2,500 to more than 100,000 bbl/day of gas-oil feed, depending on the size of the unit. Often the catalytic cracker is the tallest unit in the refinery.

The catalytic cracker consists of three major units: (1) a reactor, (2) a catalyst regenerator, and (3) a product-separation distillation column. Other processes in cracking are thermal cracking, used to split molecules of heavy components in the reduced crude feed, and catalytic reforming, by which the carbon chains in petroleum molecules are rearranged and sulfur is removed from the reforming crude.

There are also distillation columns for light gas recovery. Catalytic polymerization, a cracking process, involves the combining of two or more hydrocarbons to get larger molecules.

EFFECT OF SIZE AND OUTPUT ON COST IN REFINERIES

Large-capacity refineries, or large-size refinery equipment, are less expensive to an oil refiner than small-capacity refineries, or small-capacity equipment, if compared on a basis of cost per unit of capacity.

For example, a 1,750-ft^2 heat exchanger at $7,000 might cost only $4/ft^2, whereas a 250-ft^2 heat exchanger at $2,125 costs about $8.50/ft^2. The ratio of total costs is, therefore

$$\frac{1,750 \text{ ft}^2 \times \$4.00}{250 \text{ ft}^2 \times \$8.50} = \frac{\$7,000}{\$2,125} = 3.29 \text{ to } 1$$

which is the cost ratio between the two heat exchangers. The capacity is 7 to 1 (1,750 ft^2:250 ft^2), which means that the capacity of the larger unit is seven times that of the smaller unit. With a cost ratio of 3.29 to 1 but a capacity ratio of 7 to 1, it appears that it is more beneficial to invest in fewer, larger-size exchangers rather than in several smaller-size ones, all other things being equal.

In considering marginal costs, the changing marginal costs (MC) per barrel is the increase in total cost (TC) divided by the increase in output in barrels. The refinery experiences changing MC when it alters its output. It is the increase at the margin that changes the TC, and that therefore determines the average cost (AC), which is total cost divided by total production (TP), or TC/TP.

This cost profile should be familiar to oil management and students of the economics of oil engineering. Whenever an oil well or an oil refinery begins to produce, its average cost per barrel of output will be high, due to very high fixed costs. But as output increases, unit cost (in barrels) comes down steadily, partly because fixed cost (overhead) is now spread over more barrels, and also because factors of production used in producing oil or refining such as capital machinery are used at greater efficiency.

Eventually, however, after some point of maximum factor efficiency, the average cost per barrel begins to go up. Even though the fixed cost per barrel continues to decline, it is now so small a fraction of the cost per barrel that its decline counts for little, since the rising inefficiency of factors of production pushes up the portion of variable cost (VC) per barrel of the total cost of a barrel.

As different amounts of crude are produced, observation should be made of cost changes that occur in total cost and total output every time the number and amount of factors of production (or inputs) is altered. The reason is that it is the marginal cost curve per barrel which tells us the total cost is going up. In other words, as the oil refinery moves into higher production levels, the amount of tonnage or barrels of refined products added (marginal productivity or output) should cost considerably less than the average of all refined products processed previously. Later, when diminishing returns, or the production decline curve, begins to set in, the added marginal productivity (MP, added barrels/tons of oil) will be high-cost barrels, higher in cost than the average of all barrels produced previously.

Since the cost of marginal output always leads the cost of average output when diminishing returns begin to work, we can always understand the relationship of MC and AC curves to one another. The MC curve always cuts the AC curve at the lowest point of the AC curve. This is always so, since as long as additional barrels of oil are cheaper than the average of all (output) barrels of oil produced, the production of additional output reduces AC per barrel of oil produced. That is, as long as the MC curve is lower than the AC curve, the AC curve must be falling, which is just what we want with additional output (MP).

Conversely, as soon as additional output is more expensive than the average for all previous output, that additional production must raise average costs again. Thus, the MC curve must cross the AC curve at the minimum point of the AC curve. It is important, therefore, for the oil industry in both oil field and refinery operations to continue increasing levels of production with decreasing MCs, thus decreasing ACs per barrel.

In both oil fields and refineries, the greater part of total costs is fixed costs. These, of course, are costs incurred independently of output. In refineries, prime costs (costs directly attributable to output) of each industrial refined product produced are low. So prices for these products can fall a good deal below the level at which all costs are covered and real profits are made before the oil company will find that it should reduce its rate of production.

Thus, the oil industry is less elastic insofar as once the investments have been made, the incidence of high fixed costs makes full utilization of oil operations almost imperative. Also, unit costs, or costs per barrel or ton, rise very sharply if output is below full capacity, so the petroleum industry desires to operate at full capacity for both stability and profitability reasons.

PRORATION OF SUPPLIES

Operating at full capacity, however, is often not done by oil companies, especially in the Middle East and in the United States, where conservation of resources is deemed necessary by the producing countries. Thus proration of reserves, or supplies, acts as a deterrent to oil companies which attempt to operate efficiently by controlling unit costs through increases in production rates.

Proration means cutbacks in production. This in turn means longer payouts on new wells and tends to discourage exploration for and development of reserves. Also, prices of crudes are largely determined by costs of developing further production from oil fields; prices are not determined by low operating costs of already developed capacity. Thus, if proration does not have the effect of causing prices to increase substantially, further exploration and development of new reserves may suffer. Prices actually need to be far above long-run costs of finding crude oil.

Oil companies are convinced that aggregate demand for oil products is highly inelastic with respect to price, although not the demand for every product. So attempts are made to hold price competition to a minimum. This requires that supply be controlled. Except for special short-run circumstances, prices quoted are approximately the same for similar grades of crude. Thus, pricing policies have been an integral part of oil companies operating in world markets.

Proration, as proposed by the Organization of Petroleum Exporting Countries (OPEC), is basically to improve terms of trade for oil exporting countries by preventing erosions of realized prices to them, due to any imbalance between supply and demand.

OIL OPERATIONS IN THE MIDDLE EAST AND GULF AREAS

Most of the crude oil pumped in the Middle East is sold as crude and shipped to Western Europe and Japan where the crude is refined. In the early days of Middle East oil, more crude in percentage of total crude recovered was refined in the Gulf area. But then the developed countries began to do their own processing to meet their own needs.

Crude oil producing countries are now exerting more pressure on oil companies in their countries to refine more crude locally. The Middle East, at present, is thus mainly an exporter of crude oil as the refining activities are located mostly in the consuming developed countries. Refined products are being processed by Middle East refineries mainly for their own inland consumption and for consuming developing countries that have no refineries of their own.

There are, however, other factors that do not favor the Middle East for location of refineries. For one thing, refinery costs in the Middle East are generally high. All refinery equipment must be imported, which raises capital costs, and also other costs in chemicals and materials items. Furthermore, spare parts must be imported. Time lags between need and receipt of these items, due to the long distance between the sources of these items and the Middle East, can be weeks rather than days.

Middle East refineries also carry greater costs for economic overhead and social overhead as well. Employment and training of local personnel is expensive. Furthermore, greater costs may include the building of roads, hospitals, schools, housing, railways, ports, airports, electrical generating plants, and so on.

Other factors working against the establishing of refineries in some parts of the Middle East are

1. its long distance to markets for oil products;

2. the scarcity of much-needed water in some areas; and

3. the lack of security in certain areas of operation.

On the favorable side for Middle East refineries, those few refineries now in operation, such as those at Ras Tanura, Saudi Arabia, and at Abadan, Iran, are much larger in size than the average size of refineries in the world. Thus, the unit capital costs and unit operating costs of refineries in the Middle East are reduced substantially. Furthermore, Middle East refineries are less complex than in other regions. This reduces the value of their output, but it also cuts their capital and operating costs.

The "product mix" in Middle East refineries is usually of a lower proportion of the more valuable refined products, due to the nature of demand for their refined oil products, and also due to the less complex types of refineries in the region and the lower proportion of catalytic or reforming plants, as well as the relatively low price of crude oil.

In conclusion, the average breakdown in percent of total investment in the Middle East petroleum industry, as per World Outlook of 1971, is as follows:

Exploration for oil	9%
Development and production of crude oil	35%
Refining	17%
Transport	25%
Others, including marketing and administering of oil interests	14%
Total investment in Middle East	100%

It thus appears that the largest investments in petroleum in the Middle East now go for the development and production of crudes. However, since early in 1973, some drastic changes in the oil industry on a worldwide basis have been seen. Producing countries are now, in many cases, for the first time becoming either full or part-owners of their own oil resources. This ownership includes oil field production, refinery operations, and some marketing of oil products and crudes as well. This new look in the oligopolistic industry of petroleum, which features administrated prices by producers, is known by the oil-producing countries as "participation," not "takeover" or "nationalization," as the large consuming countries call it. It is certainly changing the structure of the oil industry in many ways.

Foreign oil companies and national oil companies, who derive their charters, capital, and powers from legislation or government decree, are engaged in joint ventures. Although these national oil companies, such as NIOC in Iran and Petromin in Saudi Arabia, are autonomous and generally operate as independent business operations and corporations, they do operate within the framework of the public sector of their own economy.

Hope for the future is bright for an inherently unstable industry. Teamwork between the two "producers," the producing country and the oil company, needs to be a natural one, since the oil industry is restricted in its operations under competitive conditions because of uncertain results of exploration, high overhead costs of operation at all stages, and high inelasticity of demand for oil and oil products, particularly in the short run. Thus, teamwork between both oil-producing countries and oil companies is necessary for success of the oil industry. The oil and oil product markets are not self-adjusting; a fall in prices might stimulate demand for crudes and oil products, but might also significantly choke off supplies, creating serious world crises for both consuming and producing countries.

Appendix A

CONVERSION FACTORS

APPROXIMATE CALORIFIC EQUIVALENTS

One million tons of oil equals approximately:

In heat units:	41 billion Btu's
	415 million therms
	10,500 teracalories
In solid fuels:	1.5 million tons of coal
	4.9 million tons of lignite
	3.3 million tons of peat

Also,

10 billion Btu's equal approximately:	0.24 million tons of oil
100 million therms equal approximately:	0.24 million tons of oil
10,000 teracalories equal approximately:	0.95 million tons of oil
1 million tons of coal equal approximately:	0.67 million tons of oil
1 million tons of lignite equal approximately:	0.20 million tons of oil
1 million tons of peat equal approximately:	0.30 million tons of oil

TABLE A.1

Approximate Conversion Factors for Crude Oil

To convert into From	Metric tons	Long tons	Short tons	Barrels	Kiloliters (cubic meters)	1,000 gallons (Imp.)	1,000 gallons (U.S.)
				Multiply by			
Metric tons	1	0.984	1.102	7.33	1.16	0.256	0.308
Long tons	1.016	1	1.120	7.45	1.18	0.261	0.313
Short tons	0.907	0.893	1	6.65	1.05	0.233	0.279
Barrels	0.136	0.134	0.150	1	0.159	0.035	0.042
Kiloliters (cubic meters)	0.863	0.849	0.951	6.29	1	0.220	0.264
1,000 gallons (Imp.)	3.91	3.83	4.29	28.6	4.55	1	1.201
1,000 gallons (U.S.)	3.25	3.19	3.58	23.8	3.79	0.833	1

TABLE A.2

Approximate Weight and Volume Conversion Factors
for Crude Oil and Oil Products

To convert from	Barrels to metric tons	Metric tons to barrels	Barrels per day to tons per year	Tons per year to barrels per day
		Multiply by		
Crude oil*	0.136	7.33	49.8	0.0201
Motor spirit	0.118	8.45	43.2	0.0232
Kerosene	0.128	7.80	46.8	0.0214
Gas/diesel	0.133	7.50	48.7	0.0205
Fuel oil	0.149	6.70	54.5	0.0184

*Based on world average gravity.

Appendix B

COMPOUND-INTEREST FACTORS

1/2%

	To find F, given P: $(1 + i)^n$	To find P, given F: $\dfrac{1}{(1 + i)^n}$	To find A, given F: $\dfrac{i}{(1 + i)^n - 1}$	To find A, given P: $\dfrac{i(1 + i)^n}{(1 + i)^n - 1}$	To find F, given A: $\dfrac{(1 + i)^n - 1}{i}$	To find P, given A: $\dfrac{(1 + i)^n - 1}{i(1 + i)^n}$	
n	$(f/p)_n^{1/2}$	$(p/f)_n^{1/2}$	$(a/f)_n^{1/2}$	$(a/p)_n^{1/2}$	$(f/a)_n^{1/2}$	$(p/a)_n^{1/2}$	n
1	1.005	0.9950	1.00000	1.00500	1.000	0.995	1
2	1.010	0.9901	0.49875	0.50375	2.005	1.985	2
3	1.015	0.9851	0.33167	0.33667	3.015	2.970	3
4	1.020	0.9802	0.24183	0.25313	4.030	3.950	4
5	1.025	0.9754	0.19801	0.20301	5.050	4.926	5
6	1.030	0.9705	0.16460	0.16960	6.076	5.896	6
7	1.036	0.9657	0.14073	0.14573	7.106	6.862	7
8	1.041	0.9609	0.12283	0.12783	8.141	7.823	8
9	1.046	0.9561	0.10891	0.11391	9.182	8.779	9
10	1.051	0.9513	0.09777	0.10277	10.288	9.730	10
11	1.056	0.9466	0.08866	0.09366	11.279	10.677	11
12	1.062	0.9419	0.08107	0.08607	12.336	11.619	12
13	1.067	0.9372	0.07464	0.07964	13.397	12.556	13
14	1.072	0.9326	0.06914	0.07414	14.464	13.489	14
15	1.078	0.9279	0.06436	0.06936	15.537	14.417	15
16	1.083	0.9233	0.06019	0.06519	16.614	15.340	16
17	1.088	0.9187	0.05615	0.06151	17.697	16.259	17
18	1.094	0.9141	0.05323	0.05823	18.786	17.173	18
19	1.099	0.9096	0.05030	0.05530	19.880	18.082	19
20	1.105	0.9051	0.04767	0.05267	20.979	18.987	20
21	1.110	0.9006	0.04528	0.05028	22.084	19.888	21
22	1.116	0.8961	0.04311	0.04811	23.194	20.784	22
23	1.122	0.8916	0.04113	0.04613	24.310	21.676	23
24	1.127	0.8872	0.03932	0.04432	25.432	22.563	24
25	1.133	0.8828	0.03767	0.04265	26.559	23.446	25
26	1.138	0.8784	0.03611	0.04111	27.692	24.324	26
27	1.144	0.8740	0.03469	0.03969	28.830	25.198	27
28	1.150	0.8697	0.03336	0.03836	29.975	26.068	28
29	1.156	0.8653	0.03213	0.03713	31.124	26.933	29
30	1.161	0.8610	0.03098	0.03598	32.280	27.794	30
31	1.167	0.8567	0.02990	0.03490	33.441	28.651	31
32	1.173	0.8525	0.02889	0.03389	34.609	29.503	32
33	1.179	0.8482	0.02795	0.03295	35.782	30.352	33
34	1.185	0.8440	0.02706	0.03206	36.961	31.196	34
35	1.191	0.8398	0.02622	0.03122	38.145	32.035	35
40	1.221	0.8191	0.02265	0.02765	44.159	36.172	40
45	1.252	0.7990	0.01987	0.02487	50.324	40.207	45
50	1.283	0.7793	0.01765	0.02265	56.645	44.143	50
55	1.316	0.7601	0.01548	0.02084	63.126	47.981	55
60	1.349	0.7414	0.01433	0.01933	69.770	51.726	60
65	1.383	0.7231	0.01306	0.01806	76.582	55.377	65
70	1.418	0.7053	0.01197	0.01697	83.566	58.939	70
75	1.454	0.6879	0.01102	0.01602	90.727	62.414	75
80	1.490	0.6710	0.01020	0.01520	98.068	65.802	80
85	1.528	0.6545	0.00947	0.01447	105.594	69.108	85
90	1.567	0.6383	0.00883	0.01383	113.311	72.331	90
95	1.606	0.6226	0.00825	0.01325	121.222	75.476	95
100	1.647	0.6073	0.00773	0.01273	129.334	78.543	100

1%

	To find F, given P: $(1 + i)^n$	To find P, given F: $\dfrac{1}{(1 + i)^n}$	To find A, given F: $\dfrac{i}{(1 + i)^n - 1}$	To find A, given P: $\dfrac{i(1 + i)^n}{(1 + i)^n - 1}$	To find F, given A: $\dfrac{(1 + i)^n - 1}{i}$	To find P, given A: $\dfrac{(1 + i)^n - 1}{i(1 + i)^n}$	
n	$(f/p)_n^1$	$(p/f)_n^1$	$(a/f)_n^1$	$(a/p)_n^1$	$(f/a)_n^1$	$(p/a)_n^1$	n
1	1.010	0.9901	1.00000	1.01000	1.000	0.990	1
2	1.020	0.9803	0.49751	0.50751	2.010	1.970	2
3	1.030	0.9706	0.33002	0.34002	3.030	2.941	3
4	1.041	0.9610	0.24628	0.25628	4.060	3.902	4
5	1.051	0.9515	0.19604	0.20604	5.101	4.853	5
6	1.062	0.9420	0.16255	0.17255	6.152	5.795	6
7	1.072	0.9327	0.13863	0.14863	7.214	6.728	7
8	1.083	0.9235	0.12069	0.13069	8.286	7.652	8
9	1.094	0.9143	0.10674	0.11674	9.369	8.566	9
10	1.105	0.9053	0.09558	0.10558	10.462	9.471	10
11	1.116	0.8963	0.08645	0.09645	11.567	10.368	11
12	1.127	0.8874	0.07885	0.08885	12.683	11.255	12
13	1.138	0.8787	0.07241	0.08241	13.809	12.134	13
14	1.149	0.8700	0.06690	0.07690	14.947	13.004	14
15	1.161	0.8613	0.06212	0.07212	16.097	13.865	15
16	1.173	0.8528	0.05794	0.06794	17.258	14.718	16
17	1.184	0.8444	0.05426	0.06426	18.430	15.562	17
18	1.196	0.8360	0.05098	0.06098	19.615	16.398	18
19	1.208	0.8277	0.04805	0.05805	20.811	17.226	19
20	1.220	0.8195	0.04542	0.05542	22.019	18.046	20
21	1.232	0.8114	0.04303	0.05303	23.239	18.857	21
22	1.245	0.8034	0.04086	0.05086	24.472	19.660	22
23	1.257	0.7954	0.03889	0.04889	25.716	20.456	23
24	1.270	0.7876	0.03707	0.04707	26.973	21.243	24
25	1.282	0.7798	0.03541	0.04541	28.243	22.023	25
26	1.295	0.7720	0.03387	0.04387	29.526	22.795	26
27	1.308	0.7644	0.03245	0.04245	30.821	23.560	27
28	1.321	0.7568	0.03112	0.04112	32.129	24.316	28
29	1.335	0.7493	0.02990	0.03990	33.450	25.066	29
30	1.348	0.7419	0.02875	0.03875	34.785	25.808	30
31	1.361	0.7346	0.02768	0.03768	36.133	26.542	31
32	1.375	0.7273	0.02667	0.03667	37.494	27.270	32
33	1.391	0.7201	0.02573	0.03573	38.869	27.990	33
34	1.403	0.7130	0.02484	0.03484	40.258	28.703	34
35	1.417	0.7059	0.02400	0.03400	41.660	29.409	35
40	1.489	0.6717	0.02046	0.03046	48.886	32.835	40
45	1.565	0.6391	0.01771	0.02771	56.481	36.095	45
50	1.645	0.6080	0.01551	0.02551	64.463	39.196	50
55	1.729	0.5785	0.01373	0.02373	72.852	42.147	55
60	1.817	0.5504	0.01224	0.02224	81.670	44.955	60
65	1.909	0.5237	0.01100	0.02100	90.937	47.627	65
70	2.007	0.4983	0.00993	0.01993	100.676	50.169	70
75	2.109	0.4741	0.00902	0.01902	110.913	52.587	75
80	2.217	0.4511	0.00822	0.01822	121.672	54.888	80
85	2.330	0.4292	0.00752	0.01752	132.979	57.078	85
90	2.449	0.4084	0.00690	0.01690	144.863	59.161	90
95	2.574	0.3886	0.00636	0.01636	157.354	61.143	95
100	2.705	0.3697	0.00587	0.01587	170.481	63.029	100

1 1/2%

	To find F, given P:	To find P, given F:	To find A, given F:	To find A, given P:	To find F, given A:	To find P, given A:	
	$(1 + i)^n$	$\dfrac{1}{(1 + i)^n}$	$\dfrac{i}{(1 + i)^n - 1}$	$\dfrac{i(1 + i)^n}{(1 + i)^n - 1}$	$\dfrac{(1 + i)^n - 1}{i}$	$\dfrac{(1 + i)^n - 1}{i(1 + i)^n}$	
n	$(f/p)_n^{1\,1/2}$	$(p/f)_n^{1\,1/2}$	$(a/f)_n^{1\,1/2}$	$(a/p)_n^{1\,1/2}$	$(f/a)_n^{1\,1/2}$	$(p/a)_n^{1\,1/2}$	n
1	1.015	0.9852	1.00000	1.01500	1.000	0.985	1
2	1.030	0.9707	0.49628	0.51128	2.015	1.956	2
3	1.046	0.9563	0.32838	0.34338	3.045	2.912	3
4	1.061	0.9422	0.24444	0.25944	4.091	3.854	4
5	1.077	0.9283	0.19409	0.20909	5.152	4.783	5
6	1.093	0.9145	0.16053	0.17553	6.230	5.697	6
7	1.110	0.9010	0.13656	0.15156	7.323	6.598	7
8	1.126	0.8877	0.11858	0.13358	8.433	7.486	8
9	1.143	0.8746	0.10461	0.11961	9.559	8.361	9
10	1.161	0.8617	0.09343	0.10843	10.703	9.222	10
11	1.178	0.8489	0.08429	0.09930	11.863	10.071	11
12	1.196	0.8364	0.07668	0.09168	13.041	10.908	12
13	1.214	0.8240	0.07024	0.08524	14.237	11.732	13
14	1.232	0.8118	0.06472	0.07972	15.450	12.543	14
15	1.250	0.7999	0.05994	0.07494	16.682	13.343	15
16	1.269	0.7880	0.05577	0.07077	17.932	14.131	16
17	1.288	0.7764	0.05208	0.06708	19.201	14.908	17
18	1.307	0.7649	0.04881	0.06381	20.489	15.673	18
19	1.327	0.7536	0.04588	0.06088	21.797	16.426	19
20	1.347	0.7425	0.04325	0.05825	23.124	17.169	20
21	1.367	0.7315	0.04087	0.05587	24.471	17.900	21
22	1.388	0.7207	0.03870	0.05370	25.838	19.621	22
23	1.408	0.7100	0.03673	0.05173	27.225	19.331	23
24	1.430	0.6995	0.03492	0.04992	28.634	20.030	24
25	1.451	0.6892	0.03325	0.04826	30.063	20.720	25
26	1.473	0.6790	0.03173	0.04673	31.514	21.399	26
27	1.495	0.6690	0.03032	0.04532	32.987	22.068	27
28	1.517	0.6591	0.02900	0.04400	34.481	22.727	28
29	1.540	0.6494	0.02778	0.04278	35.999	23.376	29
30	1.563	0.6398	0.02664	0.04164	37.539	24.016	30
31	1.587	0.6303	0.02557	0.04057	39.102	24.646	31
32	1.610	0.6210	0.02458	0.03958	40.688	25.267	32
33	1.634	0.6118	0.02364	0.03864	42.229	25.879	33
34	1.659	0.6028	0.02276	0.03776	43.933	26.482	34
35	1.684	0.5939	0.02193	0.03693	45.592	27.076	35
40	1.814	0.5513	0.01834	0.03343	54.268	29.916	40
45	1.954	0.5117	0.01572	0.03072	63.614	32.552	45
50	2.105	0.4750	0.01357	0.02857	73.683	35.000	50
55	2.268	0.4409	0.01183	0.02683	84.530	37.271	55
60	2.443	0.4093	0.01039	0.02539	96.215	39.380	60
65	2.632	0.3799	0.00919	0.02419	108.803	41.338	65
70	2.835	0.3527	0.00817	0.02317	122.364	43.155	70
75	3.055	0.3274	0.00730	0.02230	136.973	44.842	75
80	3.291	0.3039	0.00655	0.02155	152.711	46.407	80
85	3.545	0.2821	0.00589	0.02089	169.665	47.861	85
90	3.819	0.2619	0.00532	0.02032	187.930	49.210	90
95	4.114	0.2431	0.00482	0.01982	207.606	50.462	95
100	4.432	0.2256	0.00437	0.01937	228.803	51.625	100

2%

	To find F, given P:	To find P, given F:	To find A, given F:	To find A, given P:	To find F, given A:	To find P, given A:	
	$(1 + i)^n$	$\dfrac{1}{(1 + i)^n}$	$\dfrac{i}{(1 + i)^n - 1}$	$\dfrac{i(1 + i)^n}{(1 + i)^n - 1}$	$\dfrac{(1 + i)^n - 1}{i}$	$\dfrac{(1 + i)^n - 1}{i(1 + i)^n}$	
n	$(f/p)_n^2$	$(p/f)_n^2$	$(a/f)_n^2$	$(a/p)_n^2$	$(f/a)_n^2$	$(p/a)_n^2$	n
1	1.020	0.9804	1.00000	1.02000	1.000	0.980	1
2	1.040	0.9612	0.49505	0.51505	2.020	1.942	2
3	1.061	0.9423	0.32675	0.34675	3.060	2.884	3
4	1.082	0.9238	0.24262	0.26262	4.122	3.808	4
5	1.104	0.9057	0.19216	0.21216	5.204	4.713	5
6	1.126	0.8880	0.15853	0.17853	6.308	5.601	6
7	1.149	0.8706	0.13451	0.15451	7.434	6.472	7
8	1.172	0.8535	0.11651	0.13651	8.583	7.325	8
9	1.195	0.8368	0.10252	0.12252	9.755	8.162	9
10	1.219	0.8203	0.09133	0.11133	10.950	8.983	10
11	1.243	0.8043	0.08216	0.10218	12.169	9.787	11
12	1.268	0.7885	0.07456	0.09456	13.412	10.575	12
13	1.294	0.7730	0.06812	0.08812	14.680	11.348	13
14	1.319	0.7579	0.06260	0.08260	15.974	12.106	14
15	1.346	0.7430	0.05783	0.07783	17.293	12.849	15
16	1.373	0.7284	0.05365	0.07365	18.639	13.578	16
17	1.400	0.7142	0.04997	0.06997	20.012	14.292	17
18	1.428	0.7002	0.04670	0.06670	21.412	14.992	18
19	1.457	0.6864	0.04378	0.06378	22.841	15.678	19
20	1.486	0.6730	0.04116	0.06116	24.297	16.351	20
21	1.516	0.6598	0.03878	0.05878	25.783	17.011	21
22	1.546	0.6468	0.03663	0.05663	27.299	17.658	22
23	1.577	0.6342	0.03467	0.05467	28.845	18.292	23
24	1.608	0.6217	0.03287	0.05287	30.422	18.914	24
25	1.641	0.6095	0.03122	0.05122	32.030	19.523	25
26	1.673	0.5976	0.02970	0.04970	33.671	20.121	26
27	1.707	0.5859	0.02829	0.04829	35.344	20.707	27
28	1.741	0.5744	0.02699	0.04699	37.051	21.281	28
29	1.776	0.5631	0.02578	0.04578	38.792	21.844	29
30	1.811	0.5521	0.02465	0.04465	40.568	22.396	30
31	1.848	0.5412	0.02360	0.04360	42.379	22.938	31
32	1.885	0.5306	0.02261	0.04261	44.227	23.468	32
33	1.922	0.5202	0.02169	0.04169	46.112	23.989	33
34	1.961	0.5100	0.02082	0.04082	48.034	24.499	34
35	2.000	0.5000	0.02000	0.04000	49.994	24.999	35
40	2.208	0.4529	0.01656	0.03656	60.402	27.355	40
45	2.438	0.4102	0.01391	0.03391	71.893	29.490	45
50	2.692	0.3715	0.01182	0.03182	84.579	31.424	50
55	2.972	0.3365	0.01014	0.03014	98.587	33.175	55
60	3.281	0.3048	0.00877	0.02877	114.052	34.761	60
65	3.623	0.2761	0.00763	0.02763	131.126	36.197	65
70	4.000	0.2500	0.00667	0.02667	149.978	37.499	70
75	4.416	0.2265	0.00586	0.02586	170.792	38.677	75
80	4.875	0.2051	0.00516	0.02516	193.772	39.745	80
85	5.383	0.1858	0.00456	0.02456	219.144	40.711	85
90	5.943	0.1683	0.00405	0.02405	247.157	41.587	90
95	6.562	0.1524	0.00360	0.02360	278.085	42.380	95
100	7.245	0.1380	0.00320	0.02320	312.232	43.098	100

2 1/2%

	To find F, given P: $(1 + i)^n$	To find P, given F: $\dfrac{1}{(1+i)^n}$	To find A, given F: $\dfrac{i}{(1+i)^n - 1}$	To find A, given P: $\dfrac{i(1+i)^n}{(1+i)^n - 1}$	To find F, given A: $\dfrac{(1+i)^n - 1}{i}$	To find P, given A: $\dfrac{(1+i)^n - 1}{i(1+i)^n}$	
n	$(f/p)^{2\,1/2}_n$	$(p/f)^{2\,1/2}_n$	$(a/f)^{2\,1/2}_n$	$(a/p)^{2\,1/2}_n$	$(f/a)^{2\,1/2}_n$	$(p/a)^{2\,1/2}_n$	n
1	1.025	0.9756	1.00000	1.02500	1.000	0.976	1
2	1.051	0.9518	0.49383	0.51883	2.025	1.927	2
3	1.077	0.9386	0.32514	0.35014	3.076	2.856	3
4	1.104	0.9060	0.24082	0.26582	4.153	3.762	4
5	1.131	0.8839	0.19025	0.21525	5.256	4.646	5
6	1.160	0.8623	0.15655	0.18155	6.388	5.508	6
7	1.189	0.8413	0.13250	0.15750	7.547	6.349	7
8	1.218	0.8207	0.11447	0.13947	8.736	7.170	8
9	1.249	0.8007	0.10046	0.12546	9.955	7.971	9
10	1.280	0.7812	0.08926	0.11426	11.203	8.752	10
11	1.312	0.7621	0.08011	0.10511	12.483	9.514	11
12	1.345	0.7436	0.07249	0.09749	13.796	10.258	12
13	1.379	0.7254	0.06605	0.09105	15.140	10.983	13
14	1.413	0.7077	0.06054	0.08554	16.519	11.691	14
15	1.448	0.6905	0.05577	0.08077	17.932	12.381	15
16	1.485	0.6736	0.05160	0.07660	19.380	13.055	16
17	1.522	0.6572	0.04793	0.07293	20.865	13.712	17
18	1.560	0.6412	0.04467	0.06967	22.386	14.353	18
19	1.599	0.6255	0.04176	0.06676	23.946	14.979	19
20	1.639	0.6103	0.03915	0.06415	25.545	15.589	20
21	1.680	0.5954	0.03679	0.06179	27.183	16.185	21
22	1.722	0.5809	0.03465	0.05965	28.863	16.765	22
23	1.765	0.5667	0.03270	0.05770	30.584	17.332	23
24	1.809	0.5529	0.03091	0.05591	32.349	17.885	24
25	1.854	0.5394	0.02928	0.05428	34.158	18.424	25
26	1.900	0.5262	0.02777	0.05277	36.012	18.951	26
27	1.948	0.5134	0.02638	0.05138	37.912	19.464	27
28	1.996	0.5009	0.02509	0.05009	39.860	19.965	28
29	2.046	0.4887	0.02389	0.04889	41.856	20.454	29
30	2.098	0.4767	0.02278	0.04778	43.903	20.930	30
31	2.150	0.4651	0.02174	0.04674	46.000	21.395	31
32	2.204	0.4538	0.02077	0.04577	48.150	21.849	32
33	2.259	0.4427	0.01986	0.04486	50.354	22.292	33
34	2.315	0.4319	0.01901	0.04401	52.613	22.724	34
35	2.373	0.4214	0.01821	0.04321	54.928	23.145	35
40	2.685	0.3724	0.01484	0.03984	67.403	25.103	40
45	3.038	0.3292	0.01227	0.03727	81.516	26.833	45
50	3.437	0.2909	0.01026	0.03526	97.484	28.362	50
55	3.889	0.2572	0.00865	0.03365	115.551	29.714	55
60	4.400	0.2273	0.00735	0.03235	135.992	30.909	60
65	4.978	0.2009	0.00628	0.03128	159.118	31.965	65
70	5.632	0.1776	0.00540	0.03040	185.284	32.898	70
75	6.372	0.1569	0.00465	0.02965	214.888	33.723	75
80	7.210	0.1387	0.00403	0.02903	248.383	34.452	80
85	8.157	0.1226	0.00349	0.02849	286.279	35.096	85
90	9.229	0.1084	0.00304	0.02804	329.154	35.666	90
95	10.442	0.0958	0.00265	0.02765	377.664	36.169	95
100	11.814	0.0846	0.00231	0.02731	432.549	36.614	100

3%

	To find F, given P:	To find P, given F:	To find A, given F:	To find A, given P:	To find F, given A:	To find P, given A:	
	$(1 + i)^n$	$\dfrac{1}{(1 + i)^n}$	$\dfrac{i}{(1 + i)^n - 1}$	$\dfrac{i(1 + i)^n}{(1 + i)^n - 1}$	$\dfrac{(1 + i)^n - 1}{i}$	$\dfrac{(1 + i)^n - 1}{i(1 + i)^n}$	
n	$(f/p)^3_n$	$(p/f)^3_n$	$(a/f)^3_n$	$(a/p)^3_n$	$(f/a)^3_n$	$(p/a)^3_n$	n
1	1.030	0.9709	1.00000	1.03000	1.000	0.971	1
2	1.061	0.9426	0.49261	0.52261	2.030	1.913	2
3	1.093	0.9151	0.32353	0.35353	3.091	2.829	3
4	1.126	0.8885	0.23903	0.26903	4.184	3.717	4
5	1.159	0.8626	0.18835	0.21835	5.309	4.580	5
6	1.194	0.8375	0.15460	0.18460	6.468	5.417	6
7	1.230	0.8131	0.13051	0.16051	7.662	6.230	7
8	1.267	0.7894	0.11246	0.14246	8.892	7.020	8
9	1.305	0.7664	0.09843	0.12843	10.159	7.786	9
10	1.344	0.7441	0.08723	0.11723	11.464	8.530	10
11	1.384	0.7224	0.07808	0.10808	12.808	9.253	11
12	1.426	0.7014	0.07046	0.10046	14.192	9.954	12
13	1.469	0.6810	0.06403	0.09403	15.618	10.635	13
14	1.513	0.6611	0.05853	0.08853	17.086	11.296	14
15	1.558	0.6419	0.05377	0.08377	18.599	11.938	15
16	1.605	0.6232	0.04961	0.07961	20.157	12.561	16
17	1.653	0.6050	0.04595	0.07595	21.762	13.166	17
18	1.702	0.5874	0.04271	0.07271	23.414	13.754	18
19	1.754	0.5703	0.03981	0.06981	25.117	14.324	19
20	1.806	0.5537	0.03722	0.06722	26.870	14.877	20
21	1.860	0.5375	0.03487	0.06487	28.676	15.415	21
22	1.916	0.5219	0.03275	0.06275	30.537	15.937	22
23	1.974	0.5067	0.03081	0.06081	32.453	16.444	23
24	2.033	0.4919	0.02905	0.05905	34.426	16.936	24
25	2.094	0.4776	0.02743	0.05743	36.459	17.413	25
26	2.157	0.4637	0.02594	0.05594	38.553	17.877	26
27	2.221	0.4502	0.02456	0.05456	40.710	18.327	27
28	2.288	0.4371	0.02329	0.05329	42.931	18.764	28
29	2.357	0.4243	0.02211	0.05211	45.219	19.188	29
30	2.427	0.4120	0.02102	0.05102	47.575	19.600	30
31	2.500	0.4000	0.02000	0.05000	50.003	20.000	31
32	2.575	0.3883	0.01905	0.04905	52.503	20.389	32
33	2.652	0.3770	0.01816	0.04816	55.078	20.766	33
34	2.732	0.3660	0.01732	0.04732	57.730	21.132	34
35	2.814	0.3554	0.01654	0.04654	60.462	21.487	35
40	3.262	0.3066	0.01326	0.04326	75.401	23.115	40
45	3.782	0.2644	0.01079	0.04079	92.720	24.519	45
50	4.384	0.2281	0.00887	0.03887	112.797	25.730	50
55	5.082	0.1968	0.00735	0.03735	136.072	26.774	55
60	5.892	0.1697	0.00613	0.03613	163.053	27.676	60
65	6.830	0.1464	0.00515	0.03515	194.333	28.453	65
70	7.918	0.1263	0.00434	0.03434	230.594	29.123	70
75	9.179	0.1089	0.00367	0.03367	272.631	29.702	75
80	10.641	0.0940	0.00311	0.03311	321.363	30.201	80
85	12.336	0.0811	0.00265	0.03265	377.857	30.631	85
90	14.300	0.0699	0.00226	0.03226	443.349	31.002	90
95	16.578	0.0603	0.00193	0.03193	519.272	31.323	95
100	19.219	0.0520	0.00165	0.03165	607.288	31.599	100

4%

	To find F, given P: $(1 + i)^n$	To find P, given F: $\dfrac{1}{(1 + i)^n}$	To find A, given F: $\dfrac{i}{(1 + i)^n - 1}$	To find A, given P: $\dfrac{i(1 + i)^n}{(1 + i)^n - 1}$	To find F, given A: $\dfrac{(1 + i)^n - 1}{i}$	To find P, given A: $\dfrac{(1 + i)^n - 1}{i(1 + i)^n}$	
n	$(f/p)_n^4$	$(p/f)_n^4$	$(a/f)_n^4$	$(a/p)_n^4$	$(f/a)_n^4$	$(p/a)_n^4$	n
1	1.040	0.9615	1.00000	1.04000	1.000	0.962	1
2	1.082	0.9246	0.49020	0.53020	2.040	1.886	2
3	1.125	0.8890	0.32035	0.36035	3.122	2.775	3
4	1.170	0.8548	0.23549	0.27549	4.246	3.630	4
5	1.217	0.8219	0.18463	0.22463	5.416	4.452	5
6	1.265	0.7903	0.15076	0.19076	6.633	5.242	6
7	1.316	0.7599	0.12661	0.16661	7.898	6.002	7
8	1.369	0.7307	0.10853	0.14853	9.214	6.733	8
9	1.423	0.7026	0.09449	0.13449	10.583	7.435	9
10	1.480	0.6756	0.08329	0.12329	12.006	8.111	10
11	1.539	0.6496	0.07415	0.11415	13.486	8.760	11
12	1.601	0.6246	0.06655	0.10655	15.026	9.385	12
13	1.665	0.6006	0.06014	0.10014	16.627	9.986	13
14	1.732	0.5775	0.05467	0.09467	18.292	10.563	14
15	1.801	0.5553	0.04994	0.08994	20.024	11.118	15
16	1.873	0.5339	0.04582	0.08582	21.825	11.652	16
17	1.948	0.5134	0.04220	0.08220	23.698	12.166	17
18	2.026	0.4936	0.03899	0.07899	25.645	12.659	18
19	2.107	0.4746	0.03614	0.07614	27.671	13.134	19
20	2.191	0.4564	0.03358	0.07358	29.778	13.590	20
21	2.279	0.4388	0.03128	0.07128	31.969	14.029	21
22	2.370	0.4220	0.02920	0.06920	34.248	14.451	22
23	2.465	0.4057	0.02731	0.06731	36.618	14.857	23
24	2.563	0.3901	0.02559	0.06559	39.083	15.247	24
25	2.666	0.3751	0.02401	0.06401	41.646	15.622	25
26	2.772	0.3607	0.02257	0.06257	44.312	15.983	26
27	2.883	0.3468	0.02124	0.06124	47.084	16.330	27
28	2.999	0.3335	0.02001	0.06001	49.968	16.663	28
29	3.119	0.3207	0.01888	0.05888	52.966	16.984	29
30	3.243	0.3083	0.01783	0.05783	56.085	17.292	30
31	3.373	0.2965	0.01686	0.05686	59.328	17.588	31
32	3.508	0.2851	0.01595	0.05595	62.701	17.874	32
33	3.648	0.2741	0.01510	0.05510	66.210	18.148	33
34	3.794	0.2636	0.01431	0.05431	69.858	18.411	34
35	3.946	0.2534	0.01358	0.05358	73.652	18.665	35
40	4.801	0.2083	0.01052	0.05052	95.026	19.793	40
45	5.841	0.1712	0.00826	0.04826	121.029	20.720	45
50	7.107	0.1407	0.00655	0.04655	152.667	21.482	50
55	8.646	0.1157	0.00523	0.04523	191.159	22.109	55
60	10.520	0.0951	0.00420	0.04420	237.991	22.623	60
65	12.799	0.0781	0.00339	0.04339	294.968	23.047	65
70	15.572	0.0642	0.00275	0.04275	364.290	23.395	70
75	18.945	0.0528	0.00223	0.04223	448.631	23.680	75
80	23.050	0.0434	0.00181	0.04181	551.245	23.915	80
85	28.044	0.0357	0.00148	0.04148	676.090	24.109	85
90	34.119	0.0293	0.00121	0.04121	827.983	24.267	90
95	41.511	0.0241	0.00099	0.04099	1012.785	24.398	95
100	50.505	0.0198	0.00081	0.04081	1237.624	24.505	100

5%

	To find F, given P: $(1 + i)^n$	To find P, given F: $\dfrac{1}{(1 + i)^n}$	To find A, given F: $\dfrac{i}{(1 + i)^n - 1}$	To find A, given P: $\dfrac{i(1 + i)^n}{(1 + i)^n - 1}$	To find F, given A: $\dfrac{(1 + i)^n - 1}{i}$	To find P, given A: $\dfrac{(1 + i)^n - 1}{i(1 + i)^n}$	
n	$(f/p)_n^5$	$(p/f)_n^5$	$(a/f)_n^5$	$(a/p)_n^5$	$(f/a)_n^5$	$(p/a)_n^5$	n
1	1.050	0.9524	1.00000	1.05000	1.000	0.952	1
2	1.103	0.9070	0.48780	0.53780	2.050	1.859	2
3	1.158	0.8638	0.31721	0.36721	3.153	2.723	3
4	1.216	0.8227	0.23201	0.28201	4.310	3.546	4
5	1.276	0.7835	0.18097	0.23097	5.526	4.329	5
6	1.340	0.7462	0.14702	0.19702	6.802	5.076	6
7	1.407	0.7107	0.12282	0.17282	8.142	5.786	7
8	1.477	0.6768	0.10472	0.15472	9.549	6.463	8
9	1.551	0.6446	0.09069	0.14069	11.027	7.108	9
10	1.629	0.6139	0.07950	0.12950	12.578	7.722	10
11	1.710	0.5847	0.07039	0.12039	14.207	8.306	11
12	1.796	0.5568	0.06283	0.11283	15.917	8.863	12
13	1.886	0.5303	0.05646	0.10646	17.713	9.394	13
14	1.980	0.5051	0.05102	0.10102	19.599	9.899	14
15	2.079	0.4810	0.04634	0.09634	21.579	10.380	15
16	2.183	0.4581	0.04227	0.09227	23.657	10.838	16
17	2.292	0.4363	0.03870	0.08870	25.840	11.274	17
18	2.407	0.4155	0.03555	0.08555	28.132	11.690	18
19	2.527	0.3957	0.03275	0.08275	30.539	12.085	19
20	2.653	0.3769	0.03024	0.08024	33.066	12.462	20
21	2.786	0.3589	0.02800	0.07800	35.719	12.821	21
22	2.925	0.3418	0.02597	0.07597	38.505	13.163	22
23	3.072	0.3256	0.02414	0.07414	41.430	13.489	23
24	3.225	0.3101	0.02247	0.07247	44.502	13.799	24
25	3.386	0.2953	0.02095	0.07095	47.727	14.094	25
26	3.556	0.2812	0.01956	0.06956	51.113	14.375	26
27	3.733	0.2678	0.01829	0.06829	54.669	14.643	27
28	3.920	0.2551	0.01712	0.06712	58.403	14.898	28
29	4.116	0.2429	0.01605	0.06605	62.323	15.141	29
30	4.322	0.2314	0.01505	0.06505	66.439	15.372	30
31	4.538	0.2204	0.01413	0.06413	70.761	15.593	31
32	4.765	0.2099	0.01328	0.06328	75.299	15.803	32
33	5.003	0.1999	0.01249	0.06249	80.064	16.003	33
34	5.253	0.1904	0.01176	0.06176	85.067	16.193	34
35	5.516	0.1813	0.01107	0.06107	90.320	16.374	35
40	7.040	0.1420	0.00828	0.05828	120.800	17.159	40
45	8.985	0.1113	0.00626	0.05626	159.700	17.774	45
50	11.467	0.0872	0.00478	0.05478	209.348	18.256	50
55	14.636	0.0683	0.00367	0.05367	272.713	18.633	55
60	18.679	0.0535	0.00283	0.05283	353.584	18.929	60
65	23.840	0.0419	0.00219	0.05219	456.798	19.161	65
70	30.426	0.0329	0.00170	0.05170	588.529	19.343	70
75	38.833	0.0258	0.00132	0.05132	756.654	19.485	75
80	49.561	0.0202	0.00103	0.05103	971.229	19.596	80
85	63.254	0.0158	0.00080	0.05080	1245.087	19.684	85
90	80.730	0.0124	0.00063	0.05063	1594.607	19.752	90
95	103.035	0.0097	0.00049	0.05049	2040.694	19.806	95
100	131.501	0.0076	0.00038	0.05038	2610.025	19.848	100

6%

	To find F, given P:	To find P, given F:	To find A, given F:	To find A, given P:	To find F, given A:	To find P, given A:	
	$(1 + i)^n$	$\dfrac{1}{(1 + i)^n}$	$\dfrac{i}{(1 + i)^n - 1}$	$\dfrac{i(1 + i)^n}{(1 + i)^n - 1}$	$\dfrac{(1 + i)^n - 1}{i}$	$\dfrac{(1 + i)^n - 1}{i(1 + i)^n}$	
n	$(f/p)_n^6$	$(p/f)_n^6$	$(a/f)_n^6$	$(a/p)_n^6$	$(f/a)_n^6$	$(p/a)_n^6$	n
1	1.060	0.9434	1.00000	1.06000	1.000	0.943	1
2	1.124	0.8900	0.48544	0.54544	2.060	1.833	2
3	1.191	0.8396	0.31411	0.37411	3.184	2.673	3
4	1.262	0.7921	0.22859	0.28859	4.375	3.465	4
5	1.338	0.7473	0.17740	0.23740	5.637	4.212	5
6	1.419	0.7050	0.14336	0.20336	6.975	4.917	6
7	1.504	0.6651	0.11914	0.17914	8.394	5.582	7
8	1.594	0.6274	0.10104	0.16104	9.897	6.210	8
9	1.689	0.5919	0.08702	0.14702	11.491	6.802	9
10	1.791	0.5584	0.07587	0.13587	13.181	7.360	10
11	1.898	0.5268	0.06679	0.12679	14.972	7.887	11
12	2.012	0.4970	0.05928	0.11928	16.870	8.384	12
13	2.133	0.4688	0.05296	0.11296	18.882	8.853	13
14	2.261	0.4423	0.04758	0.10758	21.015	9.295	14
15	2.397	0.4173	0.04296	0.10296	23.276	9.712	15
16	2.540	0.3936	0.03895	0.09895	25.673	10.106	16
17	2.693	0.3714	0.03544	0.09544	28.213	10.477	17
18	2.854	0.3503	0.03236	0.09236	30.906	10.828	18
19	3.026	0.3305	0.02962	0.08962	33.760	11.158	19
20	3.207	0.3118	0.02718	0.08718	36.786	11.470	20
21	3.400	0.2942	0.02500	0.08500	39.993	11.764	21
22	3.604	0.2775	0.02305	0.08305	43.392	12.042	22
23	3.820	0.2618	0.02128	0.08128	46.996	12.303	23
24	4.049	0.2470	0.01968	0.07968	50.816	12.550	24
25	4.292	0.2330	0.01823	0.07823	54.865	12.783	25
26	4.549	0.2198	0.01690	0.07690	59.156	13.003	26
27	4.822	0.2074	0.01570	0.07570	63.706	13.211	27
28	5.112	0.1956	0.01459	0.07459	68.528	13.406	28
29	5.418	0.1846	0.01358	0.07358	73.640	13.591	29
30	5.743	0.1741	0.01265	0.07265	79.058	13.765	30
31	6.088	0.1643	0.01179	0.07179	84.802	13.929	31
32	6.453	0.1550	0.01100	0.07100	90.890	14.084	32
33	6.841	0.1462	0.01027	0.07027	97.343	14.230	33
34	7.251	0.1379	0.00960	0.06960	104.184	14.368	34
35	7.686	0.1301	0.00897	0.06897	111.435	14.498	35
40	10.286	0.0972	0.00646	0.06646	154.762	15.046	40
45	13.765	0.0727	0.00470	0.06470	212.744	15.456	45
50	18.420	0.0543	0.00344	0.06344	290.336	15.762	50
55	24.650	0.0406	0.00254	0.06254	394.172	15.991	55
60	32.988	0.0303	0.00188	0.06188	533.128	16.161	60
65	45.145	0.0227	0.00139	0.06139	719.083	16.289	65
70	59.076	0.0169	0.00103	0.06103	967.932	16.385	70
75	79.057	0.0126	0.00077	0.06077	1300.949	16.456	75
80	105.796	0.0095	0.00057	0.06057	1746.600	16.509	80
85	141.579	0.0071	0.00043	0.06043	2342.982	16.549	85
90	189.465	0.0053	0.00032	0.06032	3141.075	16.579	90
95	253.546	0.0039	0.00024	0.06024	4209.104	16.601	95
100	339.302	0.0029	0.00018	0.06018	5638.368	16.618	100

7%

	To find F, given P:	To find P, given F:	To find A, given F:	To find A, given P:	To find F, given A:	To find P, given A:	
	$(1 + i)^n$	$\dfrac{1}{(1 + i)^n}$	$\dfrac{i}{(1 + i)^n - 1}$	$\dfrac{i(1 + i)^n}{(1 + i)^n - 1}$	$\dfrac{(1 + i)^n - 1}{i}$	$\dfrac{(1 + i)^n - 1}{i(1 + i)^n}$	
n	$(f/p)_n^i$	$(p/f)_n^i$	$(a/f)_n^i$	$(a/p)_n^i$	$(f/a)_n^i$	$(p/a)_n^i$	n
1	1.070	0.9346	1.00000	1.07000	1.000	0.935	1
2	1.145	0.8734	0.48309	0.55309	2.070	1.808	2
3	1.225	0.8163	0.31105	0.38105	3.215	2.624	3
4	1.311	0.7629	0.22523	0.29523	4.440	3.387	4
5	1.403	0.7130	0.17389	0.24389	5.751	4.100	5
6	1.501	0.6663	0.13980	0.20980	7.153	4.767	6
7	1.606	0.6227	0.11555	0.18555	8.654	5.389	7
8	1.718	0.5820	0.09747	0.16747	10.260	5.971	8
9	1.838	0.5439	0.08349	0.15349	11.978	6.515	9
10	1.967	0.5083	0.07238	0.14238	13.816	7.024	10
11	2.105	0.4751	0.06336	0.13336	15.784	7.499	11
12	2.252	0.4440	0.05590	0.12590	17.888	7.943	12
13	2.410	0.4150	0.04965	0.11965	20.141	8.358	13
14	2.579	0.3878	0.04434	0.11434	22.550	8.745	14
15	2.759	0.3624	0.03979	0.10979	25.129	9.108	15
16	2.952	0.3387	0.03586	0.10586	27.888	9.447	16
17	3.159	0.3166	0.03243	0.10243	30.840	9.763	17
18	3.380	0.2959	0.02941	0.09941	33.999	10.059	18
19	3.617	0.2765	0.02675	0.09675	37.379	10.363	19
20	3.870	0.2584	0.02439	0.09439	40.995	10.594	20
21	4.141	0.2415	0.02229	0.09229	44.865	10.836	21
22	4.430	0.2257	0.02041	0.09041	49.006	11.061	22
23	4.741	0.2109	0.01871	0.08871	53.436	11.272	23
24	5.072	0.1971	0.01719	0.08719	58.177	11.469	24
25	5.427	0.1842	0.01581	0.08581	63.249	11.654	25
26	5.807	0.1722	0.01456	0.08456	68.676	11.826	26
27	6.214	0.1609	0.01343	0.08343	74.484	11.987	27
28	6.649	0.1504	0.01239	0.08239	80.698	12.137	28
29	7.114	0.1406	0.01145	0.08145	87.347	12.278	29
30	7.612	0.1314	0.01059	0.08059	94.461	12.409	30
31	8.145	0.1228	0.00980	0.07980	102.073	12.532	31
32	8.715	0.1147	0.00907	0.07907	110.218	12.647	32
33	9.325	0.1072	0.00841	0.07841	118.923	12.754	33
34	9.978	0.1002	0.00780	0.07780	128.259	12.854	34
35	10.677	0.0937	0.00723	0.07723	138.237	12.948	35
40	14.974	0.0668	0.00501	0.07501	199.635	13.332	40
45	21.002	0.0476	0.00350	0.07350	285.749	13.606	45
50	29.457	0.0339	0.00246	0.07246	406.529	13.801	50
55	41.315	0.0242	0.00174	0.07174	575.929	13.940	55
60	57.946	0.0173	0.00123	0.07123	813.520	14.039	60
65	81.273	0.0123	0.00087	0.07087	1146.755	14.110	65
70	113.989	0.0088	0.00062	0.07062	1614.134	14.160	70
75	159.876	0.0063	0.00044	0.07044	2269.657	14.196	75
80	224.234	0.0045	0.00031	0.07031	3189.063	14.222	80
85	314.500	0.0032	0.00022	0.07022	4478.576	14.240	85
90	441.103	0.0023	0.00016	0.07016	6287.185	14.253	90
95	618.670	0.0016	0.00011	0.07011	8823.854	14.263	95
100	867.716	0.0012	0.00008	0.07008	12381.662	14.269	100

	To find F, given P:	To find P, given F:	To find A, given F:	To find A, given P:	To find F, given A:	To find P, given A:	
	$(1+i)^n$	$\dfrac{1}{(1+i)^n}$	$\dfrac{i}{(1+i)^n-1}$	$\dfrac{i(1+i)^n}{(1+i)^n-1}$	$\dfrac{(1+i)^n-1}{i}$	$\dfrac{(1+i)^n-1}{i(1+i)^n}$	
n	$(f/p)_n^8$	$(p/f)_n^8$	$(a/f)_n^8$	$(a/p)_n^8$	$(f/a)_n^8$	$(p/a)_n^8$	n
1	1.080	0.9259	1.00000	1.08000	1.000	0.926	1
2	1.166	0.8573	0.48077	0.56077	2.080	1.783	2
3	1.260	0.7938	0.30803	0.38803	3.246	2.577	3
4	1.360	0.7350	0.22192	0.30192	4.506	3.312	4
5	1.469	0.6806	0.17046	0.25046	5.867	3.933	5
6	1.587	0.6302	0.13632	0.21632	7.336	4.623	6
7	1.714	0.5835	0.11207	0.19207	8.923	5.206	7
8	1.851	0.5403	0.09401	0.17401	10.637	5.747	8
9	1.999	0.5002	0.08008	0.16008	12.488	6.247	9
10	2.159	0.4632	0.06903	0.14903	14.487	6.710	10
11	2.332	0.4289	0.06008	0.14008	16.645	7.139	11
12	2.518	0.3971	0.05270	0.13270	18.977	7.536	12
13	2.720	0.3677	0.04652	0.12652	21.495	7.904	13
14	2.937	0.3405	0.04130	0.12130	24.215	8.244	14
15	3.172	0.3152	0.03683	0.11683	27.152	8.559	15
16	3.426	0.2919	0.03298	0.11298	30.324	8.851	16
17	3.700	0.2703	0.02963	0.10963	33.750	9.122	17
18	3.996	0.2502	0.02670	0.10670	37.450	9.372	18
19	4.316	0.2317	0.02413	0.10413	41.446	9.604	19
20	4.661	0.2145	0.02185	0.10185	45.762	9.818	20
21	5.034	0.1987	0.01983	0.09983	50.423	10.017	21
22	5.437	0.1839	0.01803	0.09803	55.457	10.201	22
23	5.781	0.1703	0.01642	0.09642	60.893	10.371	23
24	6.341	0.1577	0.01498	0.09498	66.765	10.529	24
25	6.848	0.1460	0.01368	0.09368	73.106	10.675	25
26	7.396	0.1352	0.01251	0.09251	79.954	10.810	26
27	7.988	0.1252	0.01145	0.09145	87.351	10.935	27
28	8.627	0.1159	0.01049	0.09049	95.339	11.051	28
29	9.317	0.1073	0.00962	0.08962	103.966	11.158	29
30	10.063	0.0994	0.00883	0.08883	113.283	11.258	30
31	10.868	0.0920	0.00811	0.08811	123.346	11.350	31
32	11.737	0.0852	0.00745	0.08745	134.214	11.435	32
33	12.676	0.0789	0.00685	0.08685	145.951	11.514	33
34	13.690	0.0730	0.00630	0.08630	158.627	11.587	34
35	14.785	0.0676	0.00580	0.08580	172.317	11.655	35
40	21.725	0.0460	0.00386	0.08386	259.057	11.925	40
45	31.920	0.0313	0.00259	0.08259	386.506	12.108	45
50	46.902	0.0213	0.00174	0.08174	573.770	12.233	50
55	68.914	0.0145	0.00118	0.08118	848.923	12.319	55
60	101.257	0.0099	0.00080	0.08080	1253.213	12.377	60
65	148.780	0.0067	0.00054	0.08054	1847.248	12.416	65
70	218.606	0.0046	0.00037	0.08037	2720.080	12.443	70
75	321.205	0.0031	0.00025	0.08025	4002.557	12.461	75
80	471.955	0.0021	0.00017	0.08017	5886.935	12.474	80
85	693.456	0.0014	0.00012	0.08012	8655.706	12.482	85
90	1018.915	0.0010	0.00008	0.08008	12723.939	12.488	90
95	1497.121	0.0007	0.00005	0.08005	18701.507	12.492	95
100	2199.761	0.0005	0.00004	0.08004	27484.516	12.494	100

9%

	To find F, given P: $(1 + i)^n$	To find P, given F: $\dfrac{1}{(1 + i)^n}$	To find A, given F: $\dfrac{i}{(1 + i)^n - 1}$	To find A, given P: $\dfrac{i(1 + i)^n}{(1 + i)^n - 1}$	To find F, given A: $\dfrac{(1 + i)^n - 1}{i}$	To find P, given A: $\dfrac{(1 + i)^n - 1}{i(1 + i)^n}$	
n	$(f/p)_n^9$	$(p/f)_n^9$	$(a/f)_n^9$	$(a/p)_n^9$	$(f/a)_n^9$	$(p/a)_n^9$	n
1	1.090	0.9174	1.00000	1.09000	1.000	0.917	1
2	1.188	0.8417	0.47847	0.56847	2.090	1.759	2
3	1.295	0.7722	0.30505	0.39505	3.278	2.531	3
4	1.412	0.7084	0.21867	0.30867	4.573	3.240	4
5	1.539	0.6499	0.16709	0.25709	5.985	3.890	5
6	1.677	0.5963	0.13292	0.22292	7.523	4.486	6
7	1.828	0.5470	0.10869	0.19869	9.200	5.033	7
8	1.993	0.5019	0.09067	0.18067	11.028	5.535	8
9	2.172	0.4604	0.07680	0.16680	13.021	5.995	9
10	2.367	0.4224	0.06582	0.15582	15.193	6.418	10
11	2.580	0.3875	0.05695	0.14695	17.560	6.805	11
12	2.813	0.3555	0.04965	0.13965	20.141	7.161	12
13	3.066	0.3262	0.04357	0.13357	22.953	7.487	13
14	3.342	0.2992	0.03843	0.12843	26.019	7.786	14
15	3.642	0.2745	0.03406	0.12406	29.361	8.061	15
16	3.970	0.2519	0.03030	0.12030	33.003	8.313	16
17	4.328	0.2311	0.02705	0.11705	36.974	8.544	17
18	4.717	0.2120	0.02421	0.11421	41.301	8.756	18
19	5.142	0.1945	0.02173	0.11173	46.018	8.950	19
20	5.604	0.1784	0.01955	0.10955	51.160	9.129	20
21	6.109	0.1637	0.01762	0.10762	56.765	9.292	21
22	6.659	0.1502	0.01590	0.10590	62.873	9.442	22
23	7.258	0.1378	0.01438	0.10438	69.532	9.580	23
24	7.911	0.1264	0.01302	0.10302	76.790	9.707	24
25	8.623	0.1160	0.01180	0.10181	84.701	9.823	25
26	9.399	0.1064	0.01072	0.10072	93.324	9.929	26
27	10.245	0.0976	0.00973	0.09973	102.723	10.027	27
28	11.167	0.0895	0.00885	0.09885	112.968	10.116	28
29	12.172	0.0822	0.00806	0.09806	124.135	10.198	29
30	13.268	0.0754	0.00734	0.09734	136.308	10.274	30
31	14.462	0.0691	0.00669	0.09669	149.575	10.343	31
32	15.763	0.0634	0.00610	0.09610	164.037	10.406	32
33	17.182	0.0582	0.00556	0.09556	179.800	10.464	33
34	18.728	0.0534	0.00508	0.09508	196.982	10.518	34
35	20.414	0.0490	0.00464	0.09464	215.711	10.567	35
40	31.409	0.0318	0.00296	0.09296	337.882	10.757	40
45	48.327	0.0207	0.00190	0.09190	525.859	10.881	45
50	74.358	0.0134	0.00123	0.09123	815.084	10.962	50
55	114.408	0.0087	0.00079	0.09079	1260.092	11.014	55
60	176.031	0.0057	0.00051	0.09051	1944.792	11.048	60
65	270.864	0.0037	0.00033	0.09033	2998.288	11.070	65
70	416.730	0.0024	0.00022	0.09022	4619.223	11.084	70
75	641.191	0.0016	0.00014	0.09014	7113.232	11.094	75
80	986.552	0.0010	0.00009	0.09009	10950.556	11.100	80
85	1517.948	0.0007	0.00006	0.09006	16854.444	11.104	85
90	2335.501	0.0004	0.00004	0.09004	25939.000	11.106	90
95	3593.513	0.0003	0.00003	0.09003	39917.378	11.108	95
100	5529.089	0.0002	0.00002	0.09002	61422.544	11.109	100

10%

	To find F, given P: $(1 + i)^n$	To find P, given F: $\dfrac{1}{(1 + i)^n}$	To find A, given F: $\dfrac{i}{(1 + i)^n - 1}$	To find A, given P: $\dfrac{i(1 + i)^n}{(1 + i)^n - 1}$	To find F, given A: $\dfrac{(1 + i)^n - 1}{i}$	To find P, given A: $\dfrac{(1 + i)^n - 1}{i(1 + i)^n}$	
n	$(f/p)_n^{10}$	$(p/f)_n^{10}$	$(a/f)_n^{10}$	$(a/p)_n^{10}$	$(f/a)_n^{10}$	$(p/a)_n^{10}$	n
1	1.100	0.9091	1.00000	1.10000	1.000	0.909	1
2	1.210	0.8264	0.47619	0.57619	2.100	1.736	2
3	1.331	0.7513	0.30211	0.40211	3.310	2.487	3
4	1.464	0.6830	0.21547	0.31547	4.641	3.170	4
5	1.611	0.6209	0.16380	0.26380	6.105	3.791	5
6	1.772	0.5645	0.12961	0.22961	7.716	4.355	6
7	1.949	0.5132	0.10541	0.20541	9.487	4.868	7
8	2.144	0.4665	0.08744	0.18744	11.436	5.335	8
9	2.358	0.4241	0.07364	0.17364	13.579	5.759	9
10	2.594	0.3855	0.06275	0.16275	15.937	6.144	10
11	2.853	0.3505	0.05396	0.15396	18.531	6.495	11
12	3.138	0.3186	0.04676	0.14676	21.384	6.814	12
13	3.452	0.2897	0.04078	0.14078	24.523	7.103	13
14	3.797	0.2633	0.03575	0.13575	27.975	7.367	14
15	4.177	0.2394	0.03147	0.13147	31.772	7.606	15
16	4.595	0.2176	0.02782	0.12782	35.950	7.824	16
17	5.054	0.1978	0.02466	0.12466	40.545	8.022	17
18	5.560	0.1799	0.02193	0.12193	45.599	8.201	18
19	6.116	0.1635	0.01955	0.11955	51.159	8.363	19
20	6.727	0.1486	0.01746	0.11746	57.275	8.514	20
21	7.400	0.1351	0.01562	0.11562	64.002	8.649	21
22	8.140	0.1228	0.01401	0.11401	71.403	8.772	22
23	8.954	0.1117	0.01257	0.11257	79.543	8.883	23
24	9.850	0.1015	0.01130	0.11130	88.497	8.985	24
25	10.835	0.0923	0.01017	0.11017	98.347	9.077	25
26	11.918	0.0839	0.00916	0.10916	109.182	9.161	26
27	13.110	0.0763	0.00826	0.10826	121.100	9.237	27
28	14.421	0.0693	0.00745	0.10745	134.210	9.307	28
29	15.863	0.0630	0.00673	0.10673	148.631	9.370	29
30	17.449	0.0573	0.00608	0.10608	164.494	9.427	30
31	19.194	0.0521	0.00550	0.10550	181.943	9.479	31
32	21.114	0.0474	0.00497	0.10497	201.138	9.526	32
33	23.225	0.0431	0.00450	0.10450	222.252	9.569	33
34	25.548	0.0391	0.00407	0.10407	245.477	9.609	34
35	28.102	0.0356	0.00369	0.10369	271.024	9.644	35
40	45.259	0.0221	0.00226	0.10226	442.593	9.779	40
45	72.890	0.0137	0.00139	0.10139	718.905	9.863	45
50	117.391	0.0085	0.00086	0.10086	1163.909	9.915	50
55	189.059	0.0053	0.00053	0.10053	1880.591	9.947	55
60	304.482	0.0033	0.00033	0.10033	3034.816	9.967	60
65	490.371	0.0020	0.00020	0.10020	4893.707	9.980	65
70	789.747	0.0013	0.00013	0.10013	7887.470	9.987	70
75	1271.895	0.0008	0.00008	0.10008	12708.954	9.992	75
80	2048.400	0.0005	0.00005	0.10005	20474.002	9.995	80
85	3298.969	0.0003	0.00003	0.10003	32979.690	9.997	85
90	5313.023	0.0002	0.00002	0.10002	53120.226	9.998	90
95	8556.676	0.0001	0.00001	0.10001	85556.760	9.999	95
100	13780.612	0.0001	0.00001	0.10001	137796.123	9.999	100

12%

n	To find F, given P: $(1 + i)^n$	To find P, given F: $\dfrac{1}{(1 + i)^n}$	To find A, given F: $\dfrac{i}{(1 + i)^n - 1}$	To find A, given P: $\dfrac{i(1 + i)^n}{(1 + i)^n - 1}$	To find F, given A: $\dfrac{(1 + i)^n - 1}{i}$	To find P, given A: $\dfrac{(1 + i)^n - 1}{i(1 + i)^n}$	n
	$(f/p)_n^{12}$	$(p/f)_n^{12}$	$(a/f)_n^{12}$	$(a/p)_n^{12}$	$(f/a)_n^{12}$	$(p/a)_n^{12}$	
1	1.120	0.8929	1.00000	1.12000	1.000	0.893	1
2	1.254	0.7972	0.47170	0.59170	2.120	1.690	2
3	1.405	0.7118	0.29635	0.41635	3.374	2.402	3
4	1.574	0.6355	0.20923	0.32923	4.779	3.037	4
5	1.762	0.5674	0.15741	0.27741	6.353	3.605	5
6	1.974	0.5066	0.12323	0.24323	8.115	4.111	6
7	2.211	0.4523	0.09912	0.21912	10.089	4.564	7
8	2.476	0.4039	0.08130	0.20130	12.300	4.968	8
9	2.773	0.3606	0.06768	0.18768	14.776	5.328	9
10	3.106	0.3220	0.05698	0.17698	17.549	5.650	10
11	3.479	0.2875	0.04842	0.16842	20.655	5.938	11
12	3.896	0.2567	0.04144	0.16144	24.133	6.194	12
13	4.363	0.2292	0.03568	0.15568	28.029	6.424	13
14	4.887	0.2046	0.03087	0.15087	32.393	6.628	14
15	5.474	0.1827	0.02682	0.14682	37.280	6.811	15
16	6.130	0.1631	0.02339	0.14339	42.753	6.974	16
17	6.866	0.1456	0.02046	0.14046	48.884	7.120	17
18	7.690	0.1300	0.01794	0.13794	55.750	7.250	18
19	8.613	0.1161	0.01576	0.13576	63.440	7.366	19
20	9.646	0.1037	0.01388	0.13388	72.052	7.469	20
21	10.804	0.0926	0.01224	0.13224	81.699	7.562	21
22	12.100	0.0826	0.01081	0.13081	92.503	7.645	22
23	13.552	0.0738	0.00956	0.12956	104.603	7.718	23
24	15.179	0.0659	0.00846	0.12846	118.155	7.784	24
25	17.000	0.0588	0.00750	0.12750	133.334	7.843	25
26	19.040	0.0525	0.00665	0.12665	150.334	7.896	26
27	21.325	0.0469	0.00590	0.12590	169.374	7.943	27
28	23.884	0.0419	0.00524	0.12524	190.699	7.984	28
29	26.750	0.0374	0.00466	0.12466	214.582	8.022	29
30	29.960	0.0334	0.00414	0.12414	241.333	8.055	30
31	33.555	0.0298	0.00369	0.12369	271.292	8.085	31
32	37.582	0.0266	0.00328	0.12328	304.847	8.112	32
33	42.091	0.0238	0.00292	0.12292	342.429	8.135	33
34	47.142	0.0212	0.00260	0.12260	384.520	8.157	34
35	52.800	0.0189	0.00232	0.12232	431.663	8.176	35
40	93.051	0.0107	0.00130	0.12130	767.091	8.244	40
45	163.988	0.0061	0.00074	0.12074	1358.230	8.283	45
50	289.002	0.0035	0.00042	0.12042	2400.018	8.305	50

15%

	To find F, given P: $(1 + i)^n$	To find P, given F: $\dfrac{1}{(1 + i)^n}$	To find A, given F: $\dfrac{i}{(1 + i)^n - 1}$	To find A, given P: $\dfrac{i(1 + i)^n}{(1 + i)^n - 1}$	To find F, given A: $\dfrac{(1 + i)^n - 1}{i}$	To find P, given A: $\dfrac{(1 + i)^n - 1}{i(1 + i)^n}$	
n	$(f/p)_n^{15}$	$(p/f)_n^{15}$	$(a/f)_n^{15}$	$(a/p)_n^{15}$	$(f/a)_n^{15}$	$(p/a)_n^{15}$	n
1	1.150	0.8696	1.00000	1.15000	1.000	0.870	1
2	1.322	0.7561	0.46512	0.61512	2.150	1.626	2
3	1.521	0.6575	0.28798	0.43798	3.472	2.283	3
4	1.749	0.5718	0.20027	0.35027	4.993	2.855	4
5	2.011	0.4972	0.14832	0.29832	6.742	3.352	5
6	2.313	0.4323	0.11424	0.26424	8.754	3.784	6
7	2.660	0.3759	0.09036	0.24036	11.067	4.160	7
8	3.059	0.3269	0.07285	0.22285	13.727	4.487	8
9	3.518	0.2843	0.05957	0.20957	16.786	4.772	9
10	4.046	0.2472	0.04925	0.19925	20.304	5.019	10
11	4.652	0.2149	0.04107	0.19107	24.349	5.234	11
12	5.350	0.1869	0.03448	0.18448	29.002	5.421	12
13	6.153	0.1625	0.02911	0.17911	34.352	5.583	13
14	7.076	0.1413	0.02469	0.17469	40.505	5.724	14
15	8.137	0.1229	0.02102	0.17102	47.580	5.847	15
16	9.358	0.1069	0.01795	0.16795	55.717	5.954	16
17	10.761	0.0929	0.01537	0.16537	65.075	6.047	17
18	12.375	0.0808	0.01319	0.16319	75.836	6.128	18
19	14.232	0.0703	0.01134	0.16134	88.212	6.198	19
20	16.367	0.0611	0.00976	0.15976	102.444	6.259	20
21	18.821	0.0531	0.00842	0.15842	118.810	6.312	21
22	21.645	0.0462	0.00727	0.15727	137.631	6.359	22
23	24.891	0.0402	0.00628	0.15628	159.276	6.399	23
24	28.625	0.0349	0.00543	0.15543	184.168	6.434	24
25	32.919	0.0304	0.00470	0.15470	212.793	6.464	25
26	37.857	0.0264	0.00407	0.15407	245.711	6.491	26
27	43.535	0.0230	0.00353	0.15353	283.569	6.514	27
28	50.066	0.0200	0.00306	0.15306	327.104	6.534	28
29	57.575	0.0174	0.00265	0.15265	377.170	6.551	29
30	66.212	0.0151	0.00230	0.15230	434.745	6.566	30
31	76.143	0.0131	0.00200	0.15200	500.956	6.579	31
32	87.565	0.0114	0.00173	0.15173	577.099	6.591	32
33	100.700	0.0099	0.00150	0.15150	664.664	6.600	33
34	115.805	0.0086	0.00131	0.15131	765.364	6.609	34
35	133.176	0.0075	0.00113	0.15113	881.170	6.617	35
40	267.863	0.0037	0.00056	0.15056	1779.090	6.642	40
45	538.769	0.0019	0.00028	0.15028	3585.128	6.654	45
50	1083.657	0.0009	0.00014	0.15014	7217.716	6.661	50

20%

	To find F, given P:	To find P, given F:	To find A, given F:	To find A, given P:	To find F, given A:	To find P, given A:	
	$(1 + i)^n$	$\dfrac{1}{(1 + i)^n}$	$\dfrac{i}{(1 + i)^n - 1}$	$\dfrac{i(1 + i)^n}{(1 + i)^n - 1}$	$\dfrac{(1 + i)^n - 1}{i}$	$\dfrac{(1 + i)^n - 1}{i(1 + i)^n}$	
n	$(f/p)_n^{20}$	$(p/f)_n^{20}$	$(a/f)_n^{20}$	$(a/p)_n^{20}$	$(f/a)_n^{20}$	$(p/a)_n^{20}$	n
1	1.200	0.8333	1.00000	1.20000	1.000	0.833	1
2	1.440	0.6944	0.45455	0.65455	2.200	1.528	2
3	1.728	0.5787	0.27473	0.47473	3.640	2.106	3
4	2.074	0.4823	0.18629	0.38629	5.368	2.598	4
5	2.488	0.4019	0.13438	0.33438	7.442	2.991	5
6	2.986	0.3349	0.10071	0.30071	9.930	3.326	6
7	3.583	0.2791	0.07742	0.27742	12.916	3.605	7
8	4.300	0.2326	0.06061	0.26061	16.499	3.837	8
9	5.160	0.1938	0.04808	0.24808	20.799	4.031	9
10	6.192	0.1615	0.03852	0.23852	25.959	4.192	10
11	7.430	0.1346	0.03110	0.23110	32.150	4.327	11
12	8.916	0.1122	0.02526	0.22526	39.581	4.439	12
13	10.699	0.0935	0.02062	0.22062	48.497	4.533	13
14	12.839	0.0779	0.01689	0.21689	59.196	4.611	14
15	15.407	0.0649	0.01388	0.21388	72.035	4.675	15
16	18.488	0.0541	0.01144	0.21144	87.442	4.730	16
17	22.186	0.0451	0.00944	0.20944	105.931	4.775	17
18	26.623	0.0376	0.00781	0.20781	128.117	4.812	18
19	31.948	0.0313	0.00646	0.20646	154.740	4.843	19
20	38.338	0.0261	0.00536	0.20536	186.688	4.870	20
21	46.005	0.0217	0.00444	0.20444	225.025	4.891	21
22	55.206	0.0181	0.00369	0.20369	271.031	4.909	22
23	66.247	0.0151	0.00307	0.20307	326.237	4.925	23
24	79.497	0.0126	0.00255	0.20255	392.484	4.937	24
25	95.396	0.0105	0.00212	0.20212	471.981	4.948	25
26	114.475	0.0087	0.00176	0.20176	567.377	4.956	26
27	137.371	0.0073	0.00147	0.20147	681.853	4.964	27
28	164.845	0.0061	0.00122	0.20122	819.223	4.970	28
29	197.813	0.0051	0.00102	0.20102	984.068	4.975	29
30	237.376	0.0042	0.00085	0.20085	1181.881	4.979	30
31	284.851	0.0035	0.00070	0.20070	1419.257	4.982	31
32	341.822	0.0029	0.00059	0.20059	1704.108	4.985	32
33	410.186	0.0024	0.00049	0.20049	2045.930	4.988	33
34	492.223	0.0020	0.00041	0.20041	2456.116	4.990	34
35	590.668	0.0017	0.00034	0.20034	2948.339	4.992	35
40	1469.772	0.0007	0.00014	0.20014	7343.858	4.997	40
45	3657.258	0.0003	0.00005	0.20005	18281.331	4.999	45
50	9100.427	0.0001	0.00002	0.20002	45497.191	4.999	50

SELECTED BIBLIOGRAPHY

Chapter 1

1. Ministry of Petroleum and Minerals, Economics Department, Petrole-
    um Statistical Bulletin, Riyadh, Saudi Arabia, 1968.

Chapter 2

1. Al-Miligi, M. K., The Arab Petroleum Directory (Kuwaiti City,
    Kuwait, 1972).

2. British Petroleum Statistical Review of the World Industry, 1970,
    1971, 1972, 1973, British Petroleum Company, Ltd., London.

3. Gardner, Frank, "North Sea Today: Where Tomorrow?" Oil and Gas
    Journal, vol. 71, no. 50 (1973), p. 77.

4. Gardner, Frank, "Offshore Oil—Only the Beginning," Oil and Gas
    Journal, vol. 72, no. 18 (1974), p. 124.

5. "Going Fishing in South Vietnam," Forbes, vol. 113, no. 2 (1974),
    p. 39.

6. King, Robert E., "Big Reserve Boost Foreseen in Gulf of Mexico in
    1974," World Oil, vol. 178, no. 5 (1974), p. 72.

7. McCaslin, John, "Offshore Oil Production Soars," Oil and Gas Journal,
    vol. 72, no. 18 (1974), p. 77.

8. "Oil Basic Analysis," Standard and Poor's Industry Survey, Section II,
    July 5, 1973, p. 70.

9. "Oil Current Analysis," Standard and Poor's Industry Survey, Section
    II, April 4, 1974, p. 43.

10. "Petroleum Industry Competition: It Has Been Verified Time and Time Again," Public Affairs/Briefings, Standard Oil of California, March 1974, p. 4.

11. Rose, Standford, "Our Vast Hidden Oil Reserves," Fortune, vol. 84, no. 4 (1974), p. 107.

12. Speech by M. A. Wright, Chairman of Exxon Corporation, Wall Street Transcripts, vol. 42, no. 9 (1973), p. 35128.

13. "The Interdependence of Profits and Energy," Public Affairs/Briefings, Standard Oil of California, April 1974, p. 1.

14. World Oil, August 15, 1971.

Chapter 3

1. Abdel-Aal, H. K., Notes on Refining at University of Petroleum and Minerals, 1972.

2. American Petroleum Institute, "Facts About Oil," 1966.

3. American Society for Testing Materials' booklet, 1970.

4. Aramco Handbook, "Oil and the Middle East," American Arabian Oil Company, 1968.

5. Belayim Crude Oils, Analysis and Processing (Cairo: The Egyptian General Petroleum Corporation, 1963).

6. Beychok, Milton R., Aqueous Wastes from Petroleum and Petrochemical Plants (New York: John Wiley & Sons, 1967).

7. Bryant, H. S., "Environment Needs Guide Refinery Sulphur Recovery," Oil and Gas Journal, March 26 (1973), p. 71.

8. Cassady, Ralph, Jr., Price Making and Price Behavior in the Petroleum Industry (New Haven, Conn.: Yale University Press, 1954).

9. Frankel, I., Essentials of Petroleum (London: Case and Company Ltd., 1968).

10. Hughes, Richard V., Oil Property Valuation (New York: John Wiley & Sons, 1967).

11. Nelson, W. L., "Net Cost for Desulfurizing Residues," Oil and Gas Journal, January 19 (1970), p. 60.

12. Nelson, W. L., "Refinery-Operating Costs," Oil and Gas Journal, March 2 (1970), p. 83.

13. Nelson, Richard W. L., "Crude Evaluation," Oil and Gas Journal, November 30 (1970), p. 84.

14. Russell, Clifford S., Residuals Management in Industry: A Case
    Study of Petroleum Refining (Baltimore: Johns Hopkins University
    Press, 1973).

15. Shell Petroleum Handbook, 3rd ed., Shell Oil Company.

16. Sullivan, Robert E., Handbook of Oil and Gas Law (Englewood Cliffs,
    N.J.: Prentice-Hall, 1959).

Chapter 4

1. Alt, R. M., and Bradford, W. C., Business Economics (Homewood,
   Ill.: Richard D. Irwin, 1951).

2. Aries, Robert S., and Newton, Robert D., Chemical Engineering Cost
   Estimation (New York: McGraw-Hill Book Company, 1955).

3. Lindsay, J. R., and Sametz, A. W., Financial Management and
   Analytical Approach, rev. (Homewood, Ill.: Richard D. Irwin, 1967).

4. Peters, M. S., and Timmerhause, K. D., Plant Design and Economics for Chemical Engineers (New York: McGraw-Hill Book Company, 1968).

5. Schmelzlee, R., Notes on Financial Mathematics at University of
   Petroleum and Minerals, 1972.

6. Stephens, M. M., Spencer, O. F., and Hild, D. R., Petroleum
   Engineering Fundamentals (State College, Pa.: Pennsylvania State
   University Press, 1958).

Chapter 5

1. Brock, Horace R., Palmer, Charles E., and Archer, Fred C., Cost
   Accounting Theory and Practice (New York: McGraw-Hill Book Company, 1971).

2. Foulke, Roy A., Practical Financial Statement Analysis (New York:
   McGraw-Hill Book Company, 1961).

3. Gillespie, Cecil, Cost Accounting and Control (Englewood Cliffs, N.J.:
   Prentice-Hall, 1965).

4. Henrici, Stanley B., Standard Costs for Manufacturing (New York:
   McGraw-Hill Book Company, 1960).

5. Kurtz, Max, Engineering Economics (New York: McGraw-Hill Book
   Company, 1959).

6. Livingstone, J. Leslie, Management Planning and Control (New York:
   McGraw-Hill Book Company, 1970).

7. Mathew Bender & Company, Oil and Gas Accounting (Dallas, Texas: Southwestern Legal Foundation, 1966).

8. Mauriello, Joseph A., Accounting for the Financial Analyst (Homewood, Ill.: Richard D. Irwin, 1971).

9. Porter, Stanley, Petroleum Accounting Practice (New York: McGraw-Hill Book Company, 1965).

Chapter 6

1. "Abu Dhabi Concession," Middle East Economic Survey, Beirut, Lebanon, June 12, 1970.

2. American Petroleum Institute, "Facts About Oil," 1966.

3. Baker, William J., Petroleum Refiner, July 1959, p. 129.

4. Downs, G. F., Jr., and Tait, G. R., "Selecting Pipeline Diameter for Minimum Investment," Oil and Gas Journal, vol. 52, no. 28 (1953), p. 210.

5. "Future Growth of the World Petroleum Industry," a series of Petroleum Industry Studies, by Chase Manhattan Bank, New York, 1966.

6. Issawi, C., and Yeganeh, M., The Economics of Middle Eastern Oil (London: Faber and Faber, 1962).

7. Lenezowski, P., Oil and State in the Middle East (New York: McGraw-Hill Book Company, 1967).

8. Longingg, S. H., Oil in the Middle East, Its Discovery and Development (London: Oxford University Press, 1961).

9. Meyers, R., "Planning of Crude Oil Production, Transportation and Refining to Satisfy Optimal Distribution, Patterns of Future National Market Demands," Seventh World Petroleum Congress, Mexico, 1967.

10. Middle East Economic Survey, "Concession Agreement—in Abu Dhabi," September 1970.

11. Nelson, R., Petroleum Refinery Engineering, 4th ed., (New York: McGraw-Hill Book Company, 1949).

12. Nielsen, R. F., "Types of Production Decline Curve," Oil and Gas Journal, vol. V-51, no. 17-33 (1952).

13. Oil and Gas Journal, vol. V-66, no. 1-18 (1968).

14. "Our Industry," British Petroleum Company, Ltd., 1947.

15. Peters, Max S., Elementary Chemical Engineering (New York: McGraw-Hill Book Company, 1954).

16. Peters, M. S., and Timmerhause, K. D., Plant Design and Economics for Chemical Engineers (New York: McGraw-Hill Book Company, 1968).

17. "Pumpability of Residual Fuels," Industrial and Engineering Chemistry (Quarterly) (Washington, D. C.: American Chemical Society, June, 1954).

18. Schmelzlee, R., Notes on Oil Economics at College of Petroleum and Minerals, 1972.

19. Shcherbakov, S. G., "Progress in Pipeline Transportation," Seventh World Petroleum Congress, Mexico, 1967.

20. Struth, H. J., "An Economic Study: What It Costs to Find Oil," Petroleum Engineer, November 1960.

Chapter 7

1. Abdel-Aal, H., Notes on Refining and Plant Design at University of Petroleum and Minerals, Dhahran, Saudi Arabia, 1971.

2. American Petroleum Institute Information Bulletin, No. 11.

3. Aries, Robert S., and Newton, Robert D., Chemical Engineering Cost Estimation (New York: McGraw-Hill Book Company, 1955).

4. Dantzig, S., Linear Programming and Extensions (Princeton, N.J.: Princeton University Press, 1963).

5. Dorfman, R., Samuelson, P. A., and Solow, R. M., Linear Programming and Economic Analysis (New York: McGraw-Hill Book Company, 1958).

6. Gary, James H., and Handwerk, Glenn E., Petroleum Refining: Technology and Economics (New York: Marcel Dekker, Inc., 1975).

7. Gilmore, J. F., "Short Cut Estimating of Processes," Petroleum Refiner, vol. 32, no. 10 (1953), p. 97.

8. Happel, John, Chemical Process Economics (New York: John Wiley & Sons, 1958).

9. "How to Estimate Fractionating Column Costs," Petroleum Refiner, July 1960.

10. Kim, Chaiko, Introduction to Linear Programming (New York: Holt, Rinehart and Winston, 1971).

11. Levenspiel, O., Chemical Reactor Design (New York: John Wiley & Sons, 1962).

12. Mihm, J. C., "Optimization of Gasoline Plants Offers Growing Savings," World Oil, April 1972.

13. "Optimum ΔP Cuts Exchange Costs," World Oil, July 1959.

14. Peters, M. S., Elementary Chemical Engineering (New York: McGraw-Hill Book Company, 1954).

15. Vilbrandt, F. C., and Dryden, C. E., Chemical Engineering Plant Design (New York: McGraw-Hill Book Company, 1959).

Chapter 8

1. Cassady, R., Jr., Price Making and Price Behavior in the Petroleum Industry (New Haven, Conn.: Yale University Press, 1954).

2. DeChazeau, M., and Kahn, A. E., Integration and Competition in the Petroleum Industry, vol. 3 (New Haven, Conn.: Yale University Press, 1954).

3. Frank, H. J., Crude Oil Prices in the Middle East (New York: Praeger Publishing Company, 1966).

4. Jenkins, G. I., "Company Uncertainty and Decision-Making," in the Institute of Petroleum's Petroleum Review, June 1972.

5. Leeman, W. A., The Price of Middle East Oil (Ithaca, N.Y.: Cornell University Press, 1962).

6. Lufti, Ashraf, OPEC Oil (Beirut, Lebanon: The Middle East Research and Publishing Center, 1970).

7. Middle East Economic Survey, "Devaluation: What It Means for the Producing Countries," Beirut, Lebanon, vol. XVI, February 16, 1973.

8. Parra, Ramos, and Parra, S. A., in cooperation with Middle East Economic Survey, "International Crude Oil and Product Prices," Middle East Petroleum and Economic Publications, Beirut, Lebanon, 1972.

9. Platt's Oilgram Price Service.

10. Saudi Economic Survey, "Developments in the Oil Sector in 1970," Ashoor Public Relations Services, December 8, 1971.

11. Schmelzlee, R., Notes on Oil and Oil Product Prices at University of Petroleum and Minerals, Dhahran, Saudi Arabia.

12. Zimmerman, E. W., Conservation in the Production of Petroleum, vol. 3 (New Haven, Conn.: Yale University Press, 1954).

Chapter 9

1. Aramco Economics Department, "News Digest," Dhahran, Saudi Arabia, November 10, 1972.

2. Bes, J., Chartering and Shipping Terms (Amsterdam: Drukkery Press, 1962).

3. British Petroleum Statistical Review of the World Oil Industry, 1970, 1971, 1972, British Petroleum Company Ltd., London.

4. Drewry, H. P., "Transport Tips" (London: H. P. Drewry, Shipping Consultants, 1970).

5. Schmelzlee, R., Notes on Tankers and Tanker Rates at University of Petroleum and Minerals, Dhahran, Saudi Arabia, 1972.

6. "The Role of Combined Carriers," Petroleum Press Service, February 1972, p. 61.

7. Zanetos, Zenon, The Theory of Oil Tankship Rates (Cambridge, Mass.: M.I.T. Press, 1966).

Chapter 10

1. Kliever, G., "Industry Trends Point Toward Bright Future," World Oil, February 15, 1972.

2. Lenchamsky, G., Middle East in World Affairs (New York: McGraw-Hill Book Company, 1967).

3. Lubell, H., Middle East Oil Crisis and Western Europe's Energy Supplies (Baltimore: Johns Hopkins University Press, 1963).

4. Penrose, L., The Large International Firm in Developing Countries (London: Allen & Unwin, Ltd., 1970).

5. Sayegh, K. S., Oil and Arab Regional Development (London: Praeger Publishing Company, 1968).

INDEX

A

Abadan, 366
Abu Dhabi (Arab supplement), 30
Abu Dhabi concession agreement, 145
  arbitration, 145
  area in, 143, 145
  bonus payments, 146
  education, medicine, 150
  investment, 149, 150
  natural gas, 147
  ownership, 145
  participation, 149
  price-setting, 149
  reimbursement, 149
  relinquishments, 147
  rentals, annual, 146, 147
  royalties, 147
  taxation, 148, 149
  work obligations, 145, 146
Abu Dhabi, 276, 290, 292
  Abu Dhabi marine crude, 292
  Das Island, f.o.b. crude point, 292
  Jebel Dhana, f.o.b. crude point, 292
  Murban crudes, 292
Accretion, 132

Additional costs of output, 5
Additional costs in refineries, 221, 222, 226-227
Africa, energy consumption, per capita basis, 27
After loading time, tankers, 327
Algeria (Arab supplement), 30, 31
Alice-in-wonderland pricing structure, 284-285
Alkylation process in refining, 205
Anglo-Iranian Oil Company, 145
Angola, 290, 295
Annual cost method, investment, 82, 83, 85, 86
  problems in, 82-83, 85-86
Annual cost transport problem, depreciation and discount, 347
Annuity, 78
Annulus velocity, 178
API (American Petroleum Institute) gravity ratings, 43, 44, 46, 48, 276, 286, 290, 291-294
Appreciation of currencies, 281-282
Arab Emirates, 295
Arab oil supplement, 2, 29-39
Arab Petroleum Directory, 2, 29-39
Arabian Gulf prices, 275
Arabian heavy crude, 290, 292, 361